# PRAISE FOR *NOURISHMENT*

"In *Nourishment* a wise observer of the land and the animals becomes transformed to learn the meaning of life. Fred Provenza's reflections on his long, fruitful career in behavior-based management of landscapes include his joy in observing the natural world around him."

— **Temple Grandin**, author of *Animals in Translation*

"An absolutely fascinating topic, engagingly written. Fred Provenza will wholly up-end everything you think you know about what you are supposed to eat and why. *Nourishment* should be the first book read by every student of nutrition."

— **Nicolette Hahn Niman**, author of *Defending Beef*

"Fred Provenza's *Nourishment* is a synergy of the essence of his lifetime of research coupled with commonsense nutritional advice to aid both animals and humans. All readers will benefit from the knowledge revealed in this book. I have long thought Dr. Provenza's work should comprise a required course in veterinary and medical schools, and *Nourishment* confirms that opinion."

— **Richard J. Holliday**, DVM, holistic veterinarian;
coauthor of *A Holistic Vet's Prescription for a Healthy Herd*

"Fred Provenza is a hero of mine. His deep understanding of how animals behave on the land – why they eat what they eat, how they stay healthy, and how they teach their young – has helped generations of ranchers, farmers, wildlife managers, and university students understand and strengthen their relationships with animals in sustainable ways. We are animals, too, and in this magnificent book, Fred weaves together philosophy, nutritional science, memoir, and his humble appreciation for the natural world into an inspiring meditation on our moment on Earth."

— **Courtney White**, author of *Grass, Soil, Hope*
and *Two Percent Solutions for the Planet*

"Drawing from episodes of his years of research and the cycles of life, Dr. Provenza illuminates the fascinating mechanisms by which animals' innate knowledge of healthy eating is passed on to their offspring and how we humans can learn from the simplicity of animals in our midst. This engaging book provides proof that animals are not mindless eaters – nor should humans be."

– **Hubert Karreman**, VMD, author of *Treating Dairy Cows Naturally*

"Fred Provenza has spent his career redefining the ways we understand livestock intelligence when it comes to diet. In this compelling and engaging book, he shares many of those lessons and synthesizes a wide breadth of research and knowledge to challenge some fundamental assumptions about diet, whether it be for livestock or humans."

– **Steve Gabriel**, author of *Silvopasture* and coauthor of *Farming the Woods*

"In this path-finding and paradigm-breaking book, Provenza combines a lifetime's worth of scientific work with deep empathic insight to further extend his unique contribution to healing our ailing landscapes, our animals, and the human condition. In an era when the false pathway of economic rationalism, aligned to industrial agriculture and its food systems, is poisoning the planet and ourselves, Provenza provides a vitally new holistic perspective of our species: a perspective that provides exciting solutions to our Anthropocene and human dilemmas."

– **Charles Massy**, author of *Call of the Reed Warbler*

"Fred Provenza has a brilliant, inquisitive, and open-minded spirit, the kind that should live within all of us. *Nourishment* is the result of a lifetime of observation, experimentation, and incredible insight by a master scientist. This is a book you will read, reread, dog-ear pages in, and tell your friends about. You will be blown away, as I have been. Outstanding!"

– **Bob Budd**, executive director,
Wyoming Wildlife and Natural Resource Trust

"Once again, Fred Provenza shows he is one of the freshest and most insightful thinkers about the role of herbivores (and omnivorous humans!) in our food and land management systems. This book is a landmark contribution."

– **Gary Paul Nabhan**, author of *Mesquite*
and *Food from the Radical Center*

"Part mirror and part compass springing from a bedrock of science, *Nourishment* takes us on an extraordinary journey through bodies of wisdom.

This book sets a new bar for how to think about the intersection of food, nature, and ourselves."

— **Anne Biklé**, coauthor of *The Hidden Half of Nature*

"Through Fred Provenza's eyes, a simple meal becomes a wondrous journey into the world, into our bodies, and into our deepest selves. *Nourishment* will change the way you eat and the way you think."

— **Mark Schatzker**, author of *Steak* and *The Dorito Effect*

"Throughout his illustrious research career, Fred Provenza has taught us how animals — many of which we choose to eat — learn what they themselves should consume to maintain optimal health. He has now taken these lessons and applied them to the human situation in insightful and clever ways, concluding that we ignore our body's wisdom to our nutritional detriment. But he goes further, tying his metaphysical ideas about the lives of plants, the nature of food, and the consciousness of cells to his fascinating life adventure. This book is sure to engage, educate, and provoke the reader."

— **Gary K. Beauchamp**, director and president emeritus,
Monell Chemical Senses Center

"I could describe this book as simply an enlightening treatise on the philosophy of life, but that fails to do it justice. Fred Provenza uses his life's experiences as student, rancher, fisherman, researcher, teacher, hiker, cancer patient, and retiree to illustrate vividly the fundamental principles of nutrition not only of animals (including humans) but also of plants and soil. In *Nourishment* Provenza also offers a rich exposition of his outstanding all-embracing views of life."

— **J. Michael Forbes**, professor emeritus, University of Leeds;
author of *Voluntary Food Intake and Diet Selection of Farm Animals*

"Fred Provenza spent his distinguished academic career becoming a leading authority on the food habits of rangeland animals. In this thought-provoking book, he uses his deep understanding of his four-legged study subjects to cast light on our own nutritional condition in a world where we have become increasingly remote — philosophically and physically — from our food supply."

— **Stephen J. Simpson**, PhD, academic director,
Charles Perkins Centre, University of Sydney

"This book sets a new bar for how we think about the intersection of food, nature, and [illegible]."

—Anne Biklé, coauthor of *The Hidden Half of Nature*

"Through Fred Provenza's eyes, a simple meal becomes a wondrous journey into the world, into our bodies, and into our deepest selves. *Nourishment* will change the way you eat and the way you think."

—Mark Schatzker, author of *Steak* and *The Dorito Effect*

"Throughout his illustrious research career, Fred Provenza has taught us how animals—many of which we choose to eat—learn what they themselves should consume to maintain optimal health. He has now taken these lessons and applied them to the human situation in insightful and clever ways, reminding us that we forsake our body's wisdom to our nutritional detriment. But he goes further, tying his metaphysical ideas about [illegible] of food and the [illegible] of cells to [illegible] fascinating [illegible] and [illegible] provoke the reader."

—[illegible] L. Beauchamp, director and president emeritus, [illegible]

"[illegible] moving treatise on [illegible] supply [illegible] Fred Provenza uses his life experiences as a student, rancher, fisherman, researcher, teacher, father, cancer patient, and retiree to illustrate vividly the fundamental principles of nutrition not only of animals (including humans) but also of plants and soil. In *Nourishment* Provenza also offers a rich exposition of his outstanding all-embracing views of life."

—J. Michael Forbes, professor emeritus, University of Leeds; author of *Voluntary Food Intake and Diet Selection in Farm Animals*

"Fred Provenza spent his distinguished academic career becoming a leading authority on the food habits of rangeland animals. In this thought-provoking book, he uses his deep understanding of his four-legged study subjects to cast light on our own nutritional condition in a world where we have become increasingly remote—philosophically and physically—from our food supply."

—Stephen J. Simpson, PhD, academic director, Charles Perkins Centre, University of Sydney

# Nourishment

## What Animals Can Teach Us About Rediscovering Our Nutritional Wisdom

FRED PROVENZA

CHELSEA GREEN PUBLISHING
White River Junction, Vermont
London, UK

Cover painting by Annabelle Lanfermeijer (http://www.landelijkekunst.nl).



Project Manager: Sarah Kovach
Editor: Fern Marshall Bradley
Copy Editor: Nancy N. Bailey
Proofreader: Angela Boyle
Indexer: Linda Hallinger
Designer: Melissa Jacobson

Printed in the United States of America.
First printing November, 2018.
10 9 8 7 6 5 4 3     21 22 23 24

**Library of Congress Cataloging-in-Publication Data**
Names: Provenza, Frederick D., author.
Title: Nourishment : what animals can teach us about rediscovering our nutritional wisdom / Fred Provenza.
Description: White River Junction, Vermont : Chelsea Green Publishing, [2018] | Includes bibliographical references and index.
Identifiers: LCCN 2018028053 | ISBN 9781603588027 (pbk.) | ISBN 9781603588034 (ebook)
Subjects: | MESH: Nutritional Physiological Phenomena | Diet | Animal Nutritional Physiological Phenomena
Classification: LCC QP141 | NLM QU 145 | DDC 613.2 – dc23
LC record available at https://lccn.loc.gov/2018028053

Chelsea Green Publishing
85 North Main Street, Suite 120
White River Junction, VT 05001
(802) 295-6300
www.chelseagreen.com

*To Those Who Visit Earth*

*Tat tvam asi*

*If you realize that all things change,*
*there is nothing you will try to hold on to.*
*If you aren't afraid of dying,*
*there is nothing you can't achieve.*

*Trying to control the future*
*is like trying to take the master carpenter's place.*
*When you handle the master carpenter's tools,*
*chances are that you'll cut yourself.*

*from the* Tao Te Ching *by Laozi*

# CONTENTS

## PART V
## Fading into Mystery

## POINT OF DEPARTURE

# Transforming

Things change. No assertion could be more banal. Yet, within that statement lies a profound truth: From birth, all things – individuals, social groups, nations, species, galaxies, and universes – carry the seeds for their dissolution.

Things never change. Within that assertion, too, is a profound truth: What we call "death" is endless transformation – the never-ending dance of energy and matter.

The universe is a "restaurant" consuming itself. On Earth, life forms complex food webs as organisms consume one another. The sun that provides the energy to maintain life on Earth will one day grow into a red giant and consume the planet it once sustained.

Humankind has made an art form of dining. *Dining* conjures up images of plants and animals nicely arranged on platters and served in fine restaurants, but *eating* is participating in endless transformation as plants and animals are grown, killed, cooked, and consumed. As I eat, the energy and matter in plants and animals become this entity I call "me" – which will in the flicker of a cosmic eye return to plants and animals. Every act of eating is an act of creating.

During the blink of an eye we call a lifetime, change transforms us. Change forces us to see Earth and our place in the cosmos in new ways. Some changes we embrace: Birth brings great joy. Others are unwelcome: Partings and death are sad, but sorrow is part of being alive.

I imagine life on Earth as an interesting experiment: Isolate a species (*Homo sapiens*) in a place (planet Earth) and see if they can figure out where they are; how they got there; how to survive there; and what, if any, purpose "there" serves. Is it any wonder *Homo sapiens* have come up with such a bewildering array of beliefs about things physical and metaphysical?

We are born at a particular time and place on a planet orbiting a star in a galaxy. We are taught what and how to believe and behave, based on what has worked historically in our culture. Culture and place become our context. Buckminster Fuller wrote in *A Fuller View*, "I live on earth at present, and I don't know what I am. I know that I am not a category. I am not a thing – a noun. I seem to be a verb, an evolutionary process – an integral function of the universe."

This "verb" called *Fred* came to visit this planet humans call *Earth* in mid-September 1950. I was born in mid-May of 1951. I was raised in a small town in the Rocky Mountains of Colorado. As a child, I was captivated by streams, ponds, mountains, and creatures wild and free – bees, moths, butterflies, dragonflies; pollywogs, frogs; all manner of trout; ducks, geese; grouse, ptarmigan; deer, elk, mountain goats. As a young man, I worked on a ranch for seven years, during four of which I was majoring in wildlife biology at Colorado State University. During that time, I met the woman who would become my wife on the ski slopes in Colorado when we were members of a racing team. After Sue and I married, we moved to Utah. There, we raised a family while I worked as a professor in the Department of Wildland Resources at Utah State University. My fascination with things wild and free led to a life of research on the behavioral ecology of plants and herbivores. Thirty-five years – a family and a career – later, we moved back to Colorado, where we live in the tranquility of the backwoods.

Unlike most songbirds who live here seasonally, Sue and I make our nest in this backwoods of Colorado year-round. Our nest is located at 9,500 feet, at the junction of forest and park. From our place, we gaze across twenty miles of South Park, one of four vast parks – North Park, Middle Park, South Park, and the San Luis Valley – that run from the northern to the southern borders across central Colorado. The rolling parkland transitions into aspen- and conifer-cloaked highlands rising to 14,000-foot alpine peaks that make up the Mosquito Range to the north, the Sawatch Range to the West, and the Sangre de Cristo Range to the south. The potpourri of plants that blanket South Park is a hodgepodge of grasses including squirreltail, muhley, and blue gramma; forbs such as paintbrush, penstemon, and locoweed; shrubs the likes of sagebrush, rabbitbrush, and ground juniper; and trees such as aspen, ponderosa pine, and Douglas fir. Their abundances and appearances change daily, seasonally, and annually – a kaleidoscope of forms and colors – in the ebb and flow of life that sustains insects, birds, and mammals in this place we call home for what we know is a stolen moment.

Before sunrise this morning, brilliant orange clouds in the east transformed into pink highlights on blue-gray clouds in the west. As the first rays of the sun's light shimmered over the mountains, the eastern clouds morphed from burnt orange to luminous gold, brilliant silver, and snowy white, while those in the west intensified from subtle pinks, grays, and steely blues to bright whites. Clouds give form and elegance to the sun's rays as matter gives form and grace to energy. During the past 4.54 billion years, Earth has experienced nearly two trillion sunrises, no two alike. Each sunrise is a beginning as dawn heralds a boundless array of possibilities. Sunset is a time for peaceful transitions, evening a time for quiet reflection. The seeming transparency of first light gives way to the mystery and opacity of twilight.

As the sun's rays warmed yesterday morning's landscape, Sue and I meandered from the treeless top of the great hill, the backdrop to the south of our nest, down through the aspen and pine woods that blanket the hill. Nuthatches, juncos, and chickadees fluttered from tree to tree, some dining on insects in the bark of pine and fir, others plucking seeds from Douglas fir cones. Elk foraged along the edges of grass-forb parkland, near the aspen-conifer forest.

From the hilltop in the evening, we gazed in wonder as two coyotes, a dozen ravens, and a golden eagle began feasting on a live antelope, its leg caught between strands in a barbed wire fence. It was a wake, nature-style, and the participants were celebrating a life.

From the hilltop at sunrise this morning, we could see no sign of the antelope, so we strolled down the grassy knoll to examine the scene of last night's buffet. Nothing remained, not even the grasses and forbs that yesterday resided inside the paunch and intestines of the antelope. The only evidence of the meal was a little blood on the grass, small pieces of bone, and the scrap of hide three coyotes and two ravens were enjoying for breakfast.

Pondering the banquet, I recalled author William Holston's essay about the diet of mountain men.[1] Bison was their main food. When game was plentiful, they ate the choice parts. When they had been without meat for several days, all traces of civility vanished: The bison's skull was split open, and the men devoured the brains in bloody chunks. The liver, torn from the body cavity, bloody and unwashed, was seasoned with gunpowder, or juice from the gall bladder. The gall juice was drunk sometimes, for "a man could get quite a glow if he took it straight on an empty stomach." The greasy fleece was cut away from the ribs and eaten raw. Other strips of raw meat or fat were consumed if the men still were not satiated.

To live, mountain man, coyote, raven, and eagle participate in an act as old as life: energy and matter endlessly changing forms. By demand and by design, organisms consume *matter* and *energy* to maintain *form* and sustain *function*. From time immemorial, life didn't ponder what or how to eat. Organisms don't know energy, protein, minerals, or vitamins as we discern them from lab analyses. They have no concerns about eating red meat or brains, internal organs or greasy fleece, or ratios of omega-3 to omega-6 polyunsaturated fatty acids. Those matters were beside the point for mountain men and equally irrelevant for the coyotes, ravens, and golden eagle assembled on that grassy knoll. Nature, in the form of an antelope, provided a feast, and they savored every morsel on that clear, cold winter night. Soon, coyotes, ravens, and eagles will become food for soil micro- and macro-fauna, which will become plant, which will become herbivore, omnivore, and carnivore yet again, which will become soil, in endless transformation.

I recall listening to Joseph Campbell's stories in *The Power of Myth* about the mystery of life living by consuming itself. As he stressed, society is a part of a larger organism, the landscape, in which society is rooted. The function of myth and ritual is to link individuals with this larger morphological structure. In the process, the consciousness of plants and animals becomes the collective consciousness, endlessly transforming one with the consciousness of the transcendent. Energy is matter is consciousness, ever evolving, manifest in myriad ways.

But times change. We have very little contact with the plants and animals that give us life. As Campbell lamented,

> *The animal envoys of the Unseen Power no longer serve, as in primeval times, to teach and to guide mankind. Bears, lions, elephants, ibexes, and gazelles are in cages in our zoos. Man is no longer the newcomer in a world of unexplored plains and forests, and our immediate neighbors are not wild beasts but other human beings, contending for goods and space on a planet that is whirling without end around the fireball of a star. Neither in body nor in mind do we inhabit the world of those hunting races of the Paleolithic millennia, to whose lives and life ways*

*we nevertheless owe the very forms of our bodies and structures of our minds. Memories of their animal envoys still must sleep, somewhere, within us . . .*

Perhaps if hunting for meat and gathering fruits and vegetables were part of our daily lives, we would be more aware of the enigma that plants and animals die so that we may live.

Tonight, as I bask in the warm orange glow of a fire dancing off the logs, through the glass window of the soapstone stove, up the log bannister, and into the loft, I'm captivated by feelings of transformation. I'm pondering what I've read about the births and deaths of stars as the fire consumes the aspen logs whose elements came from the stars. The logs we cut and split last fall warm me as a December blizzard tosses water in the form of snowflakes outside our windows. The logs have taken many forms – from star to soil to seedling to tree to the fuel now being consumed by the fire, which is producing light to kindle my thoughts and heat to warm my soul. If I don't linger here to add more logs, the fire, too, will consume itself.

It is a good evening to reflect on the mysteries of transformation and the guises within which we struggle in vain against relentless change. Plants convert sunlight, water, and nutrients into roots, stems, leaves, flowers, and fruits. Herbivores, omnivores, and carnivores can consume daily sustenance and enjoy life above and below ground because plants convert sunlight, water, and nutrients into roots, stems, leaves, flowers, and fruits. In the process, life lives on life – from death comes life and endless transformation. Yet, we seldom stop to reflect on the experience of this flow of energy and matter from plants and animals through us and back into earth.

If we enhance our awareness of these dynamic forces, we can experience and embrace a universe where energy and each particle of matter are changing at every instant. In his book *Now*, physicist Richard Muller asserts the expanding universe is not only creating space, which did not exist prior to the big bang, but it is also creating time. The expanding edge of space-time is what we refer to as *now*. One of the most important consequences of Muller's assertion is that the past no longer determines the future, at least not entirely. We experience each new moment differently from preceding moments because at each new instant we can exercise choice to alter relationships in the next moment in time. By participating, we are continually creating new relationships in *now*. The spatial and temporal potentialities are virtually boundless because the universe is composed of holons within holons, with no apparent beginning or end.

The term *holon* refers to a whole that is also a part of something bigger than itself. When viewed from its constituent subsystems, a holon is a unique entity, as is a person viewed from the perspective of an organ, for instance, or an organ from the vantage of a cell. But from another perspective, a holon is an undistinguished portion of a larger system: a cell as part of an organ as part of an individual as a member of a social group as part of a landscape. A holon's behavior influences behavior at larger scales, which in turn influence behavior at smaller scales. Parts create wholes, which create parts, endlessly. Interactions among holons lead to emergent properties as self-organizing systems ever create their way into now.[2]

The Provenza yard in the Colorado backwoods self-organizes in a different way from the suburban areas where Sue and I lived for thirty-five years in Utah. While our life in the suburbs required great inputs of labor, water, fertilizer, and herbicides to maintain a dandelion-free monoculture of Kentucky bluegrass lawn, the backwoods are maintained effortlessly by Mother Nature. In comparison with Mother Nature's plantings in the motley backwoods, the manicured organization of our suburban yards was extraordinary, though the diversity of flora and fauna in suburbia was paltry at best compared with the backwoods.

The chapters in this book are more akin to plants of the backwoods than to the manicured world of suburbia. They are linked through the miscellany of my experiences, subject to the haze and distortions of my memories. They tell a tale of the ever-changing nature of forms and relationships. Like a brook, this tale meanders among what appear to be (but aren't) modest and tranquil alpine meadows in parts one, two, and three, and from there into what turn out to be equally enigmatic glacial cirques, peaks, and crags in parts four and five. The parts of this book are my attempt to reconcile my experiences as a child, enchanted with things wild and free, with my later years – as a ranch manager, nurturing plants and animals; as a graduate student, learning about plant and animal behavior; as a professor, studying behavioral ecology of soil, plants, herbivores, and humans; and as a dweller in the backwoods, reflecting on my visit to Earth.

Part one begins when I was young and spry. The chapters in this section recount experiences that taught me gradually to more deeply appreciate the magnificence of nature and the wisdom that all plants and animals possess, a wisdom intimately linked with natural systems and the process of life living by consuming and transforming.

Part two describes my many surprises learning about how the wisdom of the body guides creatures to select needed nutrients and medicines, given appropriate alternatives and experiences early in life, and the kinds of memories that underlie that wisdom. This section raises a question that I pursue throughout the book: What is this entity I call "me"?

Part three shows how a palate in harmony with a landscape evolves from three interrelated processes: flavor-feedback relationships (relationships of cells and organ systems and the microbiome with foods), availability of biochemically rich foods (landscape ecology), and culture (learning in utero and early in life to eat nourishing combinations of foods). To the degree that any one of these links is damaged or broken, societies become dysfunctional. When they are all broken, as they are now, entire ecosystems and cultures become dysfunctional.

Part four illustrates how contemporary culture – blinded by the influence of academic, corporate, and political "guidance" – has evolved in ways that are an affront to the wisdom of the body. Nobody has to tell a wild plant, bacteria, insect, fish, bird, or mammal what to eat to sustain health, how to self-medicate to recover from disease, or how to develop and reproduce. Ironically, people now must be told by "authorities" what to and what not to eat. Do humans lack the ability to identify and choose nourishing foods or has that ability been hijacked?

Part five begins nearer the end of my life. The chapters in this section reflect on my experiences of change, how plants and animals cope with change, and how contemporary *Homo sapiens*, for the most part, neither recognize nor embrace change. Rather, most human endeavors are failed attempts to control outcomes. Yet, despite our efforts, we never seem to be in control of the myriad events that affect our lives. In our attempts to influence outcomes, we change the playing field in ways no one anticipates. We don't know enough to foretell or govern our collective fates. This section offers an unflinching gaze at relentless change, existence in the process of consuming itself. Optimism comes not from clinging to things wild and free – everything that's transitory is but an illusion, and everything is transitory – but from lovingly embracing the moment with all its horrors, beauties, wonders, and deep mysteries.

Seventy years ago, Aldo Leopold began *A Sand County Almanac* thus: "There are some who can live without wild things, and some who cannot.

These essays are the delights and dilemmas of one who cannot." His book was a fervent account of how people have lost touch with nature and how our detachment is wreaking havoc on the landscapes we inhabit. Yet, despite Leopold's insights and eloquent pleas, people have not, for the most part, gotten back into step. Indeed, the changes mechanized humans have wrought since Leopold's death have been nothing short of breathtaking. We are all, in a sense, being consumed by changes we've wrought, most of which we didn't foresee. From antibiotic-resistant bacteria to herbicide-resistant weeds to unending wars to warming climates, these changes all seem to emerge unexpectedly. A mere 12,000 years before Leopold's visit to Earth, hunting by our ancestors helped lead to the extinction of forty species of mammals, and modern humans are continuing to do the same to less obvious small "cogs and wheels" that are the life of soil, plants, and animals.

Humans aren't alone in annihilating things wild and free. The cosmos and Mother Nature are participating through meteors and changing climates. Paleontologists tell us 99.9 percent of the species that have lived on Earth are now extinct. Only one in one thousand species that ever visited Earth (roughly 8.7 billion) is here today (estimated 8.7 million).[3] Nor do most species stay for long – less than ten million to twenty million years.[4] Until now, the ongoing sixth mass extinction, species were wiped out in the absence of *Homo sapiens*.[5] Mother Nature, it seems, is bent on consuming herself, from life on the planet to stars and galaxies in the cosmos, and now humans are participating. As "author," she is creativity in marvelous manifestations. As "destroyer," she abolishes these splendid displays to bring forth new forms: the one wax takes many molds. As "Concealer" and "Revealer," she beckons us to ponder the mysteries of transformation.

During our moment on Earth, we dine with change, we dance with uncertainty, and we fade into mystery. We now spend huge amounts of money to prolong life a few weeks or days. Yet, if we viewed death as birth, would we embrace life as ever germinating, flowering, and setting seed; dawn as beginnings teeming with opportunities; maturity as realized potential; twilight as new beginnings? Would we understand nothing arises alone or exists independently (emptiness is form), nor does anything persist in one form (form is emptiness) – creation and destruction being mirror images? Perhaps what we view as loss isn't loss at all, but being ever creating new forms, functions, and behaviors – from quarks to universes and everything in between.

This book, then, is about the delights, dilemmas, joys, sorrows, pains, qualms, uncertainties, and deep mysteries of the experience of relentless change. "The life and times of" is the experience of change, and it hurts – if we let it. In the grand scheme of things, life flourishes with change. The demise of dinosaurs ushered in the age of mammals. Birds, in all their splendor, are descendants of dinosaurs. While we savor the beauty of the moment and the radiance of existence in myriad forms, they are fleeting. Attempting to cling to anything is futile. Letting go is creative. The challenge is to avoid illusions of stability, permanence, and control and to embrace inexorable cycles of transformation. We do that by participating creatively in now.

In the many workshops I've done around the world over three decades, people often ask me: "How do I tell others about what we've discussed?" They laugh as they say, "I'm at a loss to capture in words the depth and breadth of what we've shared about life. I can't sum it all up in an entire conversation, let alone a few words." This book is my attempt to sum it all up – to link what we like to think we know of the physical with what we like to think we know about the metaphysical. If you are enchanted by a childlike sense of mystery and wonder, if you are willing to admit none of us knows much of anything, you may find this book will warm your thoughts and lighten your spirits as you sit wrapped in the glow of an evening sunset.

# PART I

# *Dining with Change*

## CHAPTER 1

# Goats, Rats, and Clara's Kids

On July 4, 1976, while most folks in the United States were celebrating the 200th anniversary of the founding of the nation, my wife and I were driving 400 miles across Utah, from Logan in the northeast to St. George in the southwest, so we could build a fence. From July through September, Sue and I constructed two miles of fence to form six pastures on Cactus Flat, a mesa twenty-eight miles northwest of St. George and roughly two miles from the small town of Gunlock. It was no vacation. I was a graduate student, and the paddocks would be the scene of a research study that would span the next five years of my life.

The cities and towns – Ogden, Salt Lake, Provo, Nephi, Fillmore, and Cedar City – we traveled through on our way to St. George were small back then. Outside of the settled areas, we traversed long rolling stretches of arid land clothed in gray-green sagebrush and dark green juniper. As we drove over mountains and across great valleys, I tried to imagine how the landscape in the Gunlock area would look. I knew a shrub called blackbrush would be common, because my research would involve goats grazing blackbrush. I'd seen samples of blackbrush mounted on herbarium sheets, but I wasn't sure what the shrub would look like in the landscape.

I began to get a sense of that as we descended from Cedar City (5,846 feet) to St. George (2,860 feet), and the landscape changed dramatically from anything I'd ever seen. The Pine Valley Mountain Range dominated the horizon to the west, with just barely visible canyonlands to the east (Zion National Park) and the south (Grand Canyon). The black lava–strewn landscapes were attired with a diverse mix of shrubs including banana yucca, bitterbrush, broom snakeweed, creosote bush, threadleaf (old man) sagebrush, indigo bush, desert peach, and Mormon tea. It was a strange, harsh-looking land, utterly different from the Rocky Mountains where I grew up. The vista was at once foreboding and enticing. I wondered where this adventure would lead us. In retrospect, I really didn't have a clue.

We lived in a trailer in what was then the small town of St. George, and our routine was the same six days each week for twelve weeks. At 5 a.m. each morning, we began our hour-long drive to the study site on Cactus Flat. At 6 p.m. each evening, we began our hour-long trip back to St. George. In the hours between, we dug holes – four-foot-deep holes for corner posts, which we fashioned from juniper trees. Because goats can go anywhere a rat can go, we used four-foot net wire fencing held erect by the steel line posts that we pounded at sixteen-foot intervals into the rock-hard caliche soil, all the time baking in the unrelenting rays of the desert sun, which typically elevated air temperatures to as much as 110°F. Nor did temperatures moderate much at night, as asphalt and concrete that baked during the day radiated heat into the trailer at night. Building the fence on Cactus Flat and living in St. George can best be described as "hell on Earth."

We used the pastures on Cactus Flat to evaluate the effectiveness of goats as mobile pruning machines to rejuvenate landscapes dominated by blackbrush. Our goal was to evaluate the effectiveness of goats browsing at different densities (fifteen goats per pasture in 2.5-acre, 5-acre, and 10-acre pastures) to improve the quality of foraging land in the future for both wildlife and cattle that spend the winter in those landscapes. In practice, this would then be done on unfenced land. And, indeed, many people, nowadays referred to as ecological doctors, are using shepherding with goats and sheep to rejuvenate landscapes. At the time we did the research, few people were contemplating using livestock for such purposes.

We built the fence so we could manage the goats' foraging behavior during winter and observe the results. When old twigs are removed by pruning (grazing) during winter, blackbrush produces a flush of new twigs the following spring. The new growth is much higher in energy, protein, and minerals than are the old twigs. So, we reasoned, goat browsing during winter would stimulate growth of nutritious young twigs, which provide better forage for mule deer, bighorn sheep, and cattle that spend their winters foraging on blackbrush.

Blackbrush is a small shrub, eighteen to thirty-six inches in height depending on where it grows, that dominates the transition zone between cold desert to the north and hot desert to the south, forming a narrow band across southern Utah and northern Arizona. Blackbrush, a member of the rose family and the only species in the monotypic genus *Coleogyne*, was named for its gray bark, which turns jet black when it gets wet during rain- or snow-storms. The gray branches are ornamented with small evergreen leaves. Blackbrush forms vast pure stands across the desert floor and on scrubby slopes, giving the landscape a uniform dark gray color. On Cactus Flat, blackbrush shares its home with juniper trees, whose gray bark and dark green leaves brighten an otherwise blackbrush gray backdrop with a savannah of juniper green trees.

Blackbrush and juniper provide forage and construction materials for desert woodrats, mammals that have long tails, large ears, and large black eyes. Woodrats live in houses they build from branches, twigs, sticks, and other debris. The huge, beaver-dam–shaped structures may be up to four feet across. They are usually constructed in a tree or on the ground at the base of a tree or rocky ledge. On Cactus Flat, woodrats use small twigs and leaves from blackbrush and juniper to fashion large mounds at the bases of juniper trees, with juniper bark as "siding" on the outside. They build tunnels to various rooms inside their houses, which provide shelter from extremes of desert temperatures in summer and winter and protection from predators. Primarily nocturnal and vegetarian, woodrats survive on a diet of cactus, yucca pods, bark, berries, pinyon nuts, seeds, and green vegetation. Woodrat houses, the new growth of blackbrush, and the goats would become my teachers as my research study commenced.

From July through September of 1976, while we were building the fence, Sue and I observed that a juniper tree in one of the pastures had been struck by lightning the previous summer. The blackbrush shrubs around the tree had produced prodigious amounts of new growth, and we were eager to see how vigorously the goats would consume those nutritious new twigs. During the winter of 1977, we had leased ninety Angora goats from the Navajo Nation, and we put fifteen goats in each of the six pastures we'd built. Nearly as soon as the study began, we watched the goats engage in two peculiar behaviors. The first occurred early in January, just a week after we'd moved the goats to their new homes on Cactus Flats.

Near dusk one evening, while the goats were actively foraging, we slowly herded them to the spot where the blackbrush shrubs encircled the old, dead juniper. We fully expected them to chow down on the nutritious new twigs. But to our amazement, only one goat sampled the new twigs, taking one small bite. The goats simply stood there, silently gazing at us. After a while, they walked away from the dead juniper and began once again to forage enthusiastically on older twigs on other blackbrush shrubs in their pasture. I was shocked. I wondered what the goats were thinking as they watched us watching them refuse to eat the new twigs. I even tasted one of the new twigs. I remember vividly that it didn't taste bad at all!

I also recalled the words of one of my mentors: "You should never study an animal that's smarter than you are, and *you* shouldn't be studying goats." And although my mentor and I didn't know it then, the same advice was true for shrubs. At that time, ecologists were in the early stages of developing the field of chemical ecology, the study of how plants chemically mediate relationships among soil, other plants, herbivores, and human beings. Blackbrush and many other shrubs are quite good at organic chemistry and much "smarter" than I could have imagined. During the next twenty years, I learned many lessons from goats and blackbrush, lessons that transcend why goats don't like to eat the young twigs that blackbrush creates.

Their behavior that evening was not an anomaly, as we learned by observing the Angora goats through that winter and Spanish goats during the ensuing two winters. When goats have a choice between new growth and older growth, most, though not all, strongly prefer the older growth. That doesn't make any sense in the expected framework of nutrition, because the new growth is much higher in energy, protein, and minerals than the older growth.

When I told a professor of toxicology that goats prefer less nutritious older twigs over much more nutritious new twigs from blackbrush, he remarked, "That just goes to show animals lack nutritional wisdom." His comments reflected the sentiment at that time, a view sadly still in vogue today, that the animals in our care – and human beings, for that matter – seemingly lack nutritional wisdom. I didn't believe him, though I'm not sure why, and I could muster no argument to counter what he was saying. I didn't know how to prove the body of a goat, or any creature, has nutritional wisdom. People thought then, and still do now, that domesticated animals and humans simply eat foods that taste "good" and avoid those that taste "bad."

Later in January, I observed the second surprising goat feeding behavior, one with another seemingly inexplicable cause. Overall that winter, the goats didn't fare well due to the poor nutritional quality of blackbrush and their lack of familiarity with the blackbrush-dominated landscape. The goats lost weight during the three-month trial, but the amount of weight loss wasn't consistent among pastures. Goats in the 2.5- and 10-acre pastures lost more weight (more than 20 percent of total body weight) than in the 5-acre pastures (less than 10 percent).

Why the discrepancy? Had I not been living with and closely observing the goats, I might have simply called it an outlier in the data, not unlike many unexplained anomalies that increase "error" variation. However, I had noticed that goats in one of the five-acre pastures had learned to eat woodrat houses! I couldn't make sense of it because woodrat houses appear to be even less nutritious than the old, woody blackbrush twigs. But those goats lost less weight than their fellow goats that didn't partake of woodrat abodes. Another intriguing and significant detail: During all three winters of the study, out of a total of eighteen groups of goats, only that one group of goats ever learned to use woodrat houses as a source of nourishment.

Over the years, I've come to realize that the outliers – not the averages – are the most interesting part of research. The quirky behaviors of individuals interacting with each other and the environments they inhabit are how nature creates new relationships – genes expressed in organisms in uniquely evolving biophysical environments merging with chance occurrences.

But during the winter of 1977, I hadn't figured that out, and I was puzzled and intrigued. The goats were snubbing a food we thought was good for them – the new and presumably more nutritious twigs. And one group of goats was eating a "food" item – woodrat houses – that in no way resembled any forage I'd seen. We speculated, correctly, about what was happening in both cases that winter, but it would take me several years to demonstrate to skeptical scientists why goats were making these food choices. Those early observations marked the beginning of a forty-year exploration of the nutritional wisdom of wild and domestic grazing animals, and my attempts to reconcile findings from our carefully controlled studies with findings from other animals and humans. In retrospect, my PhD project initially appeared to me to be a boring study. What could be less interesting, for a wildlife biologist fascinated by wild animals and plant diversity, than domestic goats foraging on a monoculture of blackbrush? But it turned out to be a controlled field study with fascinating implications. I've often reflected on how I'd have never learned what I did about foraging behavior if I'd been able to do the study I would have preferred back then – mountain goats foraging on botanically diverse landscapes in the Rocky Mountains.

## Inept Herbivores?

Most nutritionists believe wild animals select diets to meet their nutritional needs but contend domestic animals lack such nutritional wisdom.

This belief came about during the 1960s and 1970s from studies in which researchers offered livestock a choice of different minerals or vitamins, and those animals didn't select the nutrients researchers thought they should.[1]

For example, when lactating dairy cows were offered a dicalcium phosphate supplement, they didn't "instinctively" eat recommended levels of calcium or phosphorus. Many animals never approached the source of calcium, even though they were calcium deficient; among those that did, intakes of calcium varied greatly. Some animals ate large amounts of dicalcium phosphate, even though researchers thought they didn't need either calcium or phosphorus. These results were consistent with earlier findings that sheep didn't rectify a phosphorus deficit by consuming supplemental dicalcium phosphate. And in a study that spanned a period of many weeks, dairy cows did not consistently select appropriate minerals or vitamins. Nor did lambs eat sufficient amounts of some minerals, and they overconsumed other minerals. The cumulative results of such studies led researchers to begin recommending feeding mineral mixes specifically designed to meet needs of the "average individual" for various classes of livestock.

Collectively, these studies led to the belief that domestication created animals more responsive to food flavor than to nutritive value. Scientists also hypothesized that acceptability of food flavors, rather than selective appetites or cravings for specific minerals and vitamins, influences food selection. In other words, they believed the process of domestication of livestock had removed their nutritional wisdom and inborn ability to select needed nutrients, a trait that from eons past to this day has enabled wild herbivores to eat appropriate forages and prosper.

These studies stand in surprising contrast to the findings of range and wildlife scientists who studied the foraging behavior of free-ranging animals over several decades. They did two types of studies. Some researchers followed individual animals as they roamed and grazed, attempting to collect (by handpicking) a representation of all the food items they nibbled. From their painstaking observations and laboratory analyses, scientists learned a great deal about the botanical and chemical composition of herbivore diets.

In the other type of study, veterinarians established esophageal fistulae (artificial passages) in the animals that allowed researchers to collect each bite of forage animals actually ate. A comparison of the results of these two types of studies showed that no human can pluck by hand a diet as nutritious as that actually selected by free-ranging goats, sheep, or cows. Both types of studies showed herbivores often eat twenty-five to fifty or more foods daily, though a handful of species typically make up the bulk of any one meal. What they choose changes from meal to meal, day to day, and season to season.

Collectively, these findings, which didn't fit the notion that herbivores are unable to select a nutritious diet, raised two questions. Why did livestock seem so incompetent in "nutritional wisdom trials" in confinement, but so adept at selecting a nutritious diet under free-ranging conditions with a smorgasbord of up to 100 plants from which to choose? By eating so many kinds of plants, are free-ranging animals able to select a nutritious diet merely by chance?

## Clever Rats

While observations of foraging by herbivores raised questions about nutritional wisdom, studies of rats suggested that palatability might link feeding behavior with nutritional needs. Unlike livestock, rats chose a balanced diet when allowed to self-select, even in laboratory conditions and when facing various nutritional challenges. For example, rats whose adrenal glands are removed can't retain salt, so they die due to salt deprivation within two weeks if fed a low-salt diet. However, if they are offered salt free-choice or given access to salt water and pure water, they choose either to eat salt or drink salt water and by so doing, keep themselves alive.[2]

Equally intriguing, rats that have their parathyroid glands removed die within days due to tetany, a result of calcium deficiency. Given a chance, though, rats that don't have parathyroid glands prefer to drink a solution of calcium lactate rather than water, which keeps them free of tetany and thriving.[3] (The major function of the parathyroid glands is to maintain calcium levels within a narrow range to enable nervous and muscular systems to function properly.)

Furthermore, rats fed diets deficient in various essential amino acids detect the deficiency within minutes. Given the chance, they select food that contains the specific amino acid they are lacking and thereby retain the ability to create proteins required for their proper form and function.[4] Amino acids, the building blocks of proteins, cannot be created from other compounds by the body, and the nine essential amino acids must be obtained from the foods animals eat.

Finally, rats can be rendered diabetic with an injection of streptozotocin, a compound that causes them to display all the symptoms of diabetes, including polyuria, polydipsia, and glycosuria as well as elevated fasting plasma glucose levels and glucose intolerance. Diabetic rats allowed a choice consume more protein and less carbohydrate than nondiabetic rats.[5] As a result, they lose their symptoms of diabetes: Blood sugar levels return to normal, they gain weight, they eat less food, and they drink normal amounts of water. The same happens when diabetic humans increase their intake of protein and reduce carbohydrates in their diet.[6]

Unlike with livestock, these and many other studies suggest domestication didn't do away with the nutritional wisdom of laboratory rats.[7] How were rats able to select nutritious diets when challenged with a range of deficiencies, from minerals (calcium and sodium) to amino acids to diabetes? Are human beings more akin to livestock or rats in confinement?

## Clara's Kids

In June of 1939, Clara Davis, a pediatrician from Chicago, told attendees of the 70th annual meeting of the Canadian Medical Association about findings from what is the world's longest and most involved dietary experiment on self-selection in human beings. In a six-year study, Davis became the "mother" of fifteen infants who had been put in an orphanage at ages that ranged from six to eleven months.[8] She chose infants because they had not eaten adult foods previously nor were they influenced by beliefs of older people. Davis stressed they were without preconceived prejudices and biases regarding the foods she offered. Her studies were destined to become a citation classic in the *Canadian Medical Association Journal* and a basis for argument, discussion, and reinterpretation for anyone hoping to untangle perplexing questions about how children's appetites, food choices, and health intersect.[9]

During her studies, the children selected nutritious diets when offered a wide range of foods of animal and vegetable origins. The thirty-four foods Davis offered the kids, procured fresh in the market, jointly provided requisite fats, carbohydrates, amino acids, minerals, and vitamins. Even so, the children easily could have become deficient by selecting the wrong combinations of foods. As she emphasized to her Quebec audience, no adult was allowed even to hint to the children what might be a proper choice or portion amount. The children typically ate several foods and a beverage at any one meal, including brains, raw beef, bone jelly, and bone marrow – foods that are repulsive to many adults. No two children ever selected the same foods and no child ever selected the same mix of foods from day to day. Nevertheless, their fervent individuality fashioned fifteen uniformly well-nourished, healthy children, as attested to by attending pediatricians. As Dr. Joseph Brennemann noted in an article in the *Journal of Pediatrics*, "I saw them on a number of occasions and they were the finest group of specimens from the physical and behaviour standpoint that I have even seen in children of that age."[10]

Despite these findings, many scientists don't believe humans have nutritional wisdom.[11] They cite as evidence the obesity crisis. The Centers for Disease Control and Prevention (CDC) projects that 70 percent of people in the United States today will die of diet-related diseases. Humans ostensibly can't do what wild or free-ranging domestic herbivores do without a bit of advice from dieticians. Perhaps that's why some people write – and other people read – an endless stream of articles and books that tell us what and how to eat to stay well. Now, for a few hundred dollars, you can even get personalized nutrition recommendations.[12]

The foods Davis offered included: four fruits (apples, bananas, peaches, pineapples); ten vegetables (beets, cabbage, carrots, cauliflower, lettuce, peas, potatoes, tomatoes, spinach, turnips); five grains/grain products (barley, wheat, cornmeal, oatmeal, Ry-Krisp); six red meats (beef, lamb, liver, kidneys, brains, sweetbreads); one white meat (chicken); one fish (haddock); one high-fat food (bone marrow); bone jelly (rich in collagen, gelatin, amino acids, minerals, glucosamine, chondroitin, hyaluronic acid, and fats); sea salt; and four drinks (water, orange juice, sweet milk, sour milk). Today, depending on which authority one asks, many of these foods are thought to lead to diet-related illnesses including cardiovascular disease and cancer.

Why do humans like foods that are thought to be "bad for us" such as red meat and sugary, salty, fatty foods? Why do so many humans dislike fruits and vegetables, which are reputed to be so good for us? Why did each of Davis's kids eat such different combinations of foods, and why did each child select different foods from meal to meal? Davis was concerned about culturally inherited food prejudices and biases: Have we become so maladapted culturally that we no longer know how to enable the nutritional wisdom that resides within our bodies or those of the animals in our care? Can herbivores help us rediscover nutritional wisdom?

## CHAPTER 2

# Challenges for Guests

Though I'd been raised with my mother's vegetable and flower gardens, and I'd worked in a greenhouse in high school, I'd never really "seen" a plant until my sophomore year in college when I learned to identify native plants. That opened my eyes to a world I'd never appreciated. Astonishingly, to identify plants, I first had to memorize a book that described their many and varied leaf sizes, shapes, and placements on stems. Their flower parts are even more complex – from stamens and pistils to brightly colored sepals and petals to calyces and corollas. With that knowledge I could navigate the dreaded dichotomous keys – detailed descriptions of the thousands of different species that live in Colorado. For the class, I had to collect and identify fifty species and mount each one on herbarium paper. My initial "victim" was pasque flower, one of the earliest plants to appear in spring; its scientific name is *Pulsatilla ludoviciana*.

That June I returned home to the mountains intending to fish one of my favorite streams, a brook so clear I could see every move a brown trout made as it rose to a Royal Coachman or Rio Grande King fluttering over riffles and into pools. But my lifelong fascination with fish was overcome by the brilliant displays of flowers along the stream and in the meadows. There were universes of dazzling colors – blue-eyed grass, purple and red penstemons, magenta shooting stars and geraniums, silky blue gentians, bright yellow sunflowers – and although I'd walked this stream for years, I'd never experienced the vivid presence of the plants until now.

That summer, Sue and I began collecting plant specimens. Since then, we've spent hundreds of hours collecting and identifying plants. We created a herbarium of more than 500 species, each one a work of art carefully arranged on herbarium sheets, each prepared with love and respect for the beautiful friends we were making. Even today, any time we drive a country road, stroll through a vacant lot, hike across a plain, fish a stream, or climb to a high-mountain lake, we are enchanted by what's growing and why it's there.

Plants tell stories about the relationships of herbivores, omnivores, and carnivores with landscapes. Plants are the glue that links soil with herbivores, omnivores, and carnivores – below and aboveground. Land isn't merely a network of soil, plants, and animals, though, but a cascade of energy flowing from the sun through plants into soil and animals below- and aboveground. A species is a strand in a web, linked with millions of other strands.

Plants are also the founders of the feast – all creatures ultimately eat plants. So no discourse on nutritional wisdom can be complete without considering not only how animals eat but also how plants procure the sustenance they need. And beyond that, we must also consider: How do plants manage to provide for the needs of animals and also sustain themselves?

## Challenges for Plants: The First Biochemists

Organic chemists study the structure, properties, and reactions of organic compounds – gaseous, liquid, and solid chemicals that contain carbon. Organic chemistry is the heart of materials science, organometallic and polymer chemistry, as well as the basis of biochemistry, nutrition, and medicine. Organic chemists make plastics, petrochemicals, and explosives, and from plants and plant extracts, they fashion everything from foodstuffs to medicines.

My old friend pasque flower is an organic chemist. These modest plants appear in spring, soon after snowmelt, along the edges of conifer forests. Their large flowers have colored bracts (modified leaves) that give the impression of petals. The entire plant is covered with silky hairs that give the ripe fruit the appearance of a mop head. Gazing upon these handsome flowers provides no clue of the intricacy of their inner workings, but a hint of those goings-on can be gained from considering the many ways humans use pasque flower.

Fresh plant parts of various species of pasque flower are toxic if eaten or even applied to the skin. Dried plants, on the other hand, are used medicinally in Europe and North America to treat pain during menstruation and other gynecological disorders, skin diseases, asthma, and eye infections. It is also useful as a diuretic and as an expectorant to help bring up mucus and other material from the lungs and airways. Pasque flower is widely used in homeopathic preparations, was once considered effective specifically for measles, and was also used for toothache, earache, and indigestion. Chemical analyses by humans confirm pasque flower's prowess as an organic chemist: The plant contains compounds that have antibacterial, antimalarial, and antifungal activity, and they have cytotoxic effects as well. What we don't know is, do pasque flowers understand the multifaceted benefits of these compounds they produce?

**Plant intelligence.** According to Aristotle, plants differ from animals because animals have sense perception and plants do not. Animals differ from humans, he proposed, because humans can reason, while nonhuman animals are utterly without reason, ruled merely by instincts. Thus, he reasoned, it is appropriate that animals should be used for human purposes.

Like Aristotle, mathematician and philosopher René Descartes conceived a split between humans and the natural world, which resulted in a mechanical view of nature. By this, Descartes didn't mean that plants and animals are machine-like. Rather, they are machines that react instinctively to every facet of the environment. To Descartes's way of thinking, plants and animals are unable to learn, they lack emotions, and they lack consciousness. The ability to communicate through language and to reason is what separates plants and animals from humans.

Humans describe "intelligence" in terms of our experiences of the world, not in terms of the experiences of other creatures, which we typically consider to be less intelligent. We use sophisticated instruments to measure the form (morphology), function (physiology), and behavior of plants. But no one knows what it is like to *be* a plant. Rather, only with great effort do we labor to understand intellectually the inner and outer workings of the plant world.

When chemists learn about and create organic compounds, we consider that an act of high intelligence, as any student who has taken organic chemistry will attest. What should we think, then, when a pasque flower fashions complex kinds and mixtures of compounds? What kind of intelligence is being manifest as plants create complicated and dynamic interrelationships with soil, other plants, and animals? "Dumb" weeds consistently outmaneuver clever biochemists and agribusiness, university, and farm folks who have attempted to eradicate them with chemicals, as nearly 500 herbicide-resistant weeds worldwide can attest.[1]

**Primary roles for secondary compounds.** Research during the nineteenth and twentieth centuries identified the *primary compounds* – energy, proteins, minerals, and vitamins – that plants create and disclosed their functions in growth and reproduction. During that time, researchers also identified a host of *secondary compounds* in plants, but the roles these compounds played were unknown. Some even considered them waste products of plant metabolism.

Nevertheless, studies of the past fifty years have highlighted the diverse value to plants of secondary compounds such as phenolics, terpenes, and alkaloids. Far from being waste products of plant metabolism, secondary compounds play primary roles in mediating ecological interactions. They help plants defend their turf by inhibiting survival of other plant species. They increase drought tolerance, pest resistance, tiller numbers and biomass, seed mass and numbers, and germination rates in plants. Some, like lignins and tannins, help to build organic matter and humus in soil. They act as antioxidants to protect plants from oxygen free radicals produced in photosynthesis, and they color flowers and flavor fruits to attract pollinators and fruit eaters. Still others help plants recover from injury and enhance regrowth after grazing.

Secondary compounds reduce loss of plant tissues to bacteria, fungi, insects, birds, and mammals by limiting how much of each plant an animal can eat. By limiting intake of any one plant, secondary compounds encourage animals to eat a variety of plants and to forage in a variety of different places. In so doing, secondary compounds protect plants from overuse and spread the load of herbivory across many different species in a community of plants, and they enhance the health of herbivores, omnivores, and carnivores.

**The senses of plants.** Over thirty years ago, Nobel Prize–winning botanist Barbara McClintock issued this challenge: "A goal for the future would be to determine the extent of knowledge the cell has of itself and how it uses that knowledge in a thoughtful manner when challenged."[2] In *Plant Behaviour and Intelligence*, plant physiologist and molecular biologist Anthony Trewavas develops a rich synthesis to address McClinock's challenge. With a thorough review of top-notch science, Trewavas establishes that cells and plants are conscious – they have knowledge of themselves. Cells assess environmental challenges that cause them to change form, function, and behavior. Learning and memory are vital as cells interact with one another to address environmental challenges. The net result is that the cells and organ systems of plants behave in ways that manifest environmental assessment and intelligent behavior.

Most folks can relate to the senses and behaviors of animals. Like us, animals can see, hear, touch, taste, smell, breathe, and even talk. They walk, run, jump, explore, play, and have mates. It's not obvious that plants do any of these things. But that appearance is due to our lack of ability, until recently, to properly observe plants. You may be surprised to learn that plants have as many as twenty senses, including analogs to all of our human senses, as biologist Daniel Chamovitz, dean of the George S. Wise Faculty and founder of the Manna Center Program in Food Safety and Security at Tel Aviv University, conveys in *What a Plant Knows*.

Plants "see" different wavelengths of light, which they capture in photosynthesis. As part of that process, plants "breathe" through stomata on the surface of leaves and stems. They open stomates to inhale carbon dioxide – they use carbon atoms to build primary and secondary compounds – and they exhale oxygen. Plants respond to tactile cues: Vines and roots know when they encounter objects in soil, including their own shoots and roots or those of other plants.

Roots sense and explore the environment below ground. In addition to being aware of pressure, moisture, light, gravity, volume, and hardness, roots can sense nutrients such as nitrogen, phosphorus, salt, and various secondary compounds. Roots seek water and nutrients, and they avoid potentially harmful secondary compounds produced by other plants. Roots communicate by chemical signals, which they and nearby plants produce. Roots make choices regarding food and habitat selection. They know which nutrients they need and where and how to get them below ground. Plants share resources through roots.[3] Many plants organize themselves into intricate networks using an underground web of mycorrhizal fungi that link roots of different species, enabling trees and fungi to nurture one another in symbiotic relationships.

In native communities, plants release up to 30 percent or more of the energy (carbohydrate) they sequester through photosynthesis into the soil. They do so in mutualistic associations with algae, bacteria, fungi, protozoa, mites, microarthropods, nematodes, earthworms, millipedes, ants, and other insects.[4] Plants provide energy for organisms, which in turn provide minerals for plants. Scientists once believed plants released microbe-attracting exudates passively, but plants can send chemical signals to attract bacteria and fungi that solubilize the particular mineral(s) the plant needs from the subsoil. Minerals locked up in the subsoil are thus made available through biologically and biochemically complex interrelationships that are the basis for health of soil, plants, herbivores, omnivores, and carnivores.

The language of plants is organic chemistry. Each of the estimated 400,000 species of plants on Earth can synthesize hundreds to thousands of primary and secondary compounds. Biochemical themes are common within a species, but individual plants create variations within a theme to communicate with other plants and animals. Even from an "alphabet" of as few as twenty compounds, a plant can create trillions of different "words" by varying the relative amounts of different primary and secondary compounds. That's how plants tell herbivores and omnivores when and where to touch or eat them: Think, for instance, of poison ivy or unripe versus ripe apples, bananas, strawberries, or tomatoes. Volatile compounds reveal the status of a plant, including warning other plants of would-be attackers, a message that causes undefended plants to increase their levels of defense. Plants can "smell" volatile compounds in the air, and they can "taste" them on their tissues. When attacked by caterpillars, some plants produce volatile compounds that attract "bodyguards" in the form of predatory insects such as parasitic wasps and dragonflies that delight in dining on the caterpillars that are dining on the plant.[5]

**Surviving attack.** One capacity plants don't have is the ability to run away to escape being eaten. Instead, plants have defensive mechanisms that help them survive in the face of herbivory. While some plants invest constantly in defense (constitutive defenses), others wait to initiate defense responses (induced responses) until they receive a message from a neighbor or until just after an attack. Plants also use *priming*, the term for a kind of learning and memory that involves physiological processes in which plants prepare to more quickly or aggressively respond to stressors based on past experiences.[6] Priming is a resource-saving mechanism that allows plants to hold off on fully implementing a defensive response until the response is needed.

Induced responses and priming require surveillance systems and rapid communication among plant organs. Although they lack a nervous system, plants such as Venus flytrap (*Dionaea muscipula*) and mimosa (*Mimosa pudica*) generate leaf movements using electrical signals that can travel at a rate of over an inch per second, equivalent to rates detected in the nervous system of mussels but much slower than the nearly 300 feet per second in mammals. When researchers placed Egyptian cotton leafworm larvae on thale cress leaves and recorded changes in electrical potentials using electrodes, leaf-surface potential didn't change when a larva walked on a leaf, but as soon as the leafworm began to feed, electrical signals evoked near the site of attack spread to neighboring leaves at a rate of four inches per minute.[7] Upon receipt of an electrical signal, tissues remote from the site of feeding rapidly accumulate the plant hormone jasmonate, triggering a chemical defense response.[8] The electrical transmission is through glutamate receptor–like proteins similar to what occurs in the brains of animals.[9] These and other findings imply that glutamate receptor–type proteins existed before animals and plants diverged, generating long-distance warning signals to elicit timely initiation of protective responses.

**Budgeting for survival.** While goats were teaching me about foraging, blackbrush was schooling me in the ways plants allocate resources to protect themselves under the harsh conditions of the desert Southwest. When resources are scarce, plants invest heavily in making secondary compounds to protect their tissues from being eaten. Regrowth is difficult when water and nutrients are in short supply.[10] High levels of tannins in new twigs are how blackbrush tells goats not to eat the new twigs. Younger plant parts are more nutritious than older plant parts, so they are more likely to be eaten. Because of this, younger plant parts are heavily defended with higher levels of secondary compounds.[11] These cost-benefit analyses also take into account the fact that younger tissues are more valuable than older tissues and the appreciation that secondary compounds are costly for a plant to synthesize, transport, store, and maintain.[12]

As with young and older *plant parts*, more vulnerable and valuable *juvenile plants* often are more heavily defended than *mature plants*. When healthy adults are pruned, they revert to a juvenile growth stage and again allocate more resources for defense.[13] In the shrub sagebrush, young, leafy, actively growing, nutrient-rich plants appear to be the best food for herbivores. But herbivores prefer mature plants due to the high levels of terpenes in juvenile plants.

Such cost-benefit analyses also apply to roots, leaves, and flowers. The roots of wild parsnip are unlikely to be eaten, so they have low levels of furanocoumarin, which is produced (induced) only when roots are eaten.[14] Leaves, which have a high chance of attack, have intermediate levels of constitutive and inducible furanocoumarin defense. Flowers have the highest levels of this compound, which is a permanent (constitutive) part of flowers and seeds.

Some plants reduce costs of defense by increasing production of secondary compounds only when they are attacked.[15] These induced responses can occur quickly, followed by relaxation over days or weeks. Leaves of aspens produce phenol glycosides within *minutes* of being eaten.[16] Wild tobacco begins to allocate more resources to producing alkaloids and less to growing and producing seeds within hours of being eaten by an herbivore.[17]

Commonly within populations of plants, some *morphs* (individuals) will allocate more resources to defense than to growth and reproduction.

When plants are under attack by pathogens or herbivores, those morphs have higher survival than morphs that allocate more resources to grow and reproduce. For example, heavy browsing by Sitka black-tailed deer on the islands of Haida Gwaii off the coast of British Columbia kills western red cedar plants with lower concentrations of chemical defenses. That leads to domination by western red cedar plants with high concentrations of defenses that protect them from browsing.[18]

## Challenges for Herbivores: The Next Biochemists

Their behaviors may appear to be little more than idle wanderings in search of food and a place to rest, but the challenges herbivores face are far more complex than we might imagine observing a tranquil scene of cows or bison foraging in green meadows. Herbivores must sort through hundreds of species of grasses, forbs, shrubs, and trees, each physically and biochemically unique.[19] In the process, they ingest thousands of compounds that interact with one another and with cells in complex and potentially beneficial or harmful ways. The outcome of any meal depends on an animal's age, physiological state, past experiences with a plant, and the mix and sequence in which an herbivore eats various plants in a meal. These conditions change for plants and herbivores from meal to meal, day to day, and season to season.

**Detecting nutrients.** An animal's nutritional needs vary with age and physical activity. They change throughout pregnancy. They increase when animals are infected with parasites, when they're ill, or with shifts in physical activity or a change in the weather. To meet their fluctuating needs, herbivores must sift through a biochemically dynamic foodscape. Plants constantly alter the quantity of energy, protein, minerals, and vitamins. Individuals must maneuver through these challenges, recognizing nutritional needs in themselves. Individuals who do, survive. Those who don't, won't. Can animals detect nutrients in foods, and if so, how?

**Minimizing ingestion of toxins.** As described earlier in the chapter, plants also pose a toxin challenge to herbivores because of the secondary compounds they contain, often in high concentrations. The thousands of secondary compounds each vary in biochemical structures and activities. In animals, they interfere with metabolic processes or reduce digestibility of plants. They can also cause death. Do herbivores know which plants contain which secondary compounds, and can they use secondary compounds prophylactically (in a preventive way to maintain health) and therapeutically (to self-medicate when they are ill) to enhance their health?

**Physical attributes of plants.** Herbivores also must deal with plant morphological characteristics, such as standing dead material in some grasses, spines on forbs, thorns on woody plants, and differences in plant canopy shape and structure. As plants mature, physical attributes that make foraging difficult increase while nutrients decline. Any combination of plant physical and nutritional characteristics that optimizes nutrient intake is likely to be preferred. Animals that can navigate structural and biochemical challenges can enhance their nutritional welfare. How effective are herbivores at coping with morphological defenses?

**Food on the move.** Perhaps the trickiest challenge for herbivores is the interplay of variation in nutrients, secondary compounds, and physical defenses. Primary and secondary compounds in plants growing in natural landscapes vary from morning to night, from day to day, from season to season, and from place to place. An animal's challenge is to track phytochemical changes as they occur. Can herbivores figure out where and when to dine to meet their needs for nutrients and avoid overingesting secondary compounds? If so, how do they do that?

**Animals on the move.** Changes in terrain pose another challenge. Unlike their wild counterparts, livestock often are conceived in one locale, born in another, and then moved one or more times to yet other unfamiliar settings. These practices sever transgenerational linkages with landscapes, which explains why domestic and wild herbivores placed in unfamiliar environments often suffer from predation, malnutrition, overingestion of poisonous plants, and poor reproductive performance. Change may also come about because of calamitous events such as floods and fires. Herbivores that adjust quickly to new or changed terrain can reduce nutritional stress and increase chances for survival. How well do animals adjust to new foraging environments? Do they know instinctively what to do in unfamiliar haunts?

## Challenges for Humans: Biochemistry Run Amok

For herbivores, foraging challenges come from the complex and ever-changing nature of the foodscapes nature provides. Far from a mundane activity, foraging provides insights into an age-old dilemma faced by herbivores and humans alike: How do creatures of habit survive in a world whose only habit is change? Humans still face challenges of detecting nutrients and avoiding toxins, though the forms these entities take nowadays differ from those herbivores and our ancestors faced. As humanity morphed from hunter-gatherers into industrial-scale farmers and food manufacturers, we created challenges that are different, and in many ways more multifaceted and deceptive, from those faced by herbivores or our ancestors.

**Finding nutritious foods.** Agricultural practices increased yields of crops two- to three-fold in the past two centuries. Nevertheless, high yields came at the expense of phytochemical richness, which has declined 5 to 40 percent in forty-three fruits, vegetables, and grains in the past forty years.[20] Food quality has declined for four reasons. First, plant breeders favored quantity over quality, inadvertently selecting for varieties that are less phytochemically rich than their wild ancestors. In addition, irrigating and fertilizing with off-farm sources of nitrogen, phosphorus, and potassium increase growth at the expense of phytochemical richness. Third, much produce is picked green and shipped, rather than ripened to full phytochemical richness on the vine. Finally, as atmospheric carbon dioxide increases, nitrogen (protein) concentrations decline in a wide range of plant species.[21] Lower levels of protein have been observed in leaves, stems, roots, tubers, seeds, and grains and are correlated with negative effects on human nutrition worldwide.[22] Elevated $CO_2$ is also associated with decreases in zinc and iron in grasses and legumes.

Changes in $CO_2$ can adversely affect nutrition of humans, herbivores, and insect pollinators. For humans, increasing carbohydrate while reducing protein increases risk of hypertension, lipid disorders, and coronary heart disease.[23] For herbivores, increasing levels of $CO_2$ are reducing nitrogen content and digestibility of plants in mixed grass prairie, resulting in substantial and persistent declines in forage quality.[24] Such shifts may adversely affect the rate at which herbivores gain weight in the growing season as documented in the largest remaining rangeland ecosystem in North America. For pollinators such as bees, increasing atmospheric $CO_2$ is also reducing the protein content of floral pollen essential for their nutrition and health.[25]

The challenge in agriculture is to create a balance between production (quantity) and phytochemical richness (quality). Primary and secondary compounds often increase when plants are stressed, due to less nutrients or water.[26] Conversely, these compounds often decrease when plants are supplied with growth-promoting nutrients, $CO_2$, and water. Declines in both primary and secondary compounds occur as a result of altering patterns of energy and nutrient allocation to growth and phytochemical dilution due to the increased growth. Dilution of phytochemicals due to growth is one reason most folks aren't keen to eat vegetables or fruits. In 2013, few adults in the United States met recommendations for eating vegetables (8.9 percent) or fruits (13.1 percent). Most children don't like fruits or vegetables due to lack of flavors and of experience.

Intake of vegetables can be increased long-term by repeatedly exposing toddlers and preschool children to vegetables.[27] Children's interest in eating vegetables also can be increased by hiding vegetables in prepared dishes or offering a variety of different vegetables in a meal.[28] Adding herbs and spices to vegetables also increases liking and preference for vegetables among rural high school students.[29] Coating vegetables with a familiar sweet flavor like glucose or sucrose is another way to increase kids' liking for vegetables.[30] By adding a familiar flavor to a novel food, so-called *flavor-flavor conditioning* increases liking for the unfamiliar food.

The need to hide vegetables in dishes, offer multiple vegetables, or add flavors to vegetables could be eliminated if plant breeders increased phytochemical richness / flavor of produce. In so doing, flavor once again would be functionally linked with nutritional quality. While changing farming practices can also enhance flavor, phytochemical richness, and health benefits of crops, altering practices alone doesn't ensure that fruits and vegetables are flavorful.[31]

Compared with varieties that predate the post–Second World War period of intensive breeding, modern cultivars have fewer phytochemicals that contribute to flavor. Heirloom varieties rich in phytochemicals taste better than many modern varieties.[32] In tomatoes, levels of glucose, fructose, citrate, and malate can vary several-fold, while concentrations of the more than 400 volatile compounds that affect flavor can vary thousand-fold or more among conventional and heirloom varieties. The lack of flavor-enhancing volatile compounds in contemporary varieties is one reason for the general public's increased dislike of eating fruits and vegetables; other factors contribute, too.

Freshness is also critical for phytochemical complexity and rich flavors. Fruits picked green, prior to fully ripening on the planet, have lower levels of phytochemicals that create rich flavors and robust nutrition. Flavors and phytochemicals decline after harvest, more rapidly in some fruits and vegetables than others. For example, by the time broccoli is purchased at a store, typically ten days to two weeks since harvest, it loses more than 75 to 80 percent of health-promoting phytochemicals, 50 percent of its vitamin C, and most of its sugars and antioxidants. Loss of sugars, due to respiration after picking, reduces palatability.

Volatile compounds synergize with sugar to enhance flavor in sixty-six heirloom varieties.[33] While some abundant volatiles have little impact on flavor, other less abundant volatiles distinctly improve flavor. Still other volatiles contribute to *perceived sweetness*. For instance, the tomato variety Matina is perceived as roughly *twice as sweet* as Yellow Jelly Bean, yet Matina actually has *much less* sugar. Each of seven volatiles that contribute to sweetness is at least twice as abundant in Matina as in Yellow Jelly Bean. The same is true for fruity volatiles such as citral, amyl acetate (banana), strawberry, peach, raspberry, passion fruit, and lychee.

"Complex flavor" and "sweetness" are highly rated as favorable characteristics of the ideal strawberry and tomato.[34] Ironically, in current large-scale production systems, growers are paid for yield and appearance, not for the complex flavors that come from phytochemical richness. In most cases with tomatoes, phytochemical richness and higher sugar content are associated with reduced fruit size, but consumers are willing to purchase minimally smaller fruit with superior taste.

**Challenged by overeating.** Unfortunately, as farming and plant selection practices were making the flavors of fruits and vegetables less appetizing, processed foods were being made tastier. People in the food industry learned how to combine synthetic flavors with fats and refined carbohydrates, rich sources of energy that a body requires in large amounts. Differences in flavors are distinct enough to give consumers a false sense of variety, which stimulates food intake, despite the genuine nutritional monotony. Refined, energy-rich compounds create strong preferences for processed foods, but delayed aversive effects lead to costly diseases including obesity, diabetes, heart disease, and cancer. People now crave foods and drinks with high fat and sugar, not rich arrays of phytochemicals that nourish our bodies.

Increasing intake of refined carbohydrates relative to fat began in the early 1980s in the United States, coincident with the obesity crisis.[35] We now consume an extra 500 calories each day from refined carbohydrates. The change has occurred in peoples worldwide, including the North American Indians. In the 1700s, the Pima Indians didn't suffer from obesity or diabetes.

Their diet was corn and beans, grown on fields irrigated with water from the Gila River, along with beef, chicken, wheat, melons, figs, mesquite beans, and fruit of saguaro cactus. Only after they began to eat foods high in refined carbohydrates did these diseases appear. As growth of markets has slowed in wealthy countries, food companies have expanded marketing and sales in developing nations, contributing to their obesity and health problems.[36]

While some people emphasize the adverse effects of refined fats and carbohydrates – and their concerns are certainly valid – an excess of any nutrient can be harmful. People can die from eating too much meat and fat, ingesting too many vegetables, or even drinking too much water. Drinking excess water leads to a condition called *hyponatremia*, which causes the inside of cells to flood due to unusually low levels of sodium in the bloodstream. In severe cases, water intoxication can lead to seizures, coma, and death. Eating too much meat can result in excess protein, which can cause toxicity if nitrogen waste products aren't removed from the blood. Broccoli and cauliflower have glucosinolates, spinach has oxalates, and corn has nitrates. Overingesting these and other vegetables can also cause toxicity. We now enrich and fortify foods with vitamins and minerals, and people take supplements, unaware how those practices, too, can adversely affect food choices and contribute to the obesity crisis, as I discuss in chapter 6.

**Minimizing ingestion of pesticides.** People in the United States are exposed to a range of pesticides, many with known or suspected carcinogenic or endocrine-disrupting properties.[37] Farmworkers and children are at most risk – the former due to their work with pesticides and the latter due to more vulnerable organ systems still developing. The Centers for Disease Control and Prevention reports traces of twenty-nine pesticides in a typical American body. The effects of frequently used mixtures of pesticides are unknown. Herbicides such as glyphosate, the active agent in the omnipresent herbicide Roundup, are now ubiquitous.

Shoppers worry about pesticides in foods. A *Consumer Reports* survey found pesticides are a concern for 85 percent of Americans.[38] The risk of ingesting large amounts of pesticides varies from low to high, depending on the country where it is grown and the type of produce. In the United States, eating a serving of green beans is 200 times riskier than eating a serving of broccoli. The higher frequency of occurrence of residues in fruits (75 percent) than vegetables (32 percent) may indicate higher levels of crop protection used in fruits, use of more persistent chemicals, different sprayer technologies, and pesticide applications made closer to harvest.

Prohibiting use of synthetic pesticides under organic farming standards results in a more than four-fold reduction in the number of crops with pesticide residues.[39] The pesticide residues present in 11 percent of organic crops are due to several factors that range from cross-contamination from neighboring conventional fields; the continued presence of persistent pesticides (for example, organochlorine compounds) in fields or perennial crops from past conventional management; and accidental or fraudulent use of prohibited pesticides by organic farmers.

Many people are concerned about the risks of eating GMO foods contaminated with glyphosate residues. Glyphosate is a textbook example of disruption of cell and organism homeostasis by a toxin that inhibits liver enzymes crucial in detoxification. Glyphosate enhances the damaging effects of foodborne toxic residues and environmental toxins and is considered a pathway to disease. Negative impacts are insidious and manifest slowly as inflammation damages cells. *The Lancet Oncology*, the foremost scientific journal for cancer studies, published a paper by the World Health Organization's International Agency for Research on Cancer (IARC) that classified glyphosate as a probable carcinogen. Tolerable intakes for glyphosate in the United States and European Union are based on outdated science, as I explain in chapter 16.

People are also concerned about how chronic exposure to glyphosate affects microbial populations in soil and the microbiome in the gut of humans.[40] Glyphosate was patented as an antibiotic in 2010. Glyphosate, 2,4-D, and dicamba have been linked with antibiotic resistance. Glyphosate suppresses amino-acid synthesis by gut microbes. The implications are sobering, given the importance of the gut microbiome in immune function.

**From hunters and gatherers to farmers.** For eons, *Homo sapiens* relied on a sun-driven economy, coevolving with landscapes, learning where to find foods seasonally.[41] We learned to cook to increase nutrient availability and to reduce levels of secondary compounds in plants. During the Holocene, we changed from a species whose ancient hunting and gathering traditions linked humans with landscape into a species that lived in cities and grew lawns. The invention of agriculture and the domestication of livestock, both of which followed the creation of villages, allowed populations to grow and specialize as villages became cities and cities became nations.

While not all hunters and gatherers were well fed all the time, most fared well most of the time. The transition from hunting and gathering to farming adversely affected peoples' health.[42] Anthropologists who studied human remains that spanned the change from hunter-gatherers to farmers found that nutrition and health declined in the southwest United States. The amount of meat and wild plant foods in diets decreased as maize became the mainstay. Lower quality protein in women's diets created impoverished nutrition and poor health, particularly during pregnancy and lactation. As women's nutritional status declined, so did the health of their unborn and newly born children, whose poor health was manifest as a pervasive pattern of high infant mortality, malnutrition, and disease infestation. Demand for farm labor created a need for more children, which led to shorter interbirth intervals, earlier weaning, and higher infant mortality.

In *Guns, Germs, and Steel*, ecologist, geographer, biologist, and anthropologist Jared Diamond described agriculture as "the worst mistake in the history of the human race." That "mistake" was intensified by the Green Revolution, as livestock production systems emphasized low diversity of high-producing forages over high diversity and quality of foodstuffs. People abandoned locally adapted species of plants and animals in favor of a few new varieties and breeds. Our ancestors used hundreds of plant species for food, clothing, and shelter. Modern-day humans rely on just fifteen plant species for roughly 90 percent of global food consumption.

Today, the challenges we face aren't necessarily due to lack of food, but due to lack of food quality. We've reduced concentrations of primary and secondary compounds in plants to increase yields, and we've added pesticides to protect plants now grown in monocultures. The consequence is food lacking in the phytochemicals we require for health; produce laden with herbicides and pesticides; and livestock fed additives including antibiotics and hormones, with adverse downstream effects on our health. The consequences, manifesting in diet-related diseases, are insidious, delayed in time, and as deadly as D-Con.

# CHAPTER 3

# No Two Alike

The chipmunk that is our neighbor built a home under the logs we cut and piled to warm us in winter. Each morning and evening, she forages in our backyard, where she also drinks water from a small pan we've provided for her and the many species of birds that dine with us. She sleeps, I've read, for up to fifteen hours a day. In summer and autumn, she stockpiles forages and medicinal plants, which she then eats in the comfort of her log house in winter.

Members of her species live in diverse environments, from parks and woodlots to the alpine heights that surround our nest. They are omnivores that eat a variety of plant species and their parts – from leaves, stems, and buds to seeds, nuts, and other fruits. They also eat insects, small frogs, and worms. They are opportunists that will even eat bird eggs and nestlings. In turn, chipmunks are food for various predatory birds and mammals here in the backwoods.

"Our" chipmunk is an individual involved in relationships with plants and animals in our neighborhood. They know one another. She dines on individual plants, including oat grass, paintbrush, and sagebrush. When she eats their leaves and stems, they learn and remember that: nutrients gained for the chipmunk, tissue lost by the plant. When possible, she also eats eggs of mountain bluebirds and phoebes, which is why they harass her whenever they all visit our backyard at the same time. These acts, too, are remembered by chipmunk, bluebird, and phoebe.

In short, our chipmunk isn't the same as chipmunks that make their homes down the road or in alpine peaks across the valley. She is unique in how she is built, how she functions, and how she behaves. She is an emergent property that came about as a result of interactions among genes, chipmunk, environment, and chance – all manifesting uniquely in space and time. When she has offspring, they, too, will each be unique, even though she raises them in the same environment. From bacteria to chipmunks to hominoids and all creatures in between, the genius of evolution isn't survival of the fittest. The magic is ceaselessly generating diversity.

## Uniqueness of Individuals

Variation is perhaps the most amazing aspect of life. Scientists appreciate this, but we don't necessarily like it because of the implications for our data. Variation reflects the many differences among replicates, be they individual plants or animals or humans or plots of ground. No scientist has ever accounted for all the variation recorded in a study. Even studies published in the most prestigious journals include such unexplained "error" variation. The popular press never discusses it because neither they nor their audiences understand or appreciate what variation implies. During the forty years I was conducting studies and analyzing data, however, I became more interested in error variation than any other aspect of my findings. Why do individuals differ so much and what are the implications for life?

Scientists must be able to repeat results to make recommendations. That's why we replicate studies. To see if a drug works against cancer, for instance, researchers give 1,000 people the drug (the treatment replicates) and 1,000 people a placebo (the control replicates). Why so many? Because each person will respond differently to the drug or the placebo.

Differences among replicates are partitioned into variation among treatments (drug versus placebo) and variation not accounted for in the study (error). Statistically, that's expressed as:

| Total Variation (Yij) | = | Treatment Variation (Ti) | + | Error Variation (eij) |
|---|---|---|---|---|

When the variation among treatments (Ti) is large relative to error variation (eij), the findings are said to be statistically significant. Scientists don't like error variation (eij), especially when it's large relative to treatment variation, because that makes it impossible to show statistically significant differences among different treatment groups.

Scientific findings and ensuing recommendations are based on the "average" individual in a population. The media then write accounts about the latest scientific recommendations for the "average" person in a population.

Corporations turn those recommendations into products. In life, though, no two individuals are alike, and no one is an "average" person.

Roger Williams was a biochemist who spent his visionary career at the University of Texas at Austin. In *Biochemical Individuality*, first published in 1956, he discussed implications of individuality for medicine and nutrition. Regarding the former, he noted correctly that treatments will need to be tailored to individuals. Concerning the latter, he noted insightfully the importance of "body wisdoms" and concluded, "If this wisdom were universally possessed by all members of the human family, there would be no need to study nutrition and the only deficiency diseases would be those induced when suitable foods were not available."

Individuals are unique in form, function, and behavior. Differences in form are why forensic experts can identify us by a fingerprint. Differences in function are why a bloodhound can track us by our odors. If we could look at our insides, we'd be astonished at the differences in both form and function. Our stomachs vary in size, shape, and contour as much as our ears, noses, and mouths. Our bodies also vary in how they function: A Mayo Foundation study of 5,000 people with no known stomach ailments showed that their digestive juices varied a thousand-fold in pepsin content and hydrochloric acid content. We also vary greatly in the many hormones that influence our appetite and food intake. These are just a few of the differences that help explain why we differ in our behaviors. For instance, we tend not to eat with equal frequency or in equal amounts, nor do we choose the same foods.

Echoing observations of Clara Davis with children, Williams also highlighted the great variation in food selection when animals – young rats, mice, hamsters, and baby chicks – are allowed to select their diets: meat four-fold, butter seven-fold, carrots four-fold, sugar seventeen-fold, salt four-fold, yeast forty-six-fold. He concluded by noting that he had difficulty escaping the conclusion that animals choose differently because they need differently.

That's also true for livestock, with important implications for selecting animals adapted to local conditions – ecologically, economically, and socially.[1] For instance, cattle differ more within than among breeds in tolerance of alkaloids found in larkspur, which has implications for developing herds of cattle that don't die from eating larkspur.[2] While most goats avoid eating new twigs of blackbrush, due to their high concentrations of condensed tannins, another 10 to 20 percent of goats strongly prefer new twigs.[3] Cattle and sheep of uniform sex, age, and breed vary greatly in the proportions of roughages and grains they eat in confinement, though they all grow well.[4] To enhance nutrition, production, and health of goats, sheep, and cattle, skilled herders in France design grazing circuits to create synergies among meal phases.[5] To do so, they partition landscapes into grazing sectors, which they carefully sequence within daily grazing circuits. Allowing animals to choose among different forages within and among sectors enables individuality, which is essential for enhancing production and health.

People also differ in needs. In a study of nineteen healthy men, needs for calcium varied from 222 to 1,018 millgrams per day. The same is true for vitamins A and C. On sea voyages without fruit, only certain individuals developed scurvy. In populations that eat mainly polished rice, only certain individuals develop beriberi. In some areas of the United States where the diet was poor, chiefly among those with low income, only certain individuals developed pellagra. Not all children with limited access to sunlight develop rickets. The amount of vitamin D required by the average child is estimated to be 400 IU daily, but the daily amount required to prevent rickets in more susceptible children may be 5,000 to 10,000 IU or even higher.

Williams points out that some of the most far-reaching internal differences involve the endocrine glands – thyroids, parathyroids, adrenals, sex glands, and pituitaries – that affect every facet of life: our metabolic health; our appetites for food, drink, amusement, and sex; our emotions and psychological well-being. He gives the example of people who claim that drinking caffeinated coffee doesn't keep them awake and thus believe that the effect of caffeine is all in a person's mind. This overlooks the fact that the amount of a drug necessary to bring about the same effect in different individuals may vary as much as tenfold. Alcohol is a real danger to some people, while others have almost no chance of becoming addicted to it.

Individuals also differ in how their bodies partition the energy in carbohydrates.[6] Carbohydrates are digested to glucose, an essential source of energy that can also be converted to fatty acids and stored as triglycerides in fat cells. Some individuals allocate more energy to fat cells than to muscle cells.[7] They are prone to overeating because their muscles lack the energy they need to function. Thus, they tend to be fat and sedentary. Other individuals allocate more energy to muscle cells than to fat cells. Those people tend to be lean and active because their muscles have the energy they need.

Twenty-five percent of the human population is adapted to diets high in refined carbohydrates; another 25 percent doesn't deal well with such diets. The rest of us fall along a continuum between the two extremes. The status of each individual is influenced by myriad factors including the ratio of high to low metabolically active cells; the needs of organ systems such as skeletal muscles, the brain, red blood cells, and liver; the level of physical activity and fitness; age, sex, and reproductive status; and illness.[8]

People who don't cope well with refined carbohydrates secrete large amounts of insulin while eating foods high in refined carbohydrates. They attain better weight loss on a diet that restricts refined sugars and starches (low glycemic) than on a low-fat diet. Insulin regulates the metabolism of carbohydrates and fats by promoting absorption of glucose from the blood into skeletal muscles and fat tissue; causes fat to be stored rather than used for energy; causes muscle and liver cells to store glucose in the form of glycogen; and in many cases controls cellular electrolyte balances and uptake of amino acids. Insulin inhibits conversion of glycogen to glucose by the liver. The most important signal to the insulin-producing beta cells in the pancreas is the level of glucose in blood, which is largely determined by intake of carbohydrates. Except in folks who have diabetes or metabolic syndrome, insulin is released in a constant proportion to remove excess glucose, which otherwise would be toxic, from the blood.

As Clara Davis observed, and Roger Williams predicted, researchers are now discovering that because individuals vary greatly in response to identical meals, suggesting universal dietary recommendations has little utility. Some researchers are now showing that personalized diets – created with the help of an accurate predictor of blood glucose response that integrates dietary habits, physical activity, and gut microbiota – can lower after-meal levels of blood glucose and its metabolic consequences. To do so, they monitored blood glucose levels in 800 people who collectively ate 46,898 meals during one weeklong trial.[9] They conclude that accurate personalized predictions of nutritional effects are of great practical value for integrating diet modifications into clinical decision-making processes. Their findings are consistent with the great variability in weight changes among individuals in studies of diets that differ in amounts of carbohydrate and fat.[10] Unfortunately, diets fashioned just for you – whether based on your genes, your response to glucose, your gut microbiota, or their secret formulas – don't appreciate the one expert that knows the most about you: your own body.[11]

## Uniqueness of Genomes, Proteomes, and Metabolomes

In Darwin's theory of evolution, variation among organisms results from gene mutations and recombinations that *are not directly responsive* to the immediate demands of the environment. The individual variants are tested for suitability in social and biophysical environments that came into being *independent* of that variation. This notion has come to connote rather rigid and passive means of evolving: Populations propose, environments dispose – animals are machines, genes are destiny. "In such a conceptual structure," as evolutionary biologist, mathematician, and geneticist Richard Lewontin points out in *The Triple Helix*, "the metaphor of adaptation is indeed appropriate. . . . In a curious sense the study of the organisms is really a study of the shape of the environmental space, the organisms themselves being nothing but the passive medium through which we see the shape of the external world."

Once a species arises, it persists for ten to twenty million years – a chipmunk is a chipmunk, a deer is a deer, a bear is a bear, a mountain lion is a mountain lion, and an aspen is an aspen.[12] Yet, environments change relentlessly. The last million years have been characterized by cycles of glacial and interglacial periods within a gradually deepening ice age. The growth and retreat of continental ice sheets in the Northern Hemisphere fluctuates on time scales that vary from 40,000 to 100,000 years. How, during the several million years a species exists, do individuals evolve, given vast changes in climate that alter habitats for plants and animals?

Part of the answer lies in how genes are expressed. Every human cell, for example, has roughly 20,000 genes, all identical. Yet, cells in our eyes differ from those in our skin, which differ from those in various organ systems. How does a cell know how to become a specific part of an eye as opposed to skin, a small intestine, or a brain? Though retinal and skin cells possess the same 20,000 genes, a retinal cell activates or suppresses a unique subset of the total genome, while a skin cell activates another subset. Each cell chooses a unique spectrum of genes for itself, thereby attaining its unique form, function, and behavior. But how does a retinal cell know which genes to activate and which to suppress? In part, through master regulatory genes that exercise exquisite control over the identity, growth, shape, and size of a cell. They activate genes by toggling hundreds of molecular switches that can be "turned on" or "turned off."

After the study of genes (genomics), proteomics is the next way we can explore the uniqueness of individuals. Proteomics is the study of the structures and functions of the amino acids that create the proteins that create cells. While an organism's genome is more or less fixed, the proteome differs from cell to cell and from time to time because distinct genes are expressed in different cell types as environments change. Thus, 20,000 genes can produce approximately 200,000 types of RNA. Each RNA strand then encodes for up to 200,000 proteins. That's a geometric expansion that gives forty billion possible proteins. There can be hundreds of variants of proteins in a cell. Researchers use mass spectrometers to identify pieces of proteins, but as sophisticated as these devices are, their dynamic range – or capacity to collect and read different proteins in a sample – is exceeded by the proteins' complexity by a factor of three or four.

The metabolome is yet another manifestation of the uniqueness of individuals. Metabolomics is the study of cellular processes that involve metabolites in a cell, tissue, organ, or organism. Gene expression and proteomic analyses reveal gene products in a cell, while metabolic profiling gives a snapshot of the physiology of a cell. People have known for ages that tissues and fluids reflect an individual's metabolic state. Chinese doctors used preferences of ants for sweet to tell if urine had high levels of glucose and to thus detect diabetes. In the late 1940s, Roger Williams used paper chromatography to show that "metabolic profiles" in urine and saliva are associated with diseases like schizophrenia. Today, scientists use sophisticated techniques to assess the complexities of metabolomic individuality.[13] By 2007, scientists had identified roughly 2,500 metabolites, 1,200 drugs, and 3,500 food components for humans.[14] As of September 2015, the metabolomics database METLIN contained over 240,000 metabolites.

Metabolomics, proteomics, and genomics reveal the intricacies and complexities of the cellular processes that make each individual unique. Individuality arises as plants and animals interact with the environments they inhabit. This emergence, and the plasticity to evolve as environments change over the eons a species is on Earth, is enabled by epigenetics.

## Gene Expression through Epigenetics

In the eighteenth century, soldier, biologist, academic, and naturalist Jean-Baptiste Lamarck developed a theory of inheritance of traits acquired during an individual organism's lifetime. Until recently, this theory of "soft inheritance" was ridiculed by scientists. Now, evidence for soft inheritance is arising from the field of epigenetics – the study of heritable changes in gene expression caused by mechanisms other than changes in the DNA sequence. In other words, the individual's DNA is not altered, but the environment impacts the way the DNA is expressed. The number of patterns of epigenetic expressions is likely to be 50 to 100 times greater than the 20,000 genes in the human genome. The resulting symphony is analogous to an orchestra with organisms composing, conducting, and playing a huge number of melodies. In the brief span of time that scientists have been investigating epigenetics, more than 100 cases of epigenetic inheritance have been documented in both plants and animals.[15]

As one example, plants monitor day/night and seasonal cycles to align their metabolism, growth, and development with changing environmental conditions.[16] This plasticity requires signal perception and integration, changing gene expression in response to specific signals, and maintaining the responses until conditions change yet again. Epigenetic mechanisms enable altered cell states to persist through cell divisions, and they provide a molecular memory that maintains these responses. Environmentally induced epigenetic changes in plants can be reset in egg or pollen cells, as can most epigenetic marks in animal egg and sperm cells.

One example of epigenetic change in animals is that rat pups whose mothers lick and groom them end up being calmer as adults than pups whose mothers pay little attention to them.[17] The expression of the rat pups' DNA is changed by nurturing, which turns on genes for calmness in nurtured pups. As adults, nurtured pups have fewer stress-related diseases. Studies of grooming also involve crossfostering rat pups from biological mothers who did not groom onto mothers who did groom and vice-versa. Pups reared by foster mothers who groomed them became calm. They also became groomers as mothers, as did their pups. Conversely, pups fostered by mothers who didn't groom them grew up to be nongrooming mothers.

Mothers are also linked epigenetically with their offspring through diet. Historically, studying genes for obesity and diet-related diseases was not useful because *any one gene* has such a small influence, but understanding epigenetic marks for *multiple genes* is valuable in no small part because epigenetic traits can be reversed. Factors such as nutrients, toxins, and stress interact to cause epigenetic changes that can affect the ontogeny of diseases.[18] One way such epigenetic changes occur is through DNA methylation, a process by which methyl groups are added to the DNA molecule. Methylation changes the activity of a DNA segment without changing the sequence. When smokers kick the habit, for example, their methylation patterns rapidly return to near normal. But some changes can persist for decades, which helps to explain why ex-smokers remain at a higher risk of cancer and respiratory problems years after they quit. The challenge is unraveling multiple associations across generations, given the interdependence among genes, organisms, and ever-changing environments.[19]

Those interrelationships were revealed through studies of survivors of the Dutch Hunger Winter, a period of starvation in the Netherlands during the winter of 1944–1945, the final year of the Second World War.[20] A German blockade cut off food and fuel shipments from farms to punish the Dutch, who were reluctant to aid the Nazi war efforts. During the blockade, some 4.5 million people survived on fewer than 1,000 calories – sometimes as few as 500 calories – each day. Children of women who were pregnant during the famine were smaller in size, as expected, but when they grew up and had children, their children were also smaller, which suggests the famine experienced by the mothers caused epigenetic changes passed to the next two generations.

As food became abundant after the war, epigenetic changes that resulted in a "thrifty" metabolism were not reversed, as seen in a comparison of adult weights. Fifty years later, people conceived when food was scarce weighed an average of 14 pounds more than those whose mothers were in their second or third trimester during the blockade. Their waist circumference was 1.5 inches larger, and they were three times more likely to have coronary heart disease. A gene for growth during pregnancy – one for making insulin-like growth factor 2 or IGF2 – was more active (less methylated) in people conceived during the worst days of starvation.[21] The children and grandchildren were also more susceptible to obesity, diabetes, cardiovascular disease, microalbuminuria (increase in urine albumin), and other health problems.

These findings are supported by a study of 162 obese mothers who gave birth to babies both before and after weight-loss surgery. Compared with siblings born before surgery, children born after surgery were less obese and had improved cardiometabolic risk profiles into adulthood.[22] Children born before their mothers had gastric bypass surgery showed different epigenetic marks than their siblings born after their mothers had surgery and lost weight. Differences in methylation of more than 5,500 genes were detected between "before" and "after" siblings. Genes involving diabetes, inflammation, and cardiovascular disease were most affected.

Mothers who are obese, mothers who gain excess weight in pregnancy, and mothers who become diabetic during pregnancy are more likely to have fatter babies.[23] The greater a woman's weight gain during pregnancy, the higher the risk her child will be overweight by three years of age and continue to be overweight into adolescence and adulthood. Maternal obesity and diabetes during pregnancy increase risk of obesity and type 2 diabetes in offspring.

Interest in the transgenerational epigenetic effects of phytochemicals in food was fueled by observations of changes in DNA methylation in offspring of Agouti mice. Agouti mice have yellow fur, fat bodies, and a propensity to develop obesity, diabetes, and cancer. When Agouti mice are fed a diet of soy during pregnancy, formerly obese female mice produce offspring with dark brown rather than yellow fur. These offspring are not obese, and they have a lower incidence of diet-related diseases for multiple generations.[24] Soy contains polyphenols, such as genistein, that can deactivate expression of the agouti gene in the offspring of obese females.[25] Findings such as these have implications for reversing the obesity crisis – a topic I explore in more depth in part three.

## Uniqueness of Gut Microbiota

Given the great variation among individual relationships with environments, we'd expect the microorganisms that inhabit our bodies to be equally variable. By mapping the microbial makeup of the body of healthy humans using genome sequencing techniques, researchers of the Human Microbiome Project created a database that describes microbial variation in humans.[26] They collected close to 5,000 samples from eighteen body sites, including nose, mouth, throat, skin, lower intestine (stool), and vagina in 242 healthy American volunteers (129 males and 113 females). They estimate that more than 10,000 microbial species occupy the human ecosystem, and they identified 81 to 99 percent of the genera. They found that interindividual variation in the microbiome is specific, functionally relevant, and personalized.

The bodies of humans and herbivores are colonized with vaginal microbes at birth.[27] After birth, a baby is exposed to bacteria on mother's skin as it suckles and as it begins to ingest foods in the environment. In humans, the large intestine has the richest array of microbes. The body of an "average" adult carries three to five pounds of microbes. Based largely on an analysis of the microbes in your gut, a company called Viome offers "a personalized, easy-to-follow plan with precise diet, nutrition and supplement recommendations" for a mere $399 (and a stool sample).

The service offered by companies such as Viome largely ignore the fact that microbial populations in your gut mirror changes in your diet from day to day. Conversely, ruminant nutritionists have long appreciated that the mix of primary and secondary compounds in the diet influences both the species and the numbers of different microbial populations in the gut of ruminants like cows, sheep, goats, elk, and deer.[28] Several thousand species of bacteria, protozoa, fungi, as well as some archaea (ancient microorganisms that produce methane) live in the rumen. The number of microbes in a drop of rumen fluid (fifty billion) is more than six times the number of people on Earth. Due to differences in diet daily, weekly, and seasonally, each individual has its own dynamically unique mix of microbial species.

Researchers have identified three human enterotypes, groupings of bacteria based on the gut ecosystem they inhabit. Enterotypes are not dictated by age, gender, body weight, or national divisions.[29] Rather, within enterotypes, closely related microorganisms differ broadly in their genomic structures and corresponding phenotypes – characteristics of an individual that result from the interaction of its genotype with the environment – so novel species proliferate.[30] Thus, while taxonomic similarities exist among human populations, at finer scales, gut microbiota exhibit dramatic metabolic differences tailored to suit different diets and microenvironments.

The gut microbes in humans can be regarded as an ecosystem coevolving as part of a host supraorganism within a landscape. The Hadza, for example, provide a rare example of human subsistence through hunting and gathering that persists in the same East African region where early hominids lived. Differences in gut microbes between Hadza men and women correspond to sexual division of labor and differences in diet.[31] Women selectively forage for tubers and plant foods and spend a great deal of time in camp with children, family members, and close friends. Men are highly mobile foragers and range far from the central campsite to obtain game meat and honey. Although all foods are brought back to camp and shared, men and women consume slightly more of their targeted foods from snacking throughout the day.

Gut microbes align with the nutrient acquisition strategy of the host, thus potentially buffering women from "resource gaps" that may lead to nutritional deficiencies. Women's foraging must provide nutrients for pregnancy and lactation, which is a strong adaptive pressure for gut microbes to derive the most energy from available plant foods. The ability of gut microbes of Hadza women to respond to greater intake of plants represents a marked break with traditional thinking that the human gut has only a limited ability to digest fiber.

## What Should a Human Eat?

Much of the advice about "ancestor diets" ignores the reality of "no two alike" – creatures are continually evolving in form, function, and behavior as genes are being expressed within the context of ever-changing environments. During the transition from hunting and gathering to farming grains, for instance, people adapted to an energy-rich, but initially detrimental, food supply.[32] Compared with populations that ate low-starch diets, populations that ate high-starch diets now produced more salivary amylase, an enzyme that enables digestion of starch into glucose. While salivary amylase is ineffective for digesting raw crystalline starch, cooking greatly increases the digestibility of starch to glucose, which increases energy for tissues that have high glucose demands, such as the brain, red blood cells, and a developing fetus.

Archeological and anthropological evidence, along with studies of extant hunter-gatherers, show human diets ranged from plant-based in deserts and tropical grasslands to animal-based in northern coniferous forests and tundra.[33] Hunter-gatherers had similar intake of carbohydrate – 30 to 35 percent of total energy – over a range of latitudes from 11 degrees to 40 degrees north or south of the equator. Slowly digested, the high-fiber plants they ate yielded low glycemic and insulin responses with less risk of diabetes and heart disease. Intake of carbohydrate declined from 20 to 9 percent or less of energy intake as intake of meat increased at higher latitudes from 41 degrees to above 60 degrees.

Studies of Arctic peoples suggests about 50 percent of their calories came from fat, 30 to 35 percent from protein, and 15 to 20 percent from carbohydrate, primarily from glycogen in meat.[34] Meat eaten fresh, or frozen soon after slaughter, retains much of its glycogen, which is a source of carbohydrate. Arctic peoples likely ate more carbohydrate than is normally thought from unusual sources. Ethnographic evidence shows they got carbohydrates, as well as vitamins, minerals, and secondary compounds, from eating kelp, tundra plants, and stomach contents of herbivores. This latter practice has been observed for peoples in other parts of the world, too.

Many people now promote "the diet our ancestors ate" to argue which foods we ought to eat to stay fit. Advocates with different views of ancestor diets – from herbivores to omnivores to carnivores – amass evidence in support of the diet they sanction. These views ignore marked variation among individuals in response to eating different foods and diets; overlook striking dissimilarities in foods people ate in different parts of the world at different times of the year; and disregard variation in foods available in different locations during the eons humans have been on Earth. No wonder researchers who study human food selection find it impossible to define diets, even for regimens like the so-called Mediterranean diet.[35] In their haste to tell us how to eat to be healthy, diet-book authors also ignore the fact that our ancestors came from different parts of the world and they evolved eating different diets. What foods should a child with a Hindu mother from India and a Catholic father from Argentina select to be physically and mentally healthy, given that he or she was conceived, born, and raised in the United States?

## The Sum of an Individual

If we could reset the clock to the moment you were conceived and rerun your life over and over again, you'd turn out differently every time, even if you had the same genes and were brought up in exactly the same environment. That's because some differences among individuals are due to random events during development. Richard Lewontin calls this "noisy development," and it is another way evolution hedges its bets in changing environments. Evolving relationships among genes, organisms, and environments ensure no two individuals are alike, and for good measure, nature adds a pinch of chance, increasing the odds some will survive:

Genes + Environments + Organisms + Chance = Individuals

Bacteria like *Escherichia coli* illustrate this equation. *E. coli* colonize an infant's gastrointestinal tract within forty hours of birth, arriving in food or water or from skin contact with another person. While some strains of *E. coli*, such as serotype O157:H7, can cause food poisoning, most benefit their host by producing vitamin K and preventing pathogenic bacteria from establishing in the intestines.

What's fascinating is how *E. coli* has continually evolved during the three to four billion years it's been on Earth. *E. coli* absorbs nutrients through its cell wall, typically from other cells it attacks. *E. coli* reproduces by asexual fission and can split into two daughter cells every twenty minutes. In one hour there are eight; in twelve hours, sixty-nine billion; and in fifteen hours thirty-five trillion.

In addition to rapid exponential growth, *E. coli* can exchange DNA with one another through a process called *bacterial conjugation*. Bacteria like *E. coli* can swap DNA, not only with other *E. coli*, but with other gram-negative bacteria as well. If some *E. coli* cells end up in a harsh environment inhabited by other better-adapted bacteria, they can pick up relevant DNA from the better-suited microbes. That allows an *E. coli* cell to rapidly acquire new traits, such as resistance to a novel antibiotic. This basic recipe for success – continually creating in the face of change – has kept *E. coli* thriving on Earth since the earliest days of unicellular life.

That's only part of the story science writer Carl Zimmer tells in *Microcosm*. Genetically identical *E. coli*, reared in the same environment, differ in form, function, and behavior. They differ in ability to digest lactose, physical stamina, susceptibility to viruses, and persistence. These differences arise in part because a colony of *E. coli* uses *chance* to hedge its bets at little cost. The ensuing variation is one reason why bacterial populations so easily become antibiotic resistant.

In the world of bacteria, joining the resistance movement is the norm to the extent that some scientists estimate the number of infections resistant to any form of antibiotic could cut short the lives of 300 million people and cost the global economy $100 trillion by 2050.[36] Bacteria don't just resist antibiotics, they even develop a penchant for dining on them. How? To find out, researchers isolated a number of species of soil bacteria, including one called *Pseudomonas*, and put them on a strict diet of penicillin.[37] The bacteria's response was basically "bring it on" with colonies expanding on the agar plate in close proportion to how much penicillin they were fed. Using genome sequencing, the researchers found that bacteria were able to supercharge the genes making enzymes that devour enzymes, including β-lactamase, which confers resistance to some antibiotics, when faced with a meal of penicillin.

How chance actually works can be observed in bacterial cells, as shown by Lewontin. If a continuously stirred growth medium is inoculated with a single bacterial cell, that cell will divide in a certain amount of time. But the two daughter cells will not divide simultaneously a set amount of time later; nor will the resulting four cells of that division divide simultaneously. Bacterial cultures grow continuously, not in lockstep, because each cell formed takes a slightly different time to divide. Why? Cells contain only a small number of many of the molecules that are required by organelles for metabolism. Neither molecules nor organelles occur uniformly throughout a cell, so organelles and molecules depend on movement for reactions to occur. That takes time and occupies space, so considerable variation exists from cell to cell in rates of metabolism. This variation is manifest in the time it takes cells to divide or to migrate during development.

All of the cells are growing in exactly the same conditions because the culture medium, constantly stirred, contains high concentrations of small molecules whose local concentration is effectively the same everywhere. Moreover, the cells are genetically identical because not enough time has elapsed during the few generations of division to allow mutations. Thus, the cause of their developmental asynchrony is the uneven availability of different kinds of molecules to daughter cells during division. The uneven availability is due to Brownian motion, the random movement of particles in a fluid resulting from their bombardment by fast-moving atoms or molecules in the liquid, chance bubbling up from quantum levels through to life.[38]

An analogous phenomenon occurs as multicellular organisms develop. For example, the three cells that create a sensory bristle in flies arise from two divisions of an original precursor cell. To create a bristle on the body of an adult fly, bristle-forming cells must migrate to the surface of the developing fly, a surface that is slowly but surely hardening. If the division of the original precursor cell into three takes too long and the migration is delayed, the cells will not arrive at the hardening surface in time to be included as a bristle.

As Lewontin points out, "Such random processes must underlie a great deal of the variation observed between organisms, including variation in their central nervous systems." As the human sense of smell develops, nerve cells in the brain choose among hundreds of odor receptors (neurons) in the nose. Each cell picks just one and the choice is affected by the unpredictable bursts of proteins within each neuron. Remarkably, the many other processes of neural development that occur early in life can give rise to marked differences in cognitive function that are biologically innate but neither genetic nor environmental.

## Evolving with the World

Whether or not any creature ever makes choices has been long debated, as I discuss in chapter 18. For now, the conclusion is this: If, as quantum physicists argue, chance is playing a role in the unfolding of the universe, and if "noisy development" is playing a role in creating individuals, then existence is literally creating its way into now. The cooccurrence of change and chance creates the opportunity to create relationships. Without that combination, one can argue, creatures don't make choices. Rather, we would live under the *illusion* of making choices. If change and chance do coexist, we all have the possibility of making choices, though social and biophysical environments can severely limit "free will" and the choices creatures make.

The choices each individual makes throughout life can be seen as a stream of branching points, or bifurcations, that occur in the face of varying degrees of uncertainty. At each fork in the road, the choice is intertwined in a complex web of relationships that involve the individual's cells and organ systems, including the microbiome, interacting with the physical and social environments the individual inhabits. These interactions are influenced by past experiences, beginning in utero and early in life, the alternatives available at the time, and the beliefs of the individual. The path selected at each juncture ripples through the filaments in the web of lives of all organisms, like a pebble tossed into a tranquil pool.

By participating, creatures are creating relationships with environments. As Stuart Kauffman – medical doctor, theoretical biologist, and complex systems researcher who studies the origin of life on Earth – concludes in *Investigations*, we live in a biosphere that "is doing something literally incalculable, nonalgorithmic, and outside our capacity to predict, not due to quantum uncertainty alone, nor due to deterministic chaos alone, but for a different, equally, or more profound reason: Emergent and persistent creativity in the physical universe is real."

So, life is a crapshoot, a mess, a wonderful hodgepodge of plants and animals varying uniquely in space and time, some nature and some nurture arising interdependently with chance. Sensitivity to environmental conditions leads to individuality and diversity, which we should appreciate and embrace. Life flies in the face of our attempts to categorize and generalize based on assumptions about averages – nature fills vacuums with individuals, and no two are alike. Genes, organisms, environments, and chance ever combining to affect form, function, and behaviors as landscapes evolve – organisms influencing landscapes influencing organisms in unending transformation. When we think of evolution as something that happened in the past, we miss the mark: *Transformation is unending.*

These dynamics suggest we should take our views of evolution beyond an account of how organisms developed from earlier forms during the natural history of the species to include changes during the lifetime of the individual. Organisms aren't merely adapting to environments, they are actively participating in creating relationships with them. This view recognizes that individuals are involved in the world, which enables them to evolve with the world.[39]

# PART II

# *Dancing with the Wisdom of the Body*

# CHAPTER 4

# More Than a Matter of Taste

To survive a rare lung disease, an American dancer named Claire Sylvia underwent a heart-and-lung transplant in 1988. Her chest was sawed open, her diseased organs removed, and in their place, doctors inserted the heart and lungs of a donor. Following surgery, she was certain that her harrowing journey was over. In reality, it was just beginning. New organs weren't the only thing she inherited. During her recovery in intensive care, Claire began to feel the presence of someone else inside her. Initially frightened and then captivated, she realized that some of her attitudes and behaviors had changed. In *A Change of Heart*, Claire describes how she was shocked to discover that her food preferences had changed following her organ transplant. The cells and organs from her donor had a "mind" of their own that was now influencing what she chose to eat. After an extraordinary dream, she began to search for the family of her donor.

Claire's experience raises intriguing questions. Why do we prefer to eat some foods but not others? What are we feeding? Most people don't give such questions a bit of thought, and there's really no need to do so. The wisdom of the body takes care of these matters – without a bit of thought. As a result, people really don't know why they prefer to eat certain foods or what they are feeding – stomach or maybe intestines – when they eat. Can cells and organ systems influence the food choices of a human or an herbivore? What kinds of learning and memory exist in cells and organ systems, and can cellular memory outlive physical death?

Liking for foods is typically thought to be influenced by *palatability*. Webster's dictionary defines *palatable* as pleasant or acceptable to the taste and hence fit to be eaten or drunk. Animal scientists usually explain palatability, though, as a liking influenced by a food's flavor (odor and taste) and texture, or the relish an animal shows when eating a food. Plant scientists describe palatability as attributes of plants that alter an herbivore's preference for consuming them, such as physical and chemical composition and associated plants.

## Redefining Palatability

I had begun to ponder these questions about what influences an herbivore's food choices while observing the perplexing behavior of the goats in St. George: Why didn't the goats prefer the younger more nutritious twigs of blackbrush over older, woody, less nutritious blackbrush twigs? As it turned out, the goats helped me understand their anomalous behavior.

My colleague Beth and I began with a series of trials in which we extracted and purified secondary compounds from young twigs, mixed each purified extract individually with a pelleted food, and offered the "flavored" pellets to goats one laborious trial at a time. We did the trials during fall and winter, with no sign that any of the compounds deterred feeding by the goats. By midwinter, only one compound remained to be tested – a condensed tannin plentiful in the bark of new twigs. By process of elimination, we figured, this tannin must be the feeding deterrent.

On the first morning of the final trial, the goats ate all of the tannin-infused pellets. We were surprised and bewildered. We had tested every compound that might have made the goats averse to eating new blackbrush twigs, and the goats had eaten every one. How could they be so averse to eating new growth when none of the secondary compounds we'd extracted had any effect? Not only that, but at the rate the goats ate the tannin-containing pellets the first day, we had only enough tannin-containing pellets left to conduct one more trial. We'd spent months of hard work collecting twigs and then extracting and purifying that condensed tannin. We didn't know what to do. As we pondered the situation that cold winter morning, we decided all we could do was feed the tannin-containing pellets again the next day.

Incredibly, when we offered the pellets the following day, the goats wouldn't touch them. On this second exposure to pellets high in tannins, the goats had somehow changed their preference. It wasn't a question of merely responding to flavor. If the goats had disliked the flavor of the tannin-infused pellets or innately recognized the pellets as something that would make them sick, they wouldn't have eaten them so enthusiastically on the first day. At that aha moment, we realized goats didn't innately know high-tannin pellets – or new blackbrush twigs – were bad for them. Rather, they had to learn from aversive postingestive consequences.[1] In other words, it took a queasy stomach (nausea) to teach them not to eat foods with tannins.

To confirm that hypothesis, in a subsequent trial, we supplemented goats foraging on Cactus Flats with a small amount of polyethylene glycol, a compound that binds to tannins in the gut, alleviating their aversive postingestive effects. Goats supplemented with polyethylene glycol don't experience the nauseating effects of tannins in blackbrush. With the deterrent effect neutralized, those goats were free to choose, and they preferred new to older growth twigs based on the higher energy, protein, and mineral content of the new twigs.[2]

At that time, we were also studying how lithium chloride causes food aversions in sheep. Lithium chloride – once used as a substitute for table salt and to treat manic depression in humans – in excess conditions a food aversion in animals.[3] Following the findings with blackbrush, we decided to repeat the trials with lithium chloride, but on goats as well as sheep. Sheep and goats who receive a capsule of lithium chloride acquire an aversion to any forage they ate just prior to receiving lithium chloride. Like humans, an *upset stomach* doesn't necessarily cause an aversion in goats or sheep, but *nausea* does. At the dosages we were infusing, neither the sheep nor the goats showed any overt signs of illness. Yet, the following day they avoided the food they'd eaten just prior to receiving the lithium chloride.

Though conditioned taste aversion was of key importance in psychobiology,[4] until the study with goats and blackbrush, neither we nor other scientists had a clue that secondary compounds in plants were communicating with cells and organ systems in herbivore bodies, providing feedback that changed liking for the flavor of a particular food. Rather, we had thought animals instinctively avoid foods that taste bad and choose to eat foods that taste good. During the next forty years, with this new understanding dawning, the research group I supervised carried out hundreds of studies that illustrated how *likes* and *dislikes* for the flavors of foods are caused by postingestive feedback emanating from cells, organ systems, and gut microbes.

In some studies, we worked with animals that had been made mildly deficient in primary compounds (energy, proteins, minerals, and vitamins). In our first studies, for example, we fed straw (a food with little nutritional value) to lambs deficient in energy.[5] Some of the straw was flavored with apple; some with maple. On day one, lambs in one group were given apple-flavored straw, while lambs in the other group were given maple-flavored straw. After they ate the straw, we gave all the lambs an oral drench of water directly into the gut. On day two, lambs in the group previously fed apple-flavored straw were fed maple-flavored straw, while lambs fed maple-flavored straw were fed apple-flavored straw. After the meal of straw on day two, we gave all the lambs an oral drench of energy. After several days of that protocol, the lambs were given a choice between apple- or maple-flavored straw. They strongly preferred the flavored straw that had been paired with the boost of energy delivered directly into the gut. Thus, one group preferred apple-flavored straw while the other group preferred maple-flavored straw.

We showed that feedback strongly influences preferences for flavors paired with both primary compounds and secondary compounds (phenolics, terpenes, and alkaloids).[6] We also found primary and secondary compounds interact with one another and with cells and organs to influence the choices animals make while foraging. The balance of primary and secondary compounds relative to needs strongly influences liking for flavors.

Secondary compounds set a limit on how much of any one food an animal can eat.[7] Thus, animals must eat a variety of plants that contain different secondary compounds, detoxified by different means in the gut and liver, in order to meet needs for energy and protein. Cattle who forage on high mountain pastures select from a smorgasbord of plants, including larkspur, which contains toxic alkaloids. How much larkspur a cow will eat during a meal varies from day to day. Cattle recognize when they reach a toxic threshold and they stop eating larkspur for the next few days.[8] That allows time to detoxify and eliminate those toxic alkaloids from their bodies.

Infusion studies with terpenes from sagebrush also show the sensitivity of herbivores to feedback. Terpenes give sagebrush its characteristic fragrance. Like any primary or secondary compound, in appropriate doses, terpenes are beneficial for health, but when the dose climbs too high, they become toxic. While elk, deer, cattle, and sheep use sagebrush as a nutritious forage in winter,[9] terpenes limit their intake of sagebrush in accord with the amount of terpenes these herbivores can detoxify and eliminate from their bodies.[10] When terpenes are slowly infused into the rumen or the bloodstream as sheep eat a meal, sheep stop eating before the amount of infused terpenes reaches a toxic level.[11] They resume eating only after terpenes in the body decline.

Terpenes thus affect *satiation* (processes that bring a meal to an end) and *satiety* (processes that inhibit eating between meals). Lambs reduce meal size (reach satiation sooner) and increase intervals between meals (longer satiety) when their diets are high in terpenes. When animals can eat a variety of different forages, which vary in kinds of secondary compounds, what ensues are cyclic patterns of intake of different foods from meal to meal and day to day as bodies regulate intake of foods with different kinds of secondary compounds. When animals eat small amounts of a range of plants containing secondary compounds, they expose cells and microbes in their bodies to a variety of secondary compounds beneficial for health, too.

In the early years of those studies, I was amazed that administering primary or secondary compounds directly into an animal's gut (or bloodstream) could markedly alter that animal's liking for the flavor of a food. It was counterintuitive to my experience of eating and to all I'd been taught about how taste influences preference. To further complicate matters, ruminants – including cattle, sheep, goats, elk, and deer – are walking compost heaps. They have four-chambered stomachs, and the rumen is a huge fermentation vat that contains mixed plant material being digested by thousands of species of microbes. How could signals from primary and secondary compounds not be lost in such a heap of fermentation? Over and over again, though, goats, sheep, and cattle showed us that the signals weren't lost in the rumen.

## Affective and Cognitive Processes

The body integrates information about foods through affective processes and cognitive processes.[12] Taste plays a prominent role in both processes. Receptors for taste are situated like a Janus head placed at the gateway to the body, with an "affective face" that looks at what's happening inside and a "cognitive face" that looks at what's happening outside.[13]

Affective processes integrate taste with postingestive feedback. The type of change in liking for and intake of a food item depends on whether feedback is aversive or positive. Food likes and dislikes mediated by sensations such as hunger, satiety, and nausea are how the gut tells the brain of a body's well-being – gut feelings. The result is *incentive modification*.

Cognitive processes integrate the odor and visual appearance of a food with its taste. Animals use smell and sight to select foods whose postingestive feedback has been positive; they avoid foods that have produced aversive feedback. Cognitive processes are strongly influenced by social models such as mother and other herd mates. The result is *behavior modification*.

Affective and cognitive processes are mediated by dorsal and ventral vagus nerves from the gut that converge with nerves for taste in the brain stem. From the brain stem, these nerves converge with other nerves as they relay from the brain stem to the limbic system and then to the cortex. The gut sends an enormous amount of information about the well-being of the body to the brain, including sensations that arise from digestive functions as well as stress and illness. Far more neurons ascend from the gut to the brain (afferent neurons) than descend from the brain to the gut (efferent neurons). For every message sent by the brain to the gut, about nine are sent by the gut to the brain. Because of the extensiveness of the enteric nervous system (afferent neurons), the gut has a great deal of sovereignty. As neurogastroenterology expert Michael Gershon notes in *The Second Brain*, the enteric nervous system has a mind of its own that continually informs the central nervous system about the status of the body. Following Gershon's discovery of the gut as a complex neural system, researchers have discovered that every neurotransmitter in the brain also operates in the gut. The enteric nervous system has more neurotransmitters and neuromodulators than any other part of the peripheral nervous system.

Discoveries in other fields have helped to deepen the understanding of the results we found in our research, which showed that what makes a food "taste good" to animals is not as simple as was once thought. Bodies are integrated societies of cells and organ systems, including the microbiome, each with nutritional needs. What I call "myself" is a conduit through which the cells and organ systems that make up my body meet their nutritional needs.

I can still recall how my understanding changed the day I realized that palatability and preference are not based entirely on *cognitive-rational-analytical* thought, but also on *noncognitive-emotive-synthetic* feedback that arises from within a body. The awareness that palatability is unconsciously influenced by cells and organ systems, including the microbiome, stopped me in my tracks. Until then, I'd been trained to think about foraging in a cognitive-rational-analytical sense in terms of the costs and benefits of selecting various foods.

Years later I came to appreciate that the most intriguing part of the tale reveals a dichotomy between two different kinds of rationalities. One has to do with *conscious thought* and the *rational mind*, and the other has to do with *emotional experiences* and the *wisdom body*. That's illustrated if you ask a person *why* they prefer anything from foods to mates to cars. While people can tell you *what* they prefer, we don't necessarily understand *how* their preferences originate. We don't need to think about that any more than we have to think about which enzymes to release to digest food. The body takes care of that – without a thought.

## Feedback from the Gut Microbiome

In 1968, near the Department of Defense's Dugway Proving Grounds in Utah, 6,400 sheep were killed instantly by aerially applied nerve gas that drifted in the wrong direction due to a sudden shift in winds.[14] In January 1971, over half of the sheep in a herd of 2,400 died of unknown causes in less than 24 hours near Garrison on the Utah-Nevada border. In that instance, though, nerve gas wasn't the cause. What happened near Garrison was due to a phenomenon that had been explained in an article that had been published twenty years prior. Photos in that article showed sheep carcasses strewn across the salt desert. A dead ewe lay in the foreground while another ewe nibbled an herb with bluish green leaves. That herb was halogeton (*Halogeton glomeratus*), which thrives in salt desert soil that is high in sodium and chlorine. Sodium in halogeton forms salts of oxalic acid, which is exuded as residue on leaves. That residue can be toxic to herbivores.

Ironically, sheep who have experience eating halogeton can eat up to 36 percent of their diet as halogeton without any ill effects. Poisoning occurs when hungry sheep who have never eaten the plant before (researchers use the term "naive" sheep) are forced to eat too much halogeton too quickly. Historically, that occurred when sheep were trucked long distances and put in areas dominated by halogeton.

Intoxication ensues when oxalate is absorbed faster than it can be detoxified by rumen microorganisms that degrade oxalates.[15] Sheep adapted to eating halogeton have much higher rates of oxalate degradation than do naive sheep. Successful transitions to a halogeton diet are accompanied by a ten-fold increase in the rate of oxalate metabolism by rumen microbes. Sheep need three to four days to build up an aptly large microbial population to rapidly degrade oxalate. Adapted sheep can tolerate two and a half times more oxalate than nonadapted sheep.

Scientists who study wild and domestic herbivores have long valued how diet affects the microbiome, which in turn affects the diet, as microbial ecologist Robert Hungate highlights in his classic book *The Rumen and Its Microbes*, published over fifty years ago. Ruminants and hindgut fermenters such as horses rely on bacteria to provide them with energy, protein, and vitamins. Bacteria digest the cell walls (cellulose) that give structure to plants. Volatile fatty acids such as propionate, butyrate, and acetate are produced from microbial fermentation of cellulose. They are absorbed directly through the rumen wall and provide most of the energy these animals require for activity and growth. The bodies of dead microbes provide 80 percent of the protein needs of ruminants. Bacteria in the rumen also synthesize B vitamins for ruminants.

Thus, the plants herbivores eat provide food for rumen microbes, which in turn feed herbivores. High-forage diets select for bacteria that digest cellulose, while high-grain feedlot diets select for bacteria that digest starch. Diets rich in secondary compounds select for microbes that enable herbivores to eat many otherwise potentially toxic plants such as halogeton.[16] The more diverse the diet, the greater the number of different microbial species that can thrive in the gut.

A positive feedback loop also exists between a person's food preferences, the species of microbes in their gut, and the preferences of those gut microbes for different dietary items.[17] Microbes provide feedback to the brain through nerves, neurotransmitters, peptides (short-chain amino acids), and hormones. The same is true for rodents. Studies of rats show that certain species of gut bacteria affect the levels of the appetite-stimulating hormone ghrelin, and studies of mice show that other species of gut bacteria affect the appetite-suppressing hormone leptin.[18]

The microbiome also affects gut motility and gut feelings through the hormone serotonin.[19] Serotonin regulates intestinal movements and influences appetite, sleep, and mood contributing to feelings of well-being and happiness. Serotonin is primarily found in the gastrointestinal tract (95 percent), blood platelets, and the central nervous system (5 percent) of animals, including humans. Specialized cells (enterochromaffin cells) in the gastrointestinal tract form synapses that link directly with neurons in the brain, which allows cross talk between the gut and the brain to occur in milliseconds rather than minutes. To protect the body from toxins, these cells immediately send signals to induce vomiting and diarrhea.[20] Enterochromaffin cells express chemosensory receptors for specific compounds, they are electrically excitable, and they modulate serotonin-sensitive afferent nerve fibers through synaptic connections. That enables enterochromaffin cells to detect and transduce environmental, metabolic, and homeostatic information (gut feelings) from the gut directly to the brain.

Microbes also affect emotions. For example, gamma-aminobutyric acid (GABA) is a key neurotransmitter that dampens the fear and anxiety that mammals experience when neurons in the central nervous system are over-excited. Species of gut bacteria such as *Lactobacillus rhamnosus* alter expression of GABA messenger RNA in various parts of the brain via the vagus nerve, which connects the gut with the brain.[21] Importantly, *L. rhamnosus* reduced stress-induced corticosterone and anxiety- and depression-related behavior. These findings highlight the key role of gut bacteria in communication of the gut with the brain and suggest gut bacteria may be useful therapeutic aides in stress-related disorders such as anxiety and depression.

## Learned Preferences in Herbivores

By guiding animals to eat a variety of foods, cells and organ systems in the body ensure nutritional needs are met. The body, however, can withstand departures in intake of nutrients, and thus it's not necessary to adjust intake of every nutrient at every meal. Homeostatic regulation needs only an increasing tendency, due to a steadily worsening imbalance, to generate behavior to correct a disorder. As nutritional state becomes inadequate, animals (including humans) sample novel foods readily – familiarity breeds contempt and novelty breeds contentedness. Once nutritional needs are again met, animals don't experience *cravings* from deficits or *malaise* from excesses. When nutritional state is adequate, animals sample novel foods cautiously – familiarity breeds contentedness and novelty breeds contempt.

Animals prefer foods high in energy and protein, because their bodies need large amounts of energy and protein. These nutrients strongly influence food selection when animals are given choices of foods that differ in energy and protein.[22] Lambs select diets with a higher ratio of protein to energy to meet their needs for growth, but they eat less protein as they age and their need for protein declines.[23] Ewes increase intake of protein relative to energy as their needs for protein increase with the growth of the fetus during the last trimester of gestation, or when they are infected with parasites.[24] When fed a diet with an imbalance between protein and energy, sheep choose to forage in locations with foods that rectify the imbalance.[25] Dairy cows fed protein supplements during lactation avoid eating plants (for example, legumes) and plant parts (for example, new growth) that are high in protein when they forage on grass-legume pastures.[26] They select plants and plant parts high in protein when they are fed energy-rich concentrates, such as corn grain during lactation.[27] Lambs even become averse to diets deficient in *specific amino acids* and they choose to eat foods that rectify the deficit.[28]

How do livestock maintain a balance of energy and protein in their diets? They associate the flavors of foods with nutrient-specific feedbacks. When lambs are fed a meal that consists of an "appetizer" and a "main course," they make choices that balance the ratio of energy to protein within a meal. For example, when lambs are fed a high-energy appetizer and then offered various main course choices, they prefer the flavor previously paired with protein. Conversely, lambs fed a high-protein appetizer prefer the flavor previously paired with energy for the main course.[29] They are even sensitive to the rates at which different sources of energy and protein ferment in the gut, as imbalances result in digestive and metabolic upsets.[30]

The foods that animals consume to meet their needs for energy and protein usually contain vitamins and minerals, too. If animals become deficient in vitamins or minerals, however, they will select foods to rectify the deficit. Sheep deficient in vitamin E prefer food higher in vitamin E, for example.[31] Sheep can make multiple flavor-feedback associations with minerals.[32] We designed a study in which we made lambs deficient in one of three minerals – phosphorus, calcium, or sodium – and gave them a choice of the three differently flavored foods. The lambs had previously ingested these flavored foods with one of the three minerals. Lambs preferred the flavor previously paired with repletion of the mineral – phosphorus, calcium, or sodium – they were lacking. These findings support the practice of offering a selection of individual minerals free-choice so animals can select the particular mineral(s) that are low in the forages they are consuming. The same is true in principle and practice with medicines as I discuss in chapter 7.

Unusual foraging behaviors arise as animals sample new and often strange foods in an attempt to rectify nutritional imbalances. In the wild, sheep, caribou, and red deer rectify deficits by eating lemmings, rabbits, and birds – live or dead; sheep eat arctic terns and ptarmigan eggs; white-tailed deer dine on fish; and deer gnaw on antlers.[33] Bighorn sheep use rodent middens as mineral licks.[34] Cattle deficient in phosphorus eat bones, and they stop eating bones when inorganic phosphate levels in their blood rise to normal ranges.[35] Cattle and sheep rectify mineral deficits by eating soil, licking urine patches, and eating fecal pellets and dead rabbits.

In over forty years of working with sheep, I had never seen one eat soil or feces or lick urine patches. But when I was part of a group studying sheep fed a phosphorus-deficient diet, we observed all of those behaviors. We were feeding sheep a phosphorus-deficient diet to see if they would select a food that would rectify the deficiency.[36] However, even after a couple weeks of feeding the phosphorus-deficient diet, we couldn't get blood phosphorus levels in the "deficient" sheep to drop. Why? They were sticking their heads through the wire panels to eat the feces of sheep in adjacent pens that were being fed a diet that was adequate in phosphorus. When we moved the phosphorus-deficient sheep into new pens, away from the phosphorus-replete sheep, they immediately began to eat the feces on the ground from a previous study. We eventually made them deficient, and they selected the food with phosphorus. Such perverted appetites or picas are adaptive behaviors that commence as animals begin to experience deficits.

This phenomenon also explains what happened when Angora goats of my early research study learned to eat woodrat houses.[37] Inside those houses was densely packed vegetation soaked in urine, which is high in nitrogen in the form of urea. The goats had discovered that source of nonprotein nitrogen, which rumen microbes use to synthesize aminogenic (protein) and glucogenic (energy) nutrients, and they took advantage of it to help meet their needs. But as I noted, only one group of goats ate woodrat houses the first winter, and over the ensuing two winters, out of eighteen groups of goats of different breeds and from different locations, only that one group of goats ever learned to use woodrat houses as a source of nutrition. These findings illustrate the peculiar ways herbivores can discover foods that meet their nutritional needs. By learning from mother and peers, each generation benefits, and such behaviors become part of the foraging culture – the collective nutritional wisdom – of the group.[38]

## Learned Preferences in Humans

Like herbivores, people acquire likings through flavor-feedback associations, where the flavor of food and positive consequences of nutrient ingestion lead to an acquired liking for the flavor of the food. These associations are influenced by the novelty of food, the amount of nutrients in the food, and the needs of an individual for nutrients in the food.

Unlike with herbivores, few studies have focused on flavor-nutrient learning in humans, and they don't always show changes in preference or liking after flavor-nutrient pairings.[39] Two challenges arise when researchers attempt to study humans. First, accounting for experiences in utero and early in life is virtually impossible, yet those experiences profoundly influence familiarity with flavors and foods and the choices people make. That's why young children more reliably increase preference in studies of flavor-nutrient learning. Adults show less consistent responses due to their long and unknown (to researchers) histories of experience eating different foods. With children and adults, food novelty decreases preference and intake because people prefer familiar to unfamiliar foods.[40] Second, researchers typically have little short- or long-term control over appetitive (nutritional) states in humans, yet different nutritional states strongly influence the outcomes of studies of food selection and satiation.

Most studies of food preferences in humans attempt to create an effect by using energy, but people whose daily diets are already high in energy show only weak preferences for novel-flavored, high-energy foods. They lack familiarity with the novel flavor, and they don't need more energy. That was evident in a study that compared people in the UK with Samburu people, a native group of seminomadic pastoralists in Kenya who are lean and food stressed.[41] In the study, the participants ate nearly a pound (400 g) of either an energy-dense version (1.57 kcal/g) of a novel food or a less energy-dense version (0.72 kcal/g) of the novel food.

The researchers found no evidence of flavor-nutrient learning between the two versions of the novel food. As expected, the fifty-two people from the UK reported feeling greater fullness after eating the high-energy version than after the low-energy version. Their sensitivity to the energy-dense food is a result of their high intake of energy-dense foods. Conversely, the sixty-eight Samburu people didn't report a different fullness response between the high and the low version of the food, a reflection of their overall poorer nutritional status and the relatively small differences in the energy content of the foods. The Samburu also experienced a return to premeal hunger within an hour after consuming a test food, and they reported having a capacity to consume three times more food than people in the UK. In a food-stressed environment such as the Samburu's, it makes little sense to reject food even if the food has a low-energy density.

Good examples of appetitive state-dependent learning occur when people experience nutrient deficits in real life conditions.[42] People develop cravings for fat when they are stuck eating lean-meat diets; cod liver oil when they are suffering from rickets; fruits when suffering from scurvy; calcium when vitamin D deficient; and salt when they are salt deprived. When people are deficient in minerals, they exhibit pica, as livestock do, and may crave specific, often unusual "foods" such as soil, a practice referred to as geophagy. Compelling evidence for flavor-nutrient learning in humans also comes from studies of flavor compounds in tomatoes and other fruits. The flavors people find most appealing are related with the essential nutrients – including essential amino acids, carotenoids, and omega-3 fatty acids – a body needs to function.[43] The delicious flavor of a phytochemically rich tomato is directly linked with feedback from the primary and secondary compounds that make tomatoes healthy.

These findings would not have surprised Curt Richter, America's foremost psychobiologist in the twentieth century.[44] As the first empirical scientist in the field of psychobiology, he influenced every facet of the discipline. Richter and his colleagues deprived animals of nutrients essential to survival or manipulated their hormone levels and showed that these needy states generate appetites and behaviors precisely fitting the animal's need. Richter's work was in the tradition of Claude Bernard, who emphasized regulation of the internal milieu, and Walter Cannon, who emphasized homeostasis. While both Bernard and Cannon were concerned with physiological mechanisms that underlie maintenance of the body, Richter emphasized the role of behavior. Early in Richter's career, one of his colleagues at Johns Hopkins University, noted nutritionist E. V. McCollum, showed that changes in diet invoke compensatory behavioral responses, a demonstration of Cannon's concept of "the wisdom of the body." By that time, Clara Davis had already done her studies that suggested children could maintain nutritional balance by self-selecting a diet from a variety of foods. Richter, more than anyone else, went on to show that animals could maintain homeostatic balance when offered selections of energy, protein, minerals, and vitamins, including the B-complex, A, D, and E.

Richter studied a range of behaviors, including ingestion of minerals during pregnancy and lactation. He charted mineral intake during the reproductive cycle and found that a variety of minerals essential in pregnancy and lactation were ingested in large amounts, serving both the mother and the offspring. Richter discovered many other behavioral adaptations, including coprophagia (feces ingestion) in nutritionally deprived rats, a practice that procures required nutrients. Richter also investigated bait shyness, the reluctance of rats to eat unfamiliar foods (food neophobia). He was one of the first to describe learned poison avoidance during his efforts to control rat populations in Baltimore during the Second World War. Years later, learned poison avoidance (conditioned taste aversions) became of central importance in the psychology of learning. As Richter concluded in a talk titled "Total Self-Regulatory Functions in Animals and Human Beings," "Thus, we believe that the results of our experiments indicate that in human beings and animals the effort to maintain a constant internal environment or homeostasis constitutes one of the most universal and purposeful of all behavior urges or drives."[45]

## Interactions Between Primary and Secondary Compounds

Herbivores can resist toxins as a result of morphological adaptation (not affected by a toxin) and physiological adaptation (able to detoxify the toxin). But a study of herbivorous insects' response to *Bacillus thuringiensis* (Bt) toxin indicates that diet plays a role, too. *B. thuringiensis* is a bacteria that produces proteins toxic to some insects. The toxin dissolves in the gut and then attacks the gut cells of the insect, punching holes in the lining. The Bt spores spill out of the gut and germinate in the insect, causing death within a couple days. However, when insects are given a nutritious diet higher in protein relative to carbohydrate, they can better tolerate eating genetically modified plants that contain the Bt toxin.[46]

Diet affects tolerance for toxins by large herbivores, too. Providing supplemental energy and protein enables animals to eat more of plants high in secondary compounds including terpenes, tannins, saponins, and flavones.[47] Sheep and goats double their intake of forages such as sagebrush that contain high levels of secondary compounds when their diet is more nutritious.[48] With most secondary compounds, herbivores prefer a higher ratio of protein to energy, though lambs infused with cyanide prefer foods with higher ratios of energy to protein.[49]

Why and how does a high-quality diet increase the tolerance for toxins? Animals cope with excess secondary compounds by excreting them in urine in the form of conjugated amino acids, glucuronic acid, or sulfates. Creating these compounds also produces organic acids that can disrupt acid–base balance in the body and deplete energy and protein.[50] Thus, animals require more energy, protein, and water to detoxify and eliminate these compounds.

Large herbivores also learn complementarities among secondary compounds. For example, some varieties of tall fescue, a common pasture grass, contain a fungus that produces toxic alkaloids – so-called endophyte-infected tall fescue. Eating an appetizer of plants such as birdsfoot trefoil or sainfoin, which are high in tannins, enables sheep and cattle to consume a much greater amount of endophyte-infected tall fescue.[51] The same is true for plants that are high in terpenes, such as sagebrush.[52] This is because tannins bind with alkaloids and terpenes in the animal's digestive system, rendering these compounds less harmful.

Not surprisingly, the ability of humans to cope with the adverse effects of excessive secondary compounds is also influenced by nutritional state. That is illustrated for people who eat the roots of cassava, a nutritious but potentially poisonous plant. Cassava was introduced to Africa by the Portuguese during the late sixteenth century. It was well suited for integration into the farming systems and socioeconomic conditions of Africa. With a decline in nutrition in rural areas during the twentieth century, people began to report mysterious epidemics of paralysis (konzo) in rural regions of Africa. In the mid-1980s, researchers linked the malady to cyanide in cassava.[53] Cyanide shows its effects when people suffer from malnourishment. The areas most affected by konzo are far away from markets, clinics, and paved roads – poor nutrition is frequently due to poverty, conflicts, and lack of infrastructure.

Cyanide can be eliminated when cassava is soaked in water, as demonstrated by the women of Kay Kalenge village.[54] By using the wetting method, they reduced cyanide in cassava flour to safe levels and produced tastier fufu, a staple food made by mixing and pounding equal portions of cassava and green plantain flour thoroughly with water. Less cyanide reduces thiocyanate levels in the urine of schoolchildren to safe levels, which prevents konzo. Through the wetting method, they prevented konzo for the first time since it was identified in Popokabaka Health Zone in 1938. This method is now being used in three villages in Boko Health Zone where konzo is prevalent.

With cassava we come full circle ecologically, economically, and culturally. Cassava contains cyanogenic glycosides that play a key role in defending cassava from would-be attackers. When an insect or mammalian herbivore chews on cassava, the chewing results in transformation of cyanogenic glycoside to cyanide. Hence, the plant is beautifully self-defended, and there's no need for added pesticides or engineered toxins such as Bt. Relying on naturally occurring phytochemicals in crops and forages was beneficial for humans, too, because it eliminated the need for and risks of applying pesticides to the cassava crop. Through development of a cultural practice – the wetting method – women reduced cyanide to safe levels. Such social rituals around food gathering and cooking to decrease secondary compounds and increase digestibility of fiber, now rare in cultures, were once the norm.[55]

## A Change of Heart

When Claire Sylvia was asked – while riding an exercise bike after surgery – what she wanted more than anything, she replied "a beer." As she recalled, she was "mortified that I had answered this sincere question with such a flippant response." She was even more astonished because she didn't even like beer. To her bewilderment, she was now fond of foods she hadn't liked before, especially Snickers bars and Reese's Peanut Butter Cups. She was also showing a strange new affinity for green peppers, which she had never liked. After her organ transplant, she found herself adding green peppers to every conceivable dish, although she had no idea why. "When I was finally allowed to drive again, my car practically steered itself to the nearest Kentucky Fried Chicken." Again, this made no sense at all to her because she never ate at fast-food outlets. But for some inexplicable reason, she now craved chicken nuggets. Sometime later, after she made contact with her donor's family, they informed her that she had inherited the heart and lungs of their eighteen-year-old son, who loved all of the foods she now craved.

During her years as a support-group leader, Claire listened to stories from kidney and heart recipients. The most common change they experienced was food preferences. The list is long and includes a steak eater who became a vegetarian; a milk drinker who began to hate milk; a man who always wanted to like wine but never enjoyed it until after his transplant. One man, who never liked coffee, received a kidney from his sister, a big coffee drinker, and soon he, too, was drinking coffee. A man, who received a kidney from his wife, said he craved her favorite foods; while she still liked those foods, she liked them less after she donated her kidney. One man hated Italian food, but craved spicy Italian food after his transplant. The first food he asked for after his operation was linguine. He learned later that his donor was Italian.

Historically, scientists and philosophers believed intelligence is the domain of the brain alone. They didn't accept the notion of cellular memory or emotion. As a medical doctor told Claire, "I can't blame you for feeling this way. If I were carrying somebody else's heart and lungs within me, I'd probably have all kinds of thoughts and fantasies about the donor. For centuries, poets and artists have told us that the heart is a great repository of spirit and emotion. That's a beautiful idea, but clinically there's nothing to it. The heart is just a pump."

That view changed with the research of neuroscientist and pharmacologist Candace Pert, author of *Molecules of Emotion*, who discovered that the intelligence of the body emanates from interactions among cells and organ systems mediated by molecules of emotion. Cells and organ systems communicate through peptides that attach to thousands of receptor sites and activate effector proteins residing on cell membranes. Receptors for taste occur on cell membranes throughout the body, including the tongue, gastrointestinal tract, pancreas, respiratory tract, heart, and brain. Cells lining the gut have many sensory receptors, including for taste, odor, nutrients, secondary compounds, toxins, and foreign chemicals.[56] Cells and organs have memories that influence liking and preference.

Receptor (awareness) and effector (action) proteins are embedded in the membrane of each cell. Receptor proteins – the eyes, ears, nose, and taste buds of the cell – are the basis for cellular intelligence: units of perception that create awareness of the environment through physical sensations. A cell membrane has thousands of receptors and effectors that enable a cell to engage in appropriate, life-sustaining behaviors. Cell membranes learn from interactions with the environment. The behavior of a cell can't be determined by examining any individual protein. It can only be determined by considering how all of them act in concert.

Some scientists now believe every cell in every plant and animal has a consciousness of its own. And some postulate when organs of humans are transferred from one body into another, the cells that make up the organs from the donor's body will transport and express memories in the recipient's body of their past experiences in the donor's body, as Claire Sylvia suggests. These findings illustrate the profound cellular ties that link plants and animals with the environments they inhabit. They also raise a question: What is this being I call "myself?"

During her physical and spiritual journey, Clair confirmed that the new personality cohabiting her body was actually that of her donor. Along the way, she struggled with profound and timeless questions: Where does body end and spirit begin? Is it possible for cells and organ systems to survive death? Can cellular memory outlive physical death?

# CHAPTER 5

# More Than One Kind of Memory

Years ago, I got a letter from Mick Holder, a rancher in Arizona, about an odd situation. "Gila County is mercifully deficient in poisonous plants," he wrote, "but we do have lupine and loco in small or moderate stands. In thirty years of ranching, I never had a problem with either. I leased rangeland in Apache County and moved a portion of cattle to that location during a drought. They suffered severe problems with the poisons in Apache County while the sister cattle left here in Gila County on equally poor rangeland did not have one case of loco or lupine poisoning. Did they not recognize the plants because they had been relocated 100 miles east?"

While Mick's cattle might not have recognized lupine and loco in Apache County, that's unlikely. Cattle and sheep learn new foods quickly when foraging with their mothers, and they remember those foods for years, even after only brief exposure in utero or early in life.[1] Lambs offered a new food with their mother for as little as one hour a day for five days when they are six weeks old remember the food three years later with no intermittent exposure. The same is true for cattle: They learn quickly and can remember foods for at least five years.[2]

So if Mick's cattle did recognize lupine and loco, why did they die from eating them in Apache County but not die from eating those plants in Gila County? How does a body know which foods cause which toxic postingestive consequences, given all the different foods herbivores eat within and among meals? What kinds of memories form the basis for how the cells and organ systems of herbivores and humans distinguish between safe and harmful foods?

## First Impressions Matter

In addition to the lessons goats taught me about eating woodrat houses to obtain nitrogen and avoiding new twigs of blackbrush because they are high in tannins, they taught me one way a body discerns which unfamiliar foods produce which postingestive consequences. Once I understood that goats were learning to avoid new twigs of blackbrush, I began to ponder how they learned to avoid them.[3] I'd never seen goats eat new twigs, except for the single nibble by one goat the evening Sue and I trailed them to the lightning-struck juniper tree surrounded by blackbrush. So, I reasoned, they must learn very quickly not to eat the new twigs and they must not need to eat many twigs to learn. What, I wondered, if they also eat other foods during a meal. How do they know the culprit is new blackbrush twigs and not some other item in their diet?

To figure that out, I offered goats naive to blackbrush a choice between new and old twigs, and I measured the amount of each twig type they ate in each meal throughout the day for three days. I found that as soon as a goat ate more new twigs than older twigs and ate enough new twigs to make them sick, they avoided new twigs from that meal onward.

That didn't happen in the same meal for all goats. For some, it occurred in their first meal, but for others, who didn't eat more new twigs than older twigs for several meals, the development of an aversion to new twigs happened later. But by midday of the second day, all of the goats had come to a conclusion about their taste for new twigs. And some goats went against popular opinion! Most goats (80 percent) strongly avoided new twigs, but some (20 percent) relished them, evidently because they didn't experience any postingestive distress. The goats were thus able to learn about the two twig types by eating small amounts of the twigs and letting their bodies integrate the flavor of each food with the ensuing postingestive consequences.

Bodies develop rules of association to prevent toxicity. The most important has to do with first impressions. That's when a creature assesses whether an unfamiliar (novel) food is harmful or beneficial. Animals are both curious about and cautious of anything new or different. They sample novel foods cautiously, as illustrated when the goats ate only small amounts of young and old blackbrush twigs. If the consequences are harmful, they quickly learn to avoid the food. If the consequences are beneficial, they quickly learn to eat the food. When sheep eat a meal of four familiar foods (alfalfa, barley, oats, and corn) and one novel food (rye), and then get an orally administered nauseating capsule of lithium chloride, they subsequently avoid the novel food, but not the familiar foods.[4] So they learn to associate the nausea specifically with the novel food. Conversely, if sheep experiencing a deficiency are given a similar meal and then given a capsule with the needed nutrient, they form a preference for the novel food.[5] Associating a novel food with its postingestive consequences is an effective survival strategy that is enabled because animals become familiar with foods learning from mother in utero and early in life.

Many people experience a mix of curiosity, caution, and trepidation as they search for a safe place to eat while traveling in a foreign country. Given a choice, we stick with what's familiar. That's why many folks from the United States search for the Golden Arches when traveling in foreign countries, even if they hate McDonald's burgers and fries. Creatures who aren't cautious of unfamiliar foods and places don't live long – they end up eating toxic foods.

Like people in a foreign country, cattle moved to Apache County may have stuck with eating familiar plants like loco and lupine rather than sampling potentially toxic unfamiliar plants. The strength of these responses is shown with studies of sheep, which prefer to eat a familiar food they dislike – because it had been previously paired with a dose of lithium chloride – to unfamiliar foods when they were in an unfamiliar setting.[6] The poison you know is preferable to the one you don't know.

So, what happens when people move animals to unfamiliar terrain? We override their chief defense – the familiar novel dichotomy – and they are more likely to become ill or die from overingesting poisonous plants. That's what happened when 2,400 sheep died from eating halogeton near Garrison on the Utah-Nevada border. That's also what happens when livestock producers move naive animals into areas where endophyte-infected tall fescue is the primary forage. Due to the alkaloids it contains, endophyte-infected tall fescue is toxic if ingested in overly large amounts. When animals unfamiliar with fescue are moved to fescue pastures, the conditions are ripe for creating a strong food aversion: Offer animals a novel food in a novel environment and follow food ingestion with alkaloid-induced toxicosis. There's no better way to condition a strong food aversion. To make matters worse, people expect the animals to continue eating the fescue as the bulk of their diet. No wonder food intake and animal performance are so low. Producers talk about how long it takes animals from other parts of the United States to adapt to fescue. Cattle moved from arid eastern Colorado to lush Missouri often take nine months or more to make the adjustment, and in many cases, they never do fully acclimate.

These problems can be alleviated by gradually introducing animals to fescue, so they can learn to eat small amounts without getting sick. That works best when animals have nutritious alternative forages that complement the primary and secondary compound profiles in fescue. Forages high in energy and protein and low in alkaloids facilitate detoxification of alkaloids. As previously discussed, plants such as birdsfoot trefoil, which are high in tannins, complement the alkaloids in fescue because alkaloids and tannins bind to one another when both are ingested.[7]

## Generalizing Based on Past Experiences

Animals generalize from familiar to unfamiliar circumstances. For example, sheep trained to avoid eating cinnamon-flavored rice – by following ingestion of cinnamon-flavored rice with a mild dose of lithium chloride – avoid eating any cinnamon-flavored food.[8] Flavor associations also work in the positive direction. When sheep learn to prefer a food with a distinct flavor, they generalize the preference for that familiar flavor to unfamiliar foods, and they are more likely to eat an unfamiliar food with the familiar flavor. Spraying unfamiliar foods with a known and liked substance, such as molasses or high-fructose corn syrup, encourages livestock – and humans – to eat unfamiliar foods. For livestock, this trick works well to ease the animals into eating weeds such as thistles. Moreover, the energy in molasses helps microbes in the gut degrade the nitrates in thistles, a double benefit.

The propensity to generalize is also related to Mick's question about his cattle: "Did they not recognize the plants because they had been relocated 100 miles east?" In fact, generalization may have been the Trojan Horse for lupine and loco in Apache County. Mick's cattle may have learned to eat safe amounts of lupine and loco in Gila County, but they may have generalized wrongly in Apache County, mistakenly eating amounts of these forages that were safe in Gila County, but deadly in Apache County.

The chemical features of loco and lupine likely differed in the two environments. Plants that look the same on the outside may not be the same on the inside. The amount of sunlight, nutrients, water, associated plants, and past use by herbivores all affect plant chemistry, as we discussed for crops and produce in chapter 2. The lupine or loco plants in Apache County may have been more toxic than those in Gila County. If so, the same amount of lupine or loco eaten safely in Gila County might have been quite toxic in Apache County.

We know that the mix of foods in the diet influences the toxicity of alkaloid-containing plants such as lupine and loco. We also know that when cattle and sheep eat foods high in tannins along with foods that contain alkaloids, the adverse effects of the alkaloids are greatly reduced. Apache County may have been devoid of forbs or shrubs containing tannins. In addition, well-nourished animals can tolerate a toxin dose better than poorly nourished animals. Poor nutrition could have made cattle in Apache County more susceptible to lupine and loco.

Less obvious is the notable phenomenon that a particular dose of a toxin has a greater effect in an unfamiliar than in a familiar environment. That's because environmental context prepares an individual physiologically for what's coming next, as Pavlov showed many years ago with a bell, food, and salivation in dogs. A change in context is why social drinkers are more impaired when they drink at unusual times or in different settings.[9] S. Siegel and D. W. Ellsworth tell of a cancer patient who died when injected with morphine in a different room; the patient had tolerated the same dose when injected every six hours for four weeks in a familiar room.[10] Those observations inspired their studies of rats, which showed that survival after a heroin overdose improves if heroin is taken in a familiar setting. Rats, with or without experience with heroin, were given a dose of heroin in a familiar or a novel setting. The dose was lethal for 32 percent of the experienced rats in a familiar environment, 64 percent of the experienced rats in an unfamiliar environment, and 96 percent of the inexperienced rats in an unfamiliar environment.[11]

The food industry takes advantage of our propensity to generalize from past experience to train people to eat artificially flavored foods with equally toxic, though long-delayed aversive effects. Fruits and berries are good examples. Although their flavors are similar, real and artificial strawberries are vastly different nutritionally.[12] A 284-gram (90-kcal) serving of real strawberries (cost $1.50) has 5 grams of fiber, many minerals and vitamins, and thousands of phytochemicals. A 28-gram (90-kcal) portion of Strawberry Fruit Gushers (cost $0.46, or 330 percent more) has 9 grams of sugar, 1 gram of fat, and none of the benefits of strawberries because Strawberry Fruit Gushers have no strawberries. Rather, they consist of pears (from concentrate), sugar, dried corn syrup, corn syrup, modified cornstarch, fructose, and grape juice (from concentrate). Strawberry Fruit Gushers are a good example of "empty calories" – foods with no phytochemical richness. While people tout cost as a reason to eat processed foods, the cost *per unit of nutrient richness* is far less for a strawberry than for a Strawberry Fruit Gusher.

The flavor industry is expert at mimicking flavors of fruits and spices, as Canadian science writer Mark Schatzker describes in *The Dorito Effect*. Compounds such as dimethyl anthranilate and many others, used to make grape-flavored Kool-Aid, candy, soft drinks, gums, and drugs, mimic the flavor but lack the phytochemical richness of real fruits. Man-made foods have become ever more desirable as people learned to link a multitude of synthetic flavors with feedback from energy-rich fats and carbohydrates. The variety of these flavors stimulates appetite and intake while obscuring the nutritional sameness of these foods.

Humans then generalize learned flavor-feedback relationships from high-calorie foods and beverages to similarly flavored foods and beverages that contain few or no calories – with many ill effects.[13] As the negative effects of consuming sugar-sweetened beverages have been documented, many people have turned to sweeteners such as aspartame, sucralose, and saccharin. However, consuming sweet-tasting but noncaloric or reduced-calorie food and beverages interferes with learned responses that normally contribute to glucose and energy homeostasis. Sweet tastes evoke numerous physiological responses that help to maintain energy homeostasis by signaling the imminent arrival of nutrients in the gut and by facilitating the absorption and utilization of energy contained in food. By weakening the validity of sweet taste as a signal for caloric postingestive outcomes, consumption of artificial sweeteners impairs energy and body weight regulation. Due to this interference, frequent consumption of artificial sweeteners has the counterintuitive effect of inducing metabolic imbalances. Growing evidence suggests that people who frequently consume sugar substitutes may also be at increased risk of excessive weight gain, metabolic syndrome, type 2 diabetes, and cardiovascular disease.

## Reason, Emotion, and Memory

Webster's dictionary defines *preference* as *greater liking* with the connotation of cognitive, rational thought and action. But if you look up *preference* in *You've Got to Be Believed to Be Heard*, a book by communication expert Bert Decker, you'll find it described as "acting on emotion and justifying with fact." Decker argues that to effectively communicate we must first get past another person's "gatekeeper." The gatekeeper is the gut-level emotions elicited automatically – they're noncognitive, so we don't think about them – as we interact with one another. These gut-level feelings transcend spoken words and gestures, and they make us feel at ease or suspicious of the speaker and what's being said. Basically, the gatekeeper is the "bullshit detector" everybody possesses. It operates constantly at a noncognitive gut level.

Asked which item we prefer, most of us can usually quickly choose among various alternatives – from different foods to clothes to cars to mates. Pressed further about why we prefer one over another, we can often come up with reasons for our preferences, though the more we are pressed, the less able we are to explain where the preference we have originates. At a basic level, our preferences have to do with our uniqueness in form and function. That identity, which begins to develop in the womb and early in life, fundamentally influences how we perceive the world, what we believe about the world, and how we behave in the world. Decker argues *emotion is the basis for our preferences* and the choices we make. "If the first thing to understand about communication is that we are all selling something," he writes, "then the second and even more crucial thing to understand is this: People buy on emotion and justify with fact. You may resist this statement. You may want to shout, 'No! No! No! I am a rational, cognitive human being! I make calm, considered, well-thought-out decisions! I do not buy on emotion!'" But Decker provides examples of decisions – regarding marriage and jobs, buying cars and houses, awarding sales contracts and raises – and points out that in every case, people *act on emotion and rationalize with facts*. In all cases, short-term consequences, emotionally driven, have a far greater influence on behavior than long-term possibilities, rationally conceived.

That's why creatures often don't behave rationally based on mathematical constructs such as game theory. The *emotions* of the *wisdom body* are rational, but not in a linear, predictable mathematical sense. Stock markets don't behave as economists predict they would if they were *cognitively rational* because people buy and sell based on *emotion*.

Carl Jung used the term *collective unconscious* to refer to how organ systems in organisms that have nervous systems integrate and organize experiences. He distinguished the *collective unconscious* from the *personal unconscious*. The collective unconscious assembles and organizes personal experiences in a similar manner for each member of a particular species, while the personal unconscious is a reservoir of experience unique to each individual.

Three kinds of memories link the collective unconscious with landscapes.[14] The first is the memory of a species with genetic information – fashioned by environments over millennia – for how a body in concert with the environment fashions form and function. The key is flexibility for a body to continually change form, function, and behavior based on consequences. The second kind of memory is mother, a source of transgenerational knowledge that increases efficiency and reduces risk of learning what and what not to eat, where and where not to go. The third kind of memory is acquired through experience. Postingestive feedback from nutrients (satiety) and toxins (malaise) enables individuals to experience the consequences of food ingestion and to adjust food selection commensurate with a food's utility.

Two kinds of memory underlie the personal unconscious of creatures from fruit flies to humans. *Declarative memory* provides details about objects and events. It involves recognizing specific places like your home or workplace and things like snakes or foods. *Emotional memory* links objects and events with feelings of fear or contentment, sadness or pleasure, malaise or satiety, and so on, which are integrated responses of cells and organ systems to external milieu.

In *Molecules of Emotion*, Candace Pert points out that the experiences of *emotions* and *feeling* arise when hormones, neurotransmitters, and peptides interact with and enable ongoing conversations within and among cells in a body. As she writes, "A feeling sparked in your mind will translate as a peptide being released somewhere. Peptides regulate every aspect of your body, from whether you're going to digest your food properly to whether you're going to destroy a tumor cell." What I generally refer to as "me" is ongoing interactions among cells and organ systems linked with social and biophysical environments through molecules of emotion.

It's interesting to consider most of these conversations never reach the level of conscious awareness. Creatures simply end up *feeling* certain ways about certain things, preferring one thing over another, and behaving in particular ways – little of which we comprehend. We "buy on emotion and justify with fact." So emotions aren't merely a luxury, as neuroscientist Antonio Damasio emphasizes in *Descartes' Error*, they are essential for making choices. Though they may be utterly rational beings, people who lack emotions can't make decisions.

Many scientists believe consciousness results from interactions among neurons in the brain. The brain takes in information through the senses and transforms it into a rich digital tapestry. We generally don't consider consciousness as the interrelationship of cells and organ systems in the body in concert with the biophysical environments we inhabit. Yet, this is the "thought" that underlies the wisdom of plants and the wisdom of the bodies of animals, including humans. And it is not the cognitive-rational-analytical processes we typically ascribe to intellectual matters. It is the noncognitive-intuitive-synthetic ground of our being, which humans often override by conscious thought influenced by culture.

The TED Talk "My Stroke of Insight" by neuroscientist Jill Bolte Taylor is an insightful account of the influence of the so-called left brain and right brain on how we experience life.[15] This notion, based on the fact that the brain's two hemispheres function differently, first came to light in the 1960s, a result of research by Nobel Prize–winner Roger W. Sperry. The theory that people are right-brained or left-brained states that if you are more creative or artistic, you're thought to be right-brained; if you are more analytical and methodical in your thinking, you're said to be left-brained. The left-brain is more verbal, analytical, and orderly than the right brain. Sometimes called the digital brain, it is better at tasks like reading, writing, and computations. Whether you're performing a logical or creative function, though, *you're receiving input from both sides of your brain*.[16] For example, the left brain is credited with language, but the right brain helps you understand context and tone. The left brain handles mathematical equations, but right brain helps out with comparisons and rough estimates.

On December 10, 1996, Taylor, a thirty-seven-year-old Harvard-trained brain scientist, had a stroke in the left hemisphere of her brain. As Taylor recounts in her TED Talk, she observed her mind deteriorate to the point that she could not walk, talk, read, write, or recall any of her life. During the twenty-four-hour event, she alternated between the euphoria of the intuitive and kinesthetic right brain, in which she felt a sense of complete well-being and peace, and the logical, sequential left brain, which recognized she was having a stroke and enabled her to seek help before she was completely lost. For Taylor, her stroke was a blessing and a revelation. It taught her that by "stepping to the right" of our "left brains," we can uncover emotional states of peace and well-being that are suppressed by constant "brain chatter."

Unfortunately, nowadays our right hemispheres are severely repressed. We attempt to participate in the world through the left brain, and the utter lack of balance between left and right brain isn't working well for us. We think linearly, rather than holistically. We accent ego over our linkages with others in our communities, nation, and world. The materialistic facets of life have taken over, while the spiritual parts of our nature are lost in a morass of fear, frustration, anger, and aggression. Yet, many experiences can give us access to right-hemisphere function including music, dance, shamanic rites, and rituals formerly linked with mythologies. Quiet time in meditation and in nature are effective, too.

## Back to Decker's Gatekeeper

During my years in high school in Colorado, I was a member of a ski racing team. One evening after our team took part in a race at a ski area near Glenwood Springs, we stopped for dinner at a café in Leadville. I can still recall the foods I ate that night: roast beef, green beans, and mashed potatoes with gravy. The memory is vivid because a few hours later, in the wee hours of the morning, I became dreadfully ill – nausea, vomiting, and diarrhea.

My body was attempting to rid itself of a toxin and telling me in no uncertain terms never to eat those foods again. I got the message. It was several years before I touched roast beef, green beans, or mashed potatoes. While the café is still in business, I've never again frequented that eatery, and I'm reminded of the incident each time I drive through Leadville.

I learned three lessons from the incident. First, a body links flavors of foods most recently eaten with nauseating postingestive feedback. That's why my body associated nausea, vomiting, and diarrhea with the foods I ate at the café, not with foods I ate at the ski area.

Second, a body can learn with delays of up to twelve hours between food ingestion and postingestive consequences.[17] Digestion and absorption take place slowly as cells and organ systems integrate information about a meal. That's why I formed an aversion to the foods I'd eaten at the café, even though it had been eight hours since I ate the meal.

The third lesson has to do with the staying power of food aversion. Later in the week after my awful food poisoning response, I learned that everybody who drank water at the ski area on that fateful Sunday became ill, a result of fecal contamination of the water supply. So now I knew that the cause of my illness wasn't the foods I'd eaten at the café at all.

Yet, despite all of that factual knowledge, I couldn't shake my aversion to roast beef, green beans, or mashed potatoes no matter how much I thought about it. We acquire strong aversions to foods eaten just prior to nausea, even when we later learn that contaminated water, the flu, chemotherapy, seasickness, or some other nauseating illness was responsible for the sickness rather than the foods we feel averse to eating – no amount of thinking will change that. Something similar happens when a sheep receives a dose of lithium chloride while deeply anesthetized. The sheep acquires an aversion to the food, even though it wasn't conscious while the feedback event occurred.[18] Feedback alters liking at an affective (noncognitive) level.

So, as Bert Decker argues and Candace Pert shows, our behaviors are ultimately motivated by the molecules of emotion – we "buy" on emotion and justify with facts – and only later do we rationalize with what we come to see as the evidence. That insight becomes critically important when we consider how authority can trump wisdom, how beliefs can trump authority, and how understanding can trump beliefs, issues I discuss in part four.

# CHAPTER 6

# Undermining the Wisdom Body

Island Lake is nestled deep in a glacial cirque at roughly 12,500 feet elevation in the Sawatch Range, overlooking the North Fork Drainage. The lake is named for two small islands that dot the otherwise turquoise water. The one-and-a-half mile hike to the lake follows a creek that tumbles down mountainsides and through meadows from Island Lake to North Fork Lake.

As we arrived at Island Lake today, we happened upon a small flock of ptarmigan that were foraging on the dozens of plants actively growing on the hummocks that extend into the granite crevices and up the spires that tower above the lake. The ptarmigan here are tame to the degree that a calm and careful human can nearly become a member of the flock. I watched as they plucked leaves and flowers from the smorgasbord. They ambled leisurely, except when a grasshopper jumped from their path, an ill-fated move by the grasshopper that elicited great excitement as the ptarmigan joined in a race to gobble up the unsuspecting terrestrial.

As I wandered along with the ptarmigan, I pondered different ways wild herbivores and omnivores diverge from modern humans with regard to foraging. What's available to wild animals varies seasonally. The snow-blanketed landscapes of winter offer primarily mature vegetation from last year's garden. The fresh new growth of spring develops into the flowers and fruits of summer and fall, which mature again into the dormancy of winter. Ptarmigan sequester buffers of energy and nutrients during good times, which they use during lean times.

As discussed in chapter 2, environments set physical (plant form) and biochemical (secondary compounds) limits that result in wild animals dining on a variety of phytochemically rich forages. In contrast, humans often severely limit which foods domestic livestock have on offer in pastures and feedlots. Not unlike monoculture pastures or feedlots, human foodscapes, though they appear to offer a plethora of choices, mostly lack biochemical richness. They facilitate eating energy-dense, processed foods with little or no physical or biochemical limits to moderate intake.[1] While these differences have implications for the health of humans and the animals in our care, we are mostly unaware of how the lack of appropriate choices on pastures, in feedlots, and in supermarkets affect our foraging behavior and health.

Wild and domestic herbivores learn to eat combinations of foods that nourish and satiate. Given appropriate choices, they select diets in ways that support health. In contrast, humans study and attempt to understand consequences of eating particular compounds in foods. In so doing, we are trained to fixate on how *particular* nutrients affect our health.[2] That has led to a fixation on taking supplemental nutrients, from minerals and vitamins to antioxidants.

In contrast, wild animals don't attempt to prevent deficiencies by taking supplemental nutrients. Rather, they maintain balanced diets by eating foods that rectify nutrient deficits and by limiting intake of foods that cause excesses. Despite their wise example – consistent with contradictory and often negative research findings about supplement usage – sales of supplements have exploded during the past two decades in the United States, and the number is on the rise. About half of adults in America now take dietary supplements at a cost of over $25 billion annually. In fact, 40 percent of men and women take a multivitamin daily.

These observations raise two questions. Does fortifying and enriching foods and taking supplemental minerals and vitamins adversely affect food selection, nutrition, and health of humans and herbivores? Despite taking supplemental minerals and vitamins, do humans overingest foods in an attempt to meet needs for nutrients in short supply?

The paucity of research on these questions in no way reflects the importance of these issues. Rather, it illustrates our lack of appreciation for the workings of the wisdom of the body, our emphasis in agriculture on making livestock fat rather than healthy, and our current focus in human nutrition on people who are fat. Addressing these questions has ecological, economic, and social implications for the nutrition and health of humans and the animals in our care.

## Carl's Cows

In *A Holistic Vet's Prescription for a Healthy Herd*, Richard "Doc" Holliday shares insights gained from over fifty years of working with livestock producers. He recounts one peculiar case that involved a client named Carl who had feedstuffs that looked good but, unbeknownst to Carl, were of low nutritional value. By late winter Carl consulted Doc with two seemingly unrelated problems. First, his cattle were eating nearly two pounds of a multimineral supplement per head per day. Second, about ten days before they were due to calve, Carl's heifers would abort a live calf. With care, the calf would live, but despite their best efforts, the heifer would die within three days. After the third one died, Doc sent a dying heifer to the University Vet School for autopsy. Their diagnosis – starvation – surprised Doc and Carl. Carl took good care of his animals and was feeding them all they could eat. As Doc recounts, "This diagnosis was like an insult to Carl and difficult for either of us to accept. We could have accepted a diagnosis of malnutrition because of the poor crops that year, but starvation seemed a little too harsh."

The distinction between *malnutrition* and *starvation* is critical: The animals weren't merely lacking nutrients, they were starving to death. How could that have happened? The answer may come from studies with rats conducted some fifty years ago.

Psychologists studying how rats select foods that rectify deficits showed that rats deficient in thiamine acquire an aversion to the food they are eating, which they associate with malaise due to the deficiency.[3] The aversion is manifest as a decrease in food intake, and when the aversion is severe, as with Carl's cows, it leads to starvation. As in rats, the first sign of a phosphorus deficiency in cattle, sheep, and goats is a decrease in intake of the diet. The greater the phosphorus deficiency, the greater the decrease in intake.[4]

When animals experience nutritional deficiencies, they become neophyllic, avidly sampling unfamiliar foods. That behavior increases their likelihood of encountering a food that will rectify the deficit. If and when they do, they acquire a nutritional state–dependent preference for the food. As described in chapter 5, their food choices can be bizarre. And as I had observed in my study of goats, they might even devour woodrat houses.

After the diagnosis of starvation, Doc and Carl turned their attention to the mineral consumption issue. They decided to try a cafeteria mineral program in which each mineral was fed separately based on the theory that each animal would eat what it needed to meet its needs. Carl's feeder was in the middle of his dry lot, and he had to carry each bag of mineral through the lot to the feeder. Things went well for the first few trips, but then "several of the normally docile cows suddenly surrounded him, tore a bag of mineral from his arms, chewed open the bag and greedily consumed every bit of the mineral, the bag and even some mud and muck where the mineral had spilled out . . . astounding behavior for a bunch of tame dairy cows!"

So, what was in that bag? A source of the trace mineral zinc. For the next several days, they ate numerous bags of zinc supplement while ignoring all of the other minerals. Gradually they began to taper off eating the zinc and instead ate expected amounts of the other minerals. After eating the zinc, heifers calved normally. Evidently, the forage Carl had been offering them was deficient in zinc or perhaps high in zinc antagonists. The basic multimineral mix contained a small amount of zinc, but in order to get the quantity of zinc their bodies needed, the animals had to consume extra-large amounts of the mineral mix. That meant they were also ingesting high amounts of many other minerals, including calcium. Calcium interferes with zinc absorption, which in turn increased their need for zinc. With every mouthful of mixed mineral the animals ate, they were unwittingly increasing the imbalance they were trying to rectify. Inevitably, symptoms showed up in the most vulnerable group – the young heifers who were still growing and in the last stages of pregnancy. As Doc Holliday properly concluded, "Finally, they just gave up and checked out, all for want of a few grams of zinc."

For Doc, this incident embodies the concept that, given appropriate choices, individuals can balance rations better than computers or nutritionists. Many nutritionists discount the ability of animals to balance their ration, asserting that by the time they feel the need to eat a certain item, they are in a deficient state. As Doc notes, "From their point of view, I suppose they have a point. The fallacy in their reasoning may be that they expect the animal to choose for the level of production that man desires while the animal chooses only what it needs to be healthy."

## Carl's Cows, TMRs, and Clara's Kids

The goal of animal nutritionists is to increase food intake by livestock because greater intake equals greater growth. To do so, animals confined in feedlots and dairies are fed total mixed rations (TMRs). A TMR blends feedstuffs into a complete ration that provides adequate nourishment to meet the needs of various classes of livestock. Each bite contains the required amounts of energy, protein, minerals, and vitamins to maximize growth at the least cost. The rations are made of energy concentrates (grains such as corn and barley) and roughages (forages such as corn silage and alfalfa). Minerals and vitamins are added (top-dressed) as needed.

The feedlot industry has figured out how to feed millions of animals on TMRs, but two interrelated issues may reduce efficiency and increase costs. First, as we discussed in chapter 3, there is no "average" animal. Thus, a TMR formulated to meet the needs of the average individual will exceed the needs of some animals and not meet the needs of others, even in uniform groups of animals. To compensate, individuals eat more or less of the TMR. Second, an individual that can self-select a diet from various foods may actually *eat less food* to meet its needs. Providing a variety of foods and offering supplements free-choice enables each individual to respond appropriately to its own needs, which may *decrease* food intake.

Rather than allow sheep or cattle in feedlots to choose among ingredients, livestock managers feed TMRs in part because they are afraid animals will get sick from eating too much grain offered free-choice. Grains such as corn and barley are high in rapidly fermentable carbohydrates, which cause acidosis (low rumen pH). That leads to digestive upsets that cause cattle to go off feed for an extended period of time. In other words, acidosis induces nausea and conditions food aversions.[5] Given no choice but to eat high-grain diets, sheep and cattle self-medicate on bentonite and sodium bicarbonate to alleviate acidosis.[6] When they can choose, however, calves and lambs learn to balance their intake of grain and roughages to prevent acidosis, and they learn to select combinations of foods that best meet their needs from day to day.[7] When offered a choice between alfalfa pellets and rolled barley, the "average" lamb eats 55 percent barley (grain) and 45 percent alfalfa (roughage). Individuals vary along a continuum from lambs that eat little barley to lambs that eat twice as much barley as alfalfa.[8] The average lamb eats much less grain than animals fed a TMR designed to maximize weight gain.

What would happen to food intake, weight change, and the cost for food if animals could formulate their own rations from a variety of concentrates and roughages? Conventional wisdom says animals will eat too much grain and perform poorly because they can't balance their own rations. As a result, costs would be high because the animals wouldn't put on weight well.

These questions were addressed in a two-month study.[9] Calves in one group could choose among rolled corn, rolled barley, corn silage, and alfalfa hay, while calves in the other group were fed a TMR made by grinding and mixing these four foods in proportions formulated to meet the nutritional needs of the "average" individual. Food selection varied widely among individuals offered a choice of the four foods throughout the study. Just like Clara Davis's kids, no two individuals ever selected the same diet nor did they select a diet similar in proportions to the TMR, and no animal consistently chose the same foods from day to day.

While the average ratio of protein to energy eaten during the trial was identical for calves fed the TMR and calves in the choice group, intake of energy and protein changed from day to day, as did ratios of protein to energy for individuals allowed to choose their foods. No calf in the choice group was average – each individual selected a diet consistently higher or lower in ratio of protein to energy. On twenty-one days, individuals offered a choice had protein-to-energy ratios higher than did animals fed the TMR; on two days the ratios were equal; and on forty days they had protein-to-energy ratios lower than did the animals fed the TMR.

Averaged throughout the trial, animals fed the TMR ate *more* food than animals offered a choice, but they didn't gain weight at a faster rate. All animals gained weight at the same rate, but animals given choice were more efficient at converting food into animal. Because animals offered a choice ate less, and they ate less of the more expensive grains, the daily cost per unit of weight gain was 24 percent less for the choice group than for the TMR group. Findings from similar studies suggest improved intake, rate of gain, and feed efficiency can result when lambs are allowed to select their diets from biochemically complementary foods, possibly enabling them to realize greater benefits than typically thought from inexpensive forages.[10] These assessments don't consider costs externalized to society such as taxpayer-subsidized grain production or antibiotic resistance transferred to human populations.

Trials such as these raise broader issues regarding the welfare of animals fed different rations.[11] Compared with monotonous rations, for instance, lambs fed more diverse diets eat more and experience less stress.[12] Bison in feedlots experience high stress due to social disorder and aggressive interactions, which adversely affect welfare and increase costs of production.[13] Levels of stress are higher for male bison placed in tight or loose confinement compared with free-ranging bison. Providing adequate space, offering a choice of foods, and leaving young bison on pasture prior to slaughter lowers stress, costs of gain, illness, and death, as illustrated in a long-term, ranch-scale study where bison were fed either a TMR in tight confinement, offered a choice of feeds in loose confinement, or foraged on rangelands where they were also offered a choice of grains. While the animals that were offered a choice in the loose-confinement group had higher daily gains than the other groups, the rangeland group was least costly to finish and provided the highest net returns.[14]

Collectively, these findings show that individuals can more efficiently meet their needs for nutrients when offered a choice among dietary ingredients, rather than when constrained to a single ration nutritionally balanced for the average animal. Feeding practices that allow animals to choose enable producers to capitalize on the uniqueness of individuals, enabling efficiency, improving performance, and reducing illness. That's true not only for cattle, sheep, and bison in confinement, but also for cattle, sheep, and bison foraging on monocultures as opposed to diverse mixes of plants on pastures and rangelands. That's also true when animals are offered mixed-mineral blocks formulated for the average individual as opposed to cafeteria minerals that enable individuals to choose the minerals their bodies need.

Issues with Carl's cows and TMRs – overingesting nutrients in attempts to meet needs for other nutrients – and the response of animals to supplemental nutrients raise two questions that have received scant attention with humans. First, do people who eat diets high in processed foods seek nutrients in short supply? Second, does providing supplemental primary (energy, minerals, and vitamins) or secondary (antioxidants) compounds adversely affect choices and health?

Livestock and humans will overingest energy in an attempt to meet needs for protein.[15] While most of the emphasis to control obesity in humans has been on reducing intake of fat and carbohydrates, the role of protein has generally been ignored.[16] Protein intake, which makes up only about 15 percent of dietary energy intake, has remained nearly constant within and across populations during the development of the obesity epidemic. That percentage of protein relative to energy, which is lower than most people require, can provide protein with the leverage to drive the obesity epidemic. When people are restricted to a diet that contains a lower ratio of protein to energy than they self-select, they will overingest energy to maintain their intake of protein. Protein drives overingestion of energy as people attempt to self-select a diet to meet their needs for protein.[17] Thus, human food systems, based on *food quantity* rather than on *biochemical richness*, have a maladaptive feedback loop built into them. The more energy-dense but biochemically poor foods a human eats, the more a person wants and needs to eat, in a quest to obtain nutrients such as protein and minerals present in low concentrations.[18] People can moderate these effects by eating modest amounts of more high-quality sources of protein (meat and fish) and minerals (vegetables) and less energy-dense foods high in refined carbohydrates.

While researchers who study humans and herbivores don't often consider this point, eating an array of biochemically rich dietary items is likely crucial for reducing food intake as a result of providing rich arrays of compounds that different cells and organ systems need to function. Wanting, liking, and needs are linked through biochemical richness of foods. That's why, even though eating a monotonous diet may induce satiation and help relieve obesity,[19] eating a monotonous diet is more likely to induce cravings for liked and needed foods. Eating biochemically rich foods a person likes can induce a stronger decrease in wanting to eat.[20]

In one study, for example, satiation increased faster, so people ate less when they ate a soup spiced with chili as opposed to the base soup without chili.[21] Likewise, soup with cayenne pepper, which was perceived as more spicy but equally liked to the base soup, resulted in significantly higher satiation at the end of the meal and one hour after the meal.[22] Adding cayenne pepper was associated with higher sensory satisfaction while eating and with feeling more energetic and satisfied one hour after a meal. Thus, while energy and protein are essential for health, phytochemically rich herbs and spices enhance appetite, palatability, satiation, and satiety because they are also essential for good health.[23]

## Undermining Healthy Choices in Humans

For many humans nowadays, the relationship between appetitive state and the availability of foods differs from when our ancestors hunted and gathered and were much more "lean and mean."[24] Humans today are awash in energy-dense foods that are biochemically poor. Such foods are typically altered by reducing or removing existing nutrients, replacing a nutrient with another substance, or fortifying to increase existing nutrients.[25] Even some basic foods like bread and milk are enriched with niacin and vitamin D. And people take supplements.

Scientists who study the food choice of herbivores control for appetitive (nutritional) state because excesses or deficits of energy, protein, minerals, or vitamins alter food selection. For instance, dairy cattle fed rations high in protein while they are being milked in the barn prefer to eat grass (lower in protein) rather than clover (high in protein) when they then forage on pasture. It's notable, too, that they choose to eat the less protein-rich parts of the grass.[26] Conversely, when cattle are fed concentrates high in energy relative to protein in the barn, they prefer clover to grass.

The same is true for minerals. Sheep avoid eating foods high in phosphorus, calcium, sodium, or sulfur when their needs for those minerals are met, but they select foods that rectify deficits when they are lacking those minerals.[27] When lambs have access to salt blocks, they avoid foods that contain even small amounts of sodium.[28] That's why providing salt blocks on rangelands can cause livestock to avoid eating plants such as saltbush that contain sodium.

Herbivores also satiate on secondary compounds, and this leads them to ingest a variety of forages that contain different phytochemicals.[29] That's illustrated when sheep are first given a supplemental capsule of saponins, alkaloids, or tannins and then allowed to choose among plots of alfalfa, birdsfoot trefoil, and tall fescue.[30] Sheep that get a supplement of saponins eat less alfalfa (high in saponins) and more trefoil (high in tannins) and fescue (high in alkaloids). Conversely, sheep that get a supplement of alkaloids eat less tall fescue and more alfalfa and trefoil. Sheep first given a capsule of tannins eat less trefoil and more fescue and alfalfa.

If humans respond to excesses in the way that herbivores do, then eating energy-dense foods that have been fortified and enriched (plus taking supplements) may cause us to avoid eating otherwise nutrient-rich fruits and vegetables that provide minerals, vitamins, and other phytochemicals. Would we eat more vegetables and fruits if we weren't ingesting energy-rich fortified and enriched foods and taking supplemental minerals and vitamins?

During the past three decades, people have become aware of recommendations to eat fruits and vegetables. In the United States, for example, about 8 percent of adults reported knowing they should eat at least five servings of fruits and vegetables daily in 1991. That number increased to 40 percent by 2004, and while awareness is higher today, that hasn't translated into changes in behavior.[31] In 2015, just 9 percent of adults met intake recommendations for vegetables, ranging from 6 percent in West Virginia to 12 percent in Alaska. Only 12 percent of adults met recommendations for fruit, ranging from 7 percent in West Virginia to 16 percent in Washington, DC. Consumption was lower among men, young adults, and adults living in poverty. Among adults, the primary contributor to total fruit intake was whole fruits; among adolescents, it was fruit juices. The largest single contributor to overall fruit intake among adults and adolescents was orange juice.

In the current food environment, fruits and vegetables are competing with foods and beverages rich in added sugars and nonnutritive sweeteners.[32] The more sugar-laden foods a child consumes, the less fruits and vegetables the child eats. Children's responses to sweetened foods and beverages come from sensory systems evolved to detect the once-rare calorie-rich foods that taste sweet. This sweet attraction served children well in a feast-or-famine setting, attracting them to mothers' milk and energy-rich foods required for growth, but today it makes them vulnerable to food environments replete in processed foods rich in added sugars and lacking healthy sweet foods, such as fruits. Once children become accustomed to a highly sweetened diet, they may find that foods their sensory systems evolved to eat no longer taste sweet.[33]

The glut of energy-dense foods causes children to eat less than the recommended amounts of fruits and vegetables.[34] As the serving size of energy-dense entrées increases, intake of fruits and vegetables declines. On the other hand, when served smaller portions of an energy-dense entrée, children eat more fruit and vegetables. Children also eat more vegetables when vegetables are offered as a first course, without competing foods, when children are hungry.

For sailors during the Renaissance, ocean voyages for extended periods without fresh fruits and vegetables led to scurvy, caused by a deficit of vitamin C. In 1747, a Scottish surgeon in the Royal Navy, James Lind, discovered that citrus foods prevented scurvy, though he mistakenly believed scurvy was due to faulty digestion, not dietary deficiency.[35] He proved scurvy could be treated with citrus fruit in experiments described in his 1753 book, *A Treatise of the Scurvy.* Ironically, scurvy had been described more than 2,000 years before that by Hippocrates, and cures for scurvy – including citrus fruit – have been used since antiquity.

Sailors who experienced a deficiency of vitamin C provided graphic accounts of their cravings. When they finally got the fruit they desired, they swallowed it "with emotions of the most voluptuous luxury." Upon making landfall, one British crew gorged on watercress, purslane, sorrel, turnips, and Sicilian radishes. They craved what their bodies needed, and they ate foods that met their needs. Today, those sailors could drink vitamin C–fortified orange juice. Their scurvy would disappear, along with their desire to eat fruits and vegetables.[36] Rather than ingest a rich array of phytochemicals, they would consume large quantities of sugar calories.

In addition, they could drink any of a variety of zero-calorie fruit drinks to meet their daily needs for various B vitamins, including $B_{12}$ (cobalamin), $B_6$ (pyridoxine), $B_5$ (pantothenic acid), and $B_3$ (niacin). In the early 1900s, a B-vitamin deficiency known as pellagra was common in mental asylums in the southern United States, where inmates were fed cornmeal, fatback, and molasses. Today, this diet wouldn't cause pellagra because cornmeal is fortified with niacin. Unfortunately, niacin-fortified refined grains used to make processed foods are associated with high rates of obesity.[37] Milk formulas, which have high levels of B vitamins, promote infant weight gain, especially fat mass gain, a known risk factor for children developing obesity.[38] Studies of rats, chickens, and pigs suggest the supplemental dose of niacin to maximize food intake and weight gain is roughly 60 milligrams per kilogram of body weight, a dose similar to that used to fortify wheat flour in the United States, Canada, Saudi Arabia, and Kuwait.

## Undermining the Health of Humans

One important finding from research is that livestock don't choose to eat particular foods or supplements proactively to prevent deficiencies. Rather, they respond to deficits by selecting the food or supplement with the nutrient they are lacking: If it ain't broke, they don't fix it. Only humans take supplements, allegedly to prevent deficiencies, but do they work?

Until the mid-1930s, when the first commercial yeast-extract vitamin B complex and semisynthetic vitamin C supplement tablets were sold, vitamins were obtained solely through natural foods. Seasonal changes in diet usually altered the kinds and amounts of vitamins people ingested. Intake of vitamins in vegetables and fruits was high when plants were growing in spring, summer, and fall but low during winter. Bodies coped with seasonal variation by developing mechanisms to maintain homeostasis, as illustrated with the fat-soluble vitamins.

The fat-soluble vitamins – A, D, E, and K – are obtained from sunlight (vitamin D) and by eating fruits and vegetables – especially those that are orange or dark green in color (A, E, and K). Fat in the diet is essential for absorbing and storing the fat-soluble vitamins, which are then released as needed and to some extent can be thought of as time-release micronutrients.

They can be consumed occasionally, even weeks or months apart, and still meet needs. These vitamins are stored in fat tissues and in the liver for long periods, so they usually pose less risk for deficiency than toxicity, especially if a person takes supplements.

The water-soluble vitamins – the B-complex, including $B_6$, $B_{12}$, niacin, riboflavin, and folate, and vitamin C – must dissolve in water to be absorbed in the body. The body uses vitamin C for growing and repairing tissues. It helps the body make collagen, an important protein used to make skin, cartilage, tendons, ligaments, and blood vessels. Vitamin C is also used to heal wounds and to repair and maintain bones and teeth. The human body can store only a certain amount of vitamin C, which is depleted if fresh supplies aren't consumed. The time for the onset of symptoms of scurvy in unstressed adults on a diet without any vitamin C ranges from one month to more than six months, depending on previous loading of vitamin C.[39] People rarely get too much of water-soluble vitamins from food, but taking supplements can cause problems.

That's true as well for minerals such as calcium. A ten-year follow-up of the Multi-Ethnic Study of Atherosclerosis by the National Institutes of Health (NIH) assessed the risks of calcium from diet or supplements on coronary artery calcification (a marker of atherosclerosis) and its progression among older adults.[40] Taking a calcium supplement may increase the likelihood of coronary artery calcification, although high calcium intake was associated with a decreased risk of incident atherosclerosis if that was achieved *without the use of supplements*.

Certainly, under some conditions, some individuals can develop deficiencies. In those cases, a perspective that focuses on ways to obtain needed minerals or vitamins can improve health. That's true with vitamin C for scurvy; niacin for pellagra; folate for congenital birth defects during pregnancy; vitamin $B_{12}$ for vegetarians and elderly; iron for anemia.[41]

Nutrition panels establish estimated daily average requirements (DARs) and recommended dietary allowances (RDAs). But over the past several decades, total intake of vitamins from all sources has been much higher than the DARs because foods are fortified and enriched, people take supplements, fruits and vegetables are available year-round, and the intake of meat, which is high in niacin and heme iron, has increased.

In 1992, 159 countries pledged at the FAO/WHO International Conference on Nutrition to combat deficiencies of vitamin A, iodine, and iron.[42] Daily consumption of iron per capita in the United States has surged since the

Second World War and nearly doubled over the past century due to increases in iron fortification and increased consumption of meat. After decades of fortification and supplementation, however, iron is excessive in individuals in many societies. Excess iron intake may play a role in obesity, cardiovascular disease, diabetes, and cancer.[43] However, testing potential associations between iron availability and obesity and related diseases is difficult because fortification and supplementation are now practiced so extensively.

Bodies counter nutrient deficits or excesses. To maintain homeostasis, cells can increase (up-regulate) or decrease (down-regulate) the number of receptors on the membrane, thus altering sensitivity to a compound. Deficits cause cells to up-regulate receptors to increase uptake of primary or secondary compounds, while excesses cause cells to down-regulate receptors to decrease uptake of these compounds. Feedback from cells and organ systems also encourages animals to change behavior. As animals begin to experience deficits or excesses, they reduce their intake of particular foods, and if the deficits or excesses become more extreme, they acquire aversions to the foods they are eating. Excess intake of vitamins leads to various ailments including muscle and abdominal pain, blurry vision, headaches, weakness, drowsiness, and altered mental status. They also cause nausea, vomiting, and anorexia: all ways cells and organ systems in the body are saying they can no longer deal with so much of a particular nutrient.

Still, taking supplements makes people feel better about the potentially adverse health effects of inadequate diets. Embracing supplements means the media can tell people what they want to hear and doctors have something to offer their patients. A multibillion-dollar industry is now part of our nutritional landscape, and many consumers are often misled into believing they are buying health for themselves and the animals in their care. Huge profits are a big incentive for government guidelines that in turn pave the way for an expanded market.

**General issues with supplements.** When compounds such as minerals, vitamins, or pharmaceuticals are extracted and purified, the effects of the resultant supplement or drug are amplified, typically with adverse side effects. With that silver-bullet bargain, we also do away with synergies that come about when many different compounds interact in beneficial ways to enhance health. That can be illustrated with vitamin C.[44]

In the 1930s, Albert Szent-Györgyi discovered the chemical ascorbic acid—also known as vitamin C—that enables the body to efficiently metabolize carbohydrates, fats, and proteins. He received the Nobel Prize in

Physiology or Medicine in 1937 for his explanation of the oxidation processes that vitamin C controls. Today, no vitamin supplement is more universally accepted than vitamin C, yet his research showed that vitamin C is only one part of a system of cofactors needed for health. While Szent-Györgyi was attempting to purify vitamin C, a colleague had a patient with scurvy. The scientist gave the scurvy sufferer an impure preparation of vitamin C and other compounds. The patient quickly recovered. Later, they tried a preparation of purified vitamin C on another patient with scurvy, expecting even quicker results. To their surprise, pure vitamin C was of no benefit at all, so they reverted to the impure solution. Treated with the mixture of unknown compounds and vitamin C, the second patient also recovered.

Szent-Györgyi and a colleague isolated a compound from lemon juice that served as the essential cofactor for vitamin C. During the next forty years, they learned of an electron transfer system in which vitamin C and its cofactors regulate the transfer of free radicals of oxygen. When either vitamin C or its cofactors are missing, the free radicals of oxygen produced in the normal process of respiration escaped, and the proteins became free radicals themselves. The cofactors – flavonoids and oligomeric proanthocyanidins – interact with vitamin C to protect each other from free radicals. Together, they promote vascular health and protect against cancer.

Megadosing with vitamin C has been to a large degree discredited. Modest benefits of vitamin C supplements have been shown for decreasing the length of the common cold, though not in megadoses.[45] Benefits are not greater when vitamin C intakes of more than 1,000 milligrams per day are compared to intakes between 200 and 1,000 milligrams per day.

**Issues with B vitamins.** The B vitamins play key roles in metabolism. Vitamins $B_1$ (thiamin), $B_2$ (riboflavin), $B_5$ (pantothenic acid), and $B_7$ (biotin) help convert food into energy. Vitamin $B_1$ boosts neural function and vitamin $B_2$ helps maintain good eyesight. In addition to aiding energy metabolism, vitamin $B_3$ (niacin) stimulates appetite and digestion, while vitamin $B_6$ (pyridoxine) enables synthesis of amino acid, including neurotransmitters. Pregnant and breastfeeding women need $B_6$ for their babies' brains to develop normally. Like most other B vitamins, vitamin $B_9$ (folic acid or folate) nurtures growth of red blood cells. Vitamin $B_9$ also reduces the risk of a rare birth defect (neural tube defect) during pregnancy. Vitamin $B_{12}$ (cobalamin) enables formation and growth of red blood cells and health of the nervous system.

With all these benefits, it would seem only sensible that taking the B-complex of vitamins as a supplement in pill form is just good insurance for health. Or, why not enrich foods with B vitamins? Sometimes, enriching and fortifying foods with nutrients such as B vitamins is merely a marketing decision to entice consumers. Other times it is intended to improve health in a population. Fortifying foods with vitamin $B_9$ (folic acid) was mandated in many countries to improve the folate status of pregnant women. Supplemental folic acid prevents neural tube defects, which affected 0.5 percent of births in the United States before fortification.

Folate became an example of the unpredictable consequences of "feeding" a beneficial substance in an unnatural form. Intake of folic acid from fortifying foods has been much greater than originally modeled in premandate predictions. As a result, several hundred thousand women are exposed to an excess of folic acid for each neural tube defect prevented.[46] High levels of folic acid may decrease natural killer cell cytotoxicity and reduce response to drugs used to treat malaria, rheumatoid arthritis, psoriasis, and cancer. High folate, combined with low vitamin $B_{12}$, may increase risk of cognitive impairment and anemia in the elderly, and increase risk of insulin resistance in pregnant women and obesity in their children. While folate can protect against cancer initiation, it facilitates growth of preneoplastic cells and subclinical cancers. Thus, a high folic acid intake due to fortification may be harming more people than it is helping.

In the NIH study the VITAL cohort, with more than 75,000 participants aged fifty to seventy-six, lung-cancer risk among men who took 20 milligrams of $B_6$ daily for years was twice that of men who didn't.[47] Among people who smoked, the effect of supplemental $B_6$ increased risk threefold. The risk was even higher among smokers taking $B_{12}$. Taking more than 55 micrograms daily nearly quadrupled the risk of lung cancer. There was no apparent risk in women – which is not to say it doesn't exist, only that it wasn't obvious. While these doses are excessive relative to daily US Recommended Dietary Allowances for $B_6$ (1.7 mg) and $B_{12}$ (2.4 micrograms), vendors offer supplements in even higher doses for $B_6$ (100 mg) and $B_{12}$ (5,000 micrograms).

Rather than fortify foods and take supplements, we can easily obtain B vitamins by eating foods such as cheese, milk, eggs, chicken, red meat, fish, whole grains, peanuts, citrus fruits, watermelon, beets, and dark green vegetables. The name *folic acid* is derived from the Latin word *folium*, meaning "leaf," a reference to vegetables, which are a prime source of vitamin $B_9$. Animals must obtain vitamin $B_{12}$ directly or indirectly from bacteria. The roots of plants such as carrots and potatoes – pulled from the ground and not washed – may contain $B_{12}$ acquired from the bacteria in the soil. Vitamin $B_{12}$ is also found in fish, meat, eggs, and milk.

**Issues with antioxidants.** Most plants and animals require oxygen to supply energy to cells, for chemical signaling among cells, and for detoxification processes in cells. Reactive oxygen species (ROS) are formed as a by-product of the normal metabolism of oxygen and have important roles in cell signaling and homeostasis. However, during times of environmental stress, they can increase dramatically and result in significant damage to cell structures. Thus, a paradox of metabolism is that oxidation reactions (reactive oxygen species) in cells can initiate chain reactions that can lead to premature aging, cardiovascular disease, and cancer.[48]

To counter these adverse effects, plants and animals produce a complex network of antioxidants and enzymes that collaborate to prevent oxidative damage to cellular components such as DNA, proteins, and lipids. Antioxidants remove free radical intermediates and inhibit oxidation reactions. They also help the liver detoxify noxious compounds in the environment and in foods we eat. Antioxidants can strengthen the immune system, retard aging, and reduce cancer and other chronic diseases. With age, antioxidant defenses and the ability of cells to repair oxidative damage to DNA decline, contributing to cell, organ, and whole organism senescence.

Due to their benefits when consumed in foods, antioxidants are now used as dietary supplements. Although initial studies suggested that some of these supplements may promote health, large clinical trials of antioxidant supplements such as β-carotene, vitamin A, and vitamin E alone or in combinations show increases in risk of all-cause mortality.[49] The risk is higher in developed countries, where the use of supplements is popular, even though people are more likely to meet needs for nutrients from other dietary sources.[50] High dosages of vitamins C and E appear to diminish some of the endurance training–induced adaptations in human skeletal muscles, so some suggest that high dosages of isolated antioxidants should be used with caution in individuals who are simultaneously engaged in endurance training.[51]

In the Physicians' Health Study II, a randomized, double-blind, placebo-controlled trial with 15,000 physicians aged fifty-five years and older, participants took either a basic multivitamin-mineral formulated for people older than fifty or a placebo daily for eleven years.[52] Neither antioxidant – vitamin E or vitamin C – reduced the risk of major cardiovascular events or the risk of prostate or total cancer.[53] A review of published meta-analyses regarding the value of supplemental vitamins and minerals for cardiovascular disease prevention and treatment showed little or no evidence for preventive benefits (folic acid for total cardiovascular disease, folic acid and B vitamins for stroke), no effect (multivitamins, vitamins C and D, β-carotene, calcium, and selenium), or increased risk (antioxidant mixtures and niacin [with a statin] for all-cause mortality).[54] Indeed, evidence has eroded considerably over the past several decades regarding micronutrient supplements as a viable way to prevent cancer in the general population. The outcome depends on the nutritional status of the individual taking the supplement.[55] Deficiencies of various nutrients are associated with a higher risk of cancer, but having a very high nutrient status is similarly associated with higher risk of cancer.

The precursor to vitamin A, β-carotene is an antioxidant in vegetables and fruits. Eating vegetables and fruits is linked with a reduced risk of cancer, so it seems reasonable taking β-carotene supplements might reduce risk of cancer.[56] But results of clinical trials show this is not the case. In two studies, where people were given high doses of β-carotene supplements in an attempt to prevent lung cancer and other cancers, the supplements *increased* risk of lung cancer in cigarette smokers, and a third study found neither benefit nor harm from them. What might cause these unexpected findings? While beneficial at low doses, at higher doses, antioxidants can shut down cell signaling pathways and decrease synthesis of mitochondria in new muscle cells.[57] They can also decrease production of endogenous antioxidants produced by a body.

Other studies challenge assumptions about the negative effects of reactive oxygen species on health. Rather than kill us, in low doses, they may improve our health, and the quest to neutralize them with antioxidants may do more harm than good.[58] According to this view, eating fresh fruits and vegetables promotes health, not only because doing so protects us from oxidative stress, but also because the secondary compounds they contain induce mild hormetic stress. *Hormesis* refers to a biphasic dose response characterized by a low-dose stimulation or beneficial effect and a high-dose inhibitory or toxic effect. As discussed in chapter 2, environmental stress increases concentrations of secondary compounds in plants. That increase can strengthen plants as well as the herbivores and humans who eat those plants. Among many others, phytochemicals such as resveratrol (red grapes, pistachios), sulforaphanes (cruciferous vegetables such as broccoli and cauliflower), curcumin (turmeric), capsaicin (peppers), and allicin (garlic) activate hormetic response pathways that help to strengthen cells.[59]

## If It Ain't Broke . . .

The success of research in providing knowledge about nutrients and requirements of animals has obscured the fact that animals have been nutritionists since the first cell assimilated nutrients and reproduced over two billion years ago. That's also true for birds that evolved from dinosaurs 150 million years ago. That's true for mammals, too, that have foraged well for sixty-five million years without any assistance from man. That's even true for livestock, domesticated 11,000 to 8,500 years ago, that can still select nutritious diets on rangelands with many different plant species. Neither birds (such as blackcap warblers) nor mammals (such as sheep) need access to the more than 30,000 articles on antioxidants published annually in scientific journals to respond appropriately to deficits or excesses of vitamin E.[60] The same is true for omnivorous *Homo sapiens*, who for more than 99 percent of our history lived in diverse environments and ate diets that ranged from highly carnivorous at northern latitudes to highly herbivorous in the tropics.

Our ancestors formulated their own rations, given suitable choices, even as their nutritional needs changed with age, physiological state, and environmental conditions. Nowadays, nutrition guidelines are based on clinical trials of a single nutrient or supplemental nutrient. That, in turn, encourages myopic focus on daily average requirements (DARs) and recommended dietary allowances (RDAs). Yet, the single-compound emphasis is not likely to be nearly as effective as an emphasis on the synergies created when diverse compounds interact in concert with one another. That's why supplemental nutrients aren't always health promoting; why whole foods, not nutrients, should be the fundamental unit in nutrition; why combinations and sequences of foods are key to health through nutrition; and why people should eat a diverse array of whole (unprocessed) foods. That's also why your body ought to be the final arbiter of what's fit to eat: If it ain't broke, you're probably wise not to try to fix it.

# CHAPTER 7

# Medicating in Nature's Pharmacy

A skiff of snow on the ground adds pleasing highlights to the mid-November shades of autumn – the light tans and browns of dried grasses and forbs, the beige of aspen trees, the burnt orange bark and green needles of ponderosa pine and Douglas fir trees. A recent blizzard that brought snow to the high peaks is gone. All that remains are snowcapped mountains buttressed by frosty blue conifer forests. The sky is clear. The sun's rays warm our bodies and our hearts.

Sue and I are hunting today, wandering through the woods in search of juniper berries. We are refurbishing our medicine chest for winter. We use these berries to help our bodies ward off ailments. The berries contain terpenes, secondary compounds that have strong antibiotic properties. The ones we are looking for are a dark blue-black color, which stands out against the deep green of the ground juniper plants that produced them. The prickly needles that slow our task serve as deterrents to herbivores, as do the high levels of terpenes the plants contain.

The flavor of juniper berries is sweet with more than a hint of terpenes. The seeds, however, lack all sweetness and are heavy with the essence of terpenes. "Terpenes" is a catchall term to describe their flavor, but different blends of terpenes create the unique flavors of different plants and plant parts. If I'd been a flavor taster for a living, I could offer a much more elaborate description. But my palate doesn't need words to recognize these berries or to discriminate them from the many fruits, herbs, and spices we eat that contain terpenes.

My perception of sweetness in these berries is affected by my nutritional and physiological needs.[1] They taste sweeter before a meal, when I'm hungry, than after a meal, especially if the meal contains sweets. No doubt, the berries would taste less sweet to me if our daily diet was high in sugar-sweetened foods and beverages.[2] The huge amount of sugar many people in industrialized countries now consume diminishes their perception of sweetness in fruits and vegetables, which contributes to their lack of interest in eating them. The widespread dislike of the many native fruits and vegetables that grow in natural settings is even greater because, unlike their domesticated cousins, they still pack a great deal of phytochemical punch. As we've selected for yield and transportability, the flavors of produce and meat have suffered, and our palates have been conditioned in utero and early in life to prefer bland foods high in refined carbohydrates to phytochemically rich and strongly flavored meat and produce.

As Sue and I meander from juniper to juniper, I'm wondering if this also applies to medicines. Some believe, like Canadian physician Sir William Osler – the Father of Modern Medicine – that "The desire to take medicine is perhaps the greatest feature which distinguishes man from other animals."[3] Contrary to his proposition, the first studies to display the principles that underlie self-medication were done with rats by John Garcia and colleagues beginning in 1971.[4] Their classic studies showed rats come to prefer flavors ingested prior to recuperation from illness.

Woolly bear caterpillars (*Grammia incorrupta*) also show *greater liking*, manifest as an enhanced neural response, for medicinal hemlock plants high in alkaloids.[5] Eating plants that contain alkaloids improves survival of parasitized caterpillars by conferring resistance against endoparasitic fly larvae developing in the caterpillars' guts.[6] Without access to alkaloids, most caterpillars infested with these lethal endoparasites die. But not any alkaloids will do. Given a choice between lupine (*Lupinus arboreus*) and hemlock (*Conium maculatum*), which each contain different kinds of alkaloids, healthy caterpillars prefer lupine. For self-medicating, though, parasitized caterpillars prefer the kinds of alkaloids that hemlock makes.

Monarch butterflies are equally discriminating with regard to different species of milkweed that typically grow in large patches.[7] While adult females can visit patches of different species of milkweeds to oviposit, their larvae have limited ability to move among milkweed species. Adult butterflies can't cure themselves of certain virulent protozoan parasites. To break the cycle, they lay their eggs on the most toxic milkweed plants, which reduces parasite growth and disease in their offspring and provides the first evidence of transgenerational medication.

In response to fungal infections, honeybees immunize their hives by foraging for plant resins (phenolics and terpenes), a behavior termed *social immunity*.[8] Their colony-level use of resins lessens the need for chronic elevation of an individual bee's immune response. Bees' needs for more diverse diets to maintain colony health is gaining increasing appreciation.[9] Despite low phytochemical diversity, 1.6 million colonies are annually put in monocultures of almond orchards in California, and that's reducing the life expectancy of bees. Bees experience nutritional deficits when foraging on a single pollen source, and the lack of phenolics, terpenes, and other secondary compounds is adversely affecting social immunity. Phytochemically rich diets enhance nutrition, reduce loads of parasites, and enhance detoxification of pesticides.[10]

Wandering from juniper bush to juniper bush, still picking berries, I wonder how my "civilized" upbringing in Western ways has hindered my abilities to know how to use nature's pharmacy – physiologically and behaviorally. I didn't learn as a child, and so I lacked knowledge as a young adult of how to use plants in my landscape for food or medicine. I learned about such matters only through courses and books, rather than from experiences in utero and early in life. Nor did my parents acquire this knowledge from their parents, who came from Germany and Italy where the plants growing in nature's pharmacy are different. Without familiarity with nature's medicines, we all came to rely on drugs produced by the pharmaceutical industry.

I wonder, as I pluck berries, how taking pharmaceutical drugs might affect my ability to learn to use the many plants nature provides, not only for food but also for medicine. That would be no different from how supplements and fortified foods might alter my dietary likes and dislikes. I also wonder what cells and organ systems learn and remember about which plants have which medicinal properties. Like the immune system, might cells and organ systems learn and remember which compounds they need to help them when under attack from various invaders? If so, would I then learn to like the plants I need, not unlike woolly bear caterpillars?

## First Line of Defense: Avoid

Evading parasites is the first line of defense.[11] Wild and domesticated animals avoid eggs, cysts, and larvae of parasites, which pass out of the body in feces. Primates avoid touching feces and areas where feces are deposited. Rabbits, horses, cattle, sheep, and goats avoid grazing near feces or on grass sprayed with manure fertilizer. Wild and domestic cats and dogs shun defecating near their homes. Rabbits create latrines that hygienically separate food from feces and also act as scent-marking posts. Horses exhibit a latrine behavior when grazing in highly stocked pastures, but they defecate less discriminately on rangelands where parasite risk is lower. And woodrats have urine-soaked bathrooms in their houses.

Free-ranging herds avoid feces.[12] Sheep with high parasite loads avoid parasite-rich locations in pastures to a much greater extent than do sheep with lower levels of parasites, even when those locations have forages with much higher nutrient rewards. When parasitized sheep are forced to consume contaminated pastures, they graze on leaves further from the ground, minimizing ingestion of parasite larvae, which tend to live on plant surfaces near ground level.

Pastoralists in Kenya emphasize that one of the benefits of their nomadic lifestyle is that it keeps them and their animals free of parasites. In a very real sense, they all get sick and tired if they persist in eating the same foods and foraging in the same places for too long. Food- and place-specific satiety enables health by avoiding parasites and obtaining nutritious mixtures of forages animals need to thrive. When the ability to move is thwarted, animals are more likely to need to self-medicate for various illnesses. In *Dissolving Illusions*, physician Suzanne Humphries and vaccine-skeptical parent Roman Bystrianyk argue that our emphasis on creating safer, healthier habitats, beginning in the mid-1800s – and not vaccinations – caused the subsequent drop in deaths from all infectious diseases.

## Second Line of Defense: Prophylaxis

In *Wild Health*, environmental professor Cindy Engel describes how animals keep themselves well and what we can learn from them. She nicely illustrates the many ways animals learn to evade parasites and learn to use plants therapeutically and prophylactically. For instance, wood ants (*Formica paralugubris*) mix pieces of solidified conifer resin in their nests to inhibit the growth of bacteria and fungi, protecting the ant colonies against detrimental microorganisms.[13] In a modern twist on prophylaxis, house finches that live in urban areas now add cigarette butts laced with toxic nicotine to their nests to thwart ectoparasites.[14]

Livestock that feed in diverse landscapes consume fifty or more species of plants in a day, though most species are eaten only in small amounts, so researchers have not considered them important in animal health. Yet, by eating a phytochemically rich diet, livestock are protecting themselves prophylactically against an array of maladies. Not surprisingly, with growing interest in cover crops, morbidity and mortality have decreased as stocker cattle forage on more phytochemically diverse mixtures of forages as opposed to monocultures.

When animals regularly eat medicinal plants, parasites remain at low levels. That's illustrated in the Mediterranean scrublands of Israel where Damascus and Mamber breeds of goats have different proclivities to use the high-tannin foliage of pistacia (*Pistacia lentiscus*).[15] Damascus goats frequently eat pistacia in ways that act prophylactically. Conversely, Mamber goats, which reluctantly dine on pistacia, use foliage from the plant only therapeutically. Either way, the use of pistacia alleviates nematode infection in all goats.

In the Japanese archipelago, medicinal plants are available to all populations of Japanese macaques. Troops in the north of the archipelago eat medicinal plants but the number of medicinal plants in their diet is less than in troops in the south. Researchers found an inverse correlation between the number of medicinal plants in the diet and parasite richness in a troop.[16]

Pathogens become a problem when animals are weakened by the stress that comes from being moved from familiar to unfamiliar environments. Rather than provide phytochemical diversity in their diets, we've come to rely on pharmaceutical drugs to treat animals stressed by movements. When animals are moved to a new location, their bodies haven't a clue of what's good or not good for them to ingest, and in the case of a feedlot or a monoculture pasture, they have no hope of finding natural foods that might relieve their stress. Would providing phytochemical diversity in pastures and small amounts of familiar herbs to animals in feedlots be a way to reduce dependence on antibiotics? That could also be a way to enhance the flavor and phytochemical richness of meat and thus be beneficial for the health of humans.

## Third Line of Defense: Self-Medicate

When they become ill, individuals in taxa as diverse as insects, ruminants, and primates self-medicate to help treat internal parasites, worms, external parasites, and bloat.[17]

**Internal parasites.** Parasitized chimpanzees eat bitter pith (inner stems) of *Vernonia amygdalina*, a shrub that grows in tropical Africa.[18] Parasites decrease dramatically after chimpanzees eat *V. amygdalina*, which is not surprising because *V. amygdalina* produces sesquiterpene lactones and steroid glycosides with antiparasitic activity.[19] Chimpanzees living in the wild can contract malaria, which is caused by unicellular parasites (species of *Plasmodium*). While no direct evidence exists that wild chimpanzees ingest plants specifically to treat themselves for malaria, they do consume *Trichilia rubescens*, an evergreen shrub or tree that grows in sub-Saharan Africa and tropical South America.

*T. rubescens* contains phytochemicals called limonoids (abundant in citrus fruit) that chimpanzees eat in amounts that have antimalarial medicinal effects.[20]

Comparative studies of species living in different areas suggest wild animals self-medicate to reduce internal parasites. Baboons above Awash Falls in Ethiopia don't eat the fruit of *Balanites aegyptiaca*, while those below the falls readily eat the fruit, apparently to relieve malaise from internal parasites that occur only below the falls. People living below the falls use the fruit of *Balanites* to kill schistosome-carrying snails, suggesting baboons may do likewise to reduce the impact of schistosomiasis. The berries and the leaves of *Balanites* contain the steroidal saponin diosgenin, which is active against *Schistosoma cercariae*.[21]

Anthelmintic compounds in plants expel parasitic worms (helminths) and other internal parasites from the body by either stunning or killing them, without causing significant damage to the host. Woolly spider and brown howling monkeys with access to anthelmintic plants used by Amazonian peoples in Brazil have no internal parasites, while those without access to anthelmintic plants are infested with three different species of internal parasites.[22] Other compounds in plants denature parasites. Levels of internal parasites differ in mantled howler monkeys depending on where the monkeys live in Costa Rica. Monkeys without access to fig trees have high levels of infestation compared with those who have access to fig trees. To rid themselves of worms, South Americans eat fresh fig sap, which decomposes worm proteins.

The bark of shrubs and trees is high in tannins, which protect plants and provide herbivores with many benefits, including relief from internal parasites. Asiatic two-horned rhinoceros eat so much tannin-rich bark (50 to 70 percent tannins) of the mangrove (*Ceriops candolleana*) that their urine turns dark orange.[23] Indian wild bison feed on the bark of *Holarrhena antidysenterica*, which contains tannins and an alkaloid active against an endemic amoebic dysentery protozoa. Local people use the bark to treat dysentery.

I've witnessed many times what I believe to be self-medication among elk visiting our woodlot. During winter, if an aspen is felled by wind, elk immediately strip every bit of bark from the tree, perhaps for both medicinal and nutritional benefits. Tannins are antidiarrheal, antiseptic, antibacterial, anthelmintic, and antifungal.[24] Livestock fed plants with tannins have fewer nematodes and lower fecal egg counts than livestock not fed tannins, and they gain more weight than parasitized animals. By forming insoluble complexes with plant proteins in the rumen, tannins also increase the supply of high-quality plant protein to the small intestines. This supplementary protein helps animals cope with the nutrient drain of parasites, thereby enhancing immunity and reproductive efficiency. Sifakas (*Propithecus verreauxi verreauxi*), primates that live in the Kirindy forest of western Madagascar, eat much more tannin-containing plants between pregnancy and birth compared with nonreproductive females and males.[25] Intake of tannins increases body weight and stimulates milk secretion.

Broad classifications of secondary compounds – phenolics, alkaloids, terpenes – conceal the fact that phenolic compounds such as tannins differ in their structures, which influences how they function.[26] The bark of new blackbrush twigs, which is more than 70 percent condensed tannins, has antitumor-promoting activities.[27] For most goats, though, the kinds of tannins and their high concentrations outweigh any antitumor benefits, though it would be interesting to see if goats with tumors preferred new blackbrush twigs. While tannins such as those in blackbrush are strong feeding deterrents, others bestow multiple beneficial effects on animal health.

**Mechanical expulsion.** Many animals including dogs, cats, geese, and bears eat plants that cause mechanical expulsion of worms.[28] Prior to hibernation, Alaskan brown bears eat fibrous foods that result in large dung masses composed primarily of tapeworms. Mechanical expulsion of worms has been observed in eleven different populations of chimpanzees in Africa, as well as in bonobos and eastern lowland gorillas. Animals swallow leaves with rough surface texture caused by hooklike structures called trichomes. Leaf swallowing is effective against nodular worms, trapping them in the leaves as they move freely in the large intestine searching for food and mates. In addition to trapping loose worms, rough leaves swallowed on an empty stomach stimulate diarrhea and increase gut motility, further helping shed worms from the body.

**External parasites.** One way animals combat external parasites is by personally removing them or getting others to do so on their behalf. They use mud, dust, and sunshine to make skin a less hospitable environment for pests. They also use aromatic, analgesic, and astringent plants and toxic insect secretions on their skin and in their nests to control pests. Like humans, they also use volatile secondary compounds such as citronellol, camphor, and menthol to repel insects. Though effective as insect repellents, these volatile compounds evaporate rapidly, so people have developed insect repellents such as DEET that last longer. Due to the strong toxicity of these synthetic insect repellents, however, some pharmacists and herbalists are instead recommending volatile plant compounds to deter insects.

Another way animals remove external parasites is through their diet. Phytochemicals in the diet affect blood profiles of animals, which can adversely affect external parasites, including ticks. Cardiac glycosides, neem oil, and an extract from neem oil (azadirachtin) are lethal to ticks.[29] In spite of neem's pungent odor, lambs eat food with neem extract, and at the levels of neem-treated food they consume, they show no signs of toxicity. The more neem extract in the lambs' diet, the lower the tick (*Dermacentor variabilis*) weights at detachment. The greatest tick mortality after detachment was for sheep with the highest levels of neem in their diet.[30] In a study where guinea pigs were fed fish oil, their blood lipid composition was altered in ways that adversely affected salivary gland lipid composition in ticks, inhibiting ticks from feeding on them. Impairing synthesis of type 2–series prostaglandins, believed to facilitate blood-meal acquisition by ticks, decreases the number of eggs that ticks lay.[31]

**Alleviating bloat.** Tannins in plants such as birdsfoot trefoil and sainfoin reduce the incidence of bloat because they reduce microbial activities that lead to gas production in the rumen.[32] While rumen distension is an important signal of satiety in ruminants, bloat is characterized by a rapid accumulation of fermentation gas from eating forages, such as alfalfa. Sheep acquire aversions to foods paired with extreme distention, and they prefer foods paired with relief from distension.[33] Current recommendations to prevent bloat include restricting the availability of lush legumes in pastures and feeding dry roughage or grass to reduce consumption of bloat-inducing pastures. Adding tannin-containing plants to the pharmacy can also reduce incidence of bloat.

## Different Medicines for Different Maladies

People take aspirin to relieve headaches, antacids for stomachaches, and ibuprofen for pain, but until recently no one had ever demonstrated that nonhuman animals make multiple illness-medicine associations. To do so, we conducted a series of studies to determine if sheep learn to use different medicines for different maladies. We first showed that sheep fed high-grain diets ingest sodium bicarbonate and bentonite, substances that attenuate acidosis and restore acid-base balance.[34] We then showed that sheep learn to eat grape pomace with polyethylene glycol (PEG).[35] Sheep and goats eat PEG only when they are consuming high-tannin diets, and they prefer to forage in locations with PEG when eating foods high in tannins. They also regulate the amount of PEG they eat in accord with the amount of tannin in their diet.[36]

To learn whether sheep can make multiple illness-medicine associations, we then conditioned sheep to use three medicines that diminish illnesses: sodium bicarbonate (to alleviate illness from eating too much grain), polyethylene glycol (to counteract food containing tannins), and dicalcium phosphate (to counteract food with oxalic acid).[37] We then offered sheep either grain or food with tannins or food with oxalic acid and gave them access to the three medicines. Sheep chose the medicine that rectified the malady. Demonstrating multiple malaise-medicine associations illustrates herbivores learn to rectify specific illnesses – just as they learn to rectify nutritional deficits and excesses – by selecting foods with appropriate phytochemicals.

These studies, along with those reviewed in the preceding chapters, show the remarkable ability of the wisdom body to form specific flavor-feedback relationships under a wide range of nutritional, toxicological, and disease states. As this research demonstrates, when given a choice of natural foods, livestock show an astonishingly refined palate, learning from past experiences to meet their needs for nutrients and to self-medicate, as they nibble their way through the day eating a variety of grasses, forbs, and shrubs. While we use terms such as satiety and nausea to describe a continuum from well-being to malaise, those terms don't begin to capture the nuances of flavor-feedback relationships among cells and primary or secondary compounds in forages. We still know little about the communication that occurs among cells and organ systems to enable the complex interactions that occur with primary and secondary compounds in plants – and all of that is accomplished without a bit of thought from the host.

## Social Influences on Self-Medication

Following observations that goats with high levels of internal parasites browse the antiparasitic plant *Albizia anthelmintica*, J. T. Gradé and colleagues surveyed 147 Karamojong pastoralists and healers to see if livestock self-medicate.[38] Pastoralists are the best source of information due to their constant proximity to animals, high motivation to observe and monitor disease, and years of experience with animal care, diagnosis, and treatment. Pastoralists provided 124 observations of 50 proposed self-medicating behaviors used to treat 35 different diseases. To be considered self-medicating, animals had to show signs of illness pastoralists could visually diagnose, and they had to engage in foraging behaviors that were absent or rare when they are healthy, including eating what would ordinarily be considered an unpalatable plant or plant part. After eating the plant or plant part, they had to observe improvement in symptoms that lead to the cessation of self-medication. Over 70 percent of the documented observations of self-medication led to improved symptoms and cessation of self-medication. Of the plants used by sick animals, 72 percent are used by pastoralists to treat livestock and human diseases.

Many shamans, healers, herbalists, pastoralists, and hunters have learned about nature's pharmacy from observing the foraging behaviors of healthy and sick animals.[39] Chimpanzees and gorillas eat plants that healers claim cure diseases in humans. Among the 172 species of plants ingested by chimpanzees at Mahale Mountains National Park, 21 percent of the plants are used as anthelmintics by people in Africa. Among the more than 150 plant parts wild chimpanzees eat in the Kibale National Park in Uganda, at least 35 are used in traditional medicine to treat intestinal parasites, skin infections, and reproductive and respiratory diseases. Similarly, nearly half of 260 medicinal plants used by the Kani of southwestern India are eaten by wild animals in the same habitat.[40] Japanese macaques from 10 populations across the Japanese archipelago eat 1,664 different plant parts from 694 species representing 159 plant families. These plants have a wide range of properties valuable for macaques and humans, including antibacterial, antiviral, antifungal, antiprotozoal, antiparasitic, and antimalaria properties. Thus, through observations and personal experience of being sick and recovering from an illness after ingesting an herb, people learn which plants work for which illnesses.

With time, that knowledge becomes codified as part of the local culture and tradition. Traditional Chinese Medicine includes a range of practices based on traditions that date back over 2,000 years, including acupuncture, massage, exercise, dietary therapy, and herbal medicine. Though these approaches are considered "alternative medicine" in the Western world, they are conventional in East Asia, accounting for an estimated 40 percent of all health care in China. In Taiwan, contemporary Western university hospitals such as the Kaohsiung Medical University have Chinese Medicine departments where patients have the option of a wide range of traditional treatments. In China, formulations of botanical mixtures have evolved over the last thousand years, with more than 100,000 formulas now documented.[41] Current Western pharmacopoeias such as *Pharmacopoeia Europaea* list medicinal plant combinations as ancient as Dioscorides's *De Materia Medica*, which also dates back nearly two millennia.[42] The same is true for traditional pharmacopoeias worldwide. The emerging view is that 10 to 30 percent of all higher plants are used therapeutically, regarded as medicinal, depending on the region and culture.[43]

Ayurveda, a Hindu system of medicine native to India, stresses use of hundreds of plant-based medicines, including cardamom and cinnamon. Animal products may be used as well, for example, milk, bones, and gallstones. Fats are used internally and externally. Minerals such as sulfur, arsenic, lead, copper sulfate, and gold are also consumed as prescribed. The *Sushruta Samhita* and *Charaka Samhita* encyclopedias of medicine, compiled from various sources from the middle of the first millennium BCE to about 500 CE, are among the foundational works of Ayurveda. Over the following centuries, Ayurveda practitioners developed a number of medicinal preparations and surgical procedures to treat various ailments.

Though their knowledge and practices were never codified into a system of alternative medicine, Native Americans used more than 2,800 plant species in North America for medicinal purposes.[44] They used different plant parts to treat different maladies and often combined several botanicals for specific healing purposes. They also recognized plants either as potentially toxic or medicinal, depending on the dose. The Native American model of health emphasizes balance, harmony, and beauty by aligning body, mind, and heart with the environment.

The wisdom that shamans, herbalists, and healers acquire from observing animals and experimenting with plants, as well as culturally from other humans, raises interesting questions. How many people nowadays spend time carefully observing the foraging behaviors of wild or domestic animals? Would such observations even be meaningful for domestic animals conceived in one environment, born in another, and raised in yet another? To what degree have we humans become as naive as the animals in our care? Some livestock producers are now selecting for locally adapted herds of cattle and sheep.[45] To do so, they are calving and lambing in sync with nature's cycles, as wildlife do; they are no longer feeding hay during winter; and they are no longer using vaccines. Are these practices encouraging animals to learn to better use the nutrition centers and pharmacies nature provides, provided suitable choices are available?

Not unlike findings of altered food selection for livestock and humans provided with supplementary primary or secondary compounds, animals are less inclined to self-medicate when they are provided with antiparasitic drugs. Goats treated with anthelmintic drugs that kill internal parasites eat less tannin-containing heather than do goats infected with internal parasites.[46] Likewise, parasitized sheep reduce their intake of high-tannin food when their parasite infection is terminated with ivermectin, a drug that kills internal parasites.[47] Sheep infected with *Haemonchus contortus* eat more tannin-rich *Lysiloma latisiliquum* than noninfected animals.[48] As parasite loads escalate, goats and sheep increase their intake of plants that contain tannins, which in turn decreases parasite loads.

In 1840, Illinois horse farmer John Hoxsey learned about nature's pharmacy and self-medication by observing the foraging behavior of his prize stallion.[49] Hoxsey discovered a malignant tumor on the right hock of his horse, but as a Quaker, he couldn't bear to shoot the animal, so he put it out to pasture to die. Rather than perish, though, the animal started grazing knee-deep in a corner of the pasture, eating plants not part of its normal diet. After three weeks, Hoxsey noticed the tumor had stabilized, and after three months, he observed that the tumor had dried up and began to separate from healthy tissue. At this point, he retreated to the barn, where he began to experiment with these herbs revealed to him by "horse sense."

Hoxsey devised three formulas for healing cancer: an internal tonic, an herbal-mineral red paste, and a mineral-based yellow powder for external use. Eight of the nine herbs he used have antitumor activity, five have antioxidant effects, and all nine have antimicrobial activity that may be linked to cancer-fighting effects, which are of increasing interest to scientists nowadays.[50] For instance, sesquiterpene lactones – found in many native plants in North America, including ragweed, cocklebur, burdock, sneezeweed, and sagebrush, and exotic plants such as chamomile, artichoke, sunflower, lettuce, spinach, and ginkgo biloba – have anticancer properties. Taxol, a diterpene in Pacific yew, is active against leukemia-like cancers; carrots contain β-carotene and bioactive polyacetylenes active against cancer. Vernodalin, a sesquiterpene lactone in *Vernonia amygdalina*, has cytotoxic and apoptotic effects in vivo in human breast cancer cells.[51]

Hoxsey's grandson, John C. Hoxsey, tried these medicines on people with positive results. John's son Harry Hoxsey began working with John at the age of eight. John suffered an unfortunate accident when Harry was fifteen. On his deathbed, John bestowed the formulas to Harry with a charge to treat poor people for free and to minister to all races, creeds, and religions without prejudice. Finally, he warned that the "'High Priests of Medicine' would fight him tooth and nail because he was taking money out of their pockets."[52] Indeed, Harry Hoxsey survived decades of being "hunted like a wild beast" by "authorities" in the medical establishment, only to have his clinics forcibly shut down without the scientific studies he relentlessly implored the medical establishment to conduct. In the 1950s, at the height of medicine's battle against Hoxsey Cancer Clinics, the American Medical Association voiced establishment views in editorials in the *Journal of the American Medical Association*.[53] Hoxsey Cancer Clinics attempted in vain to have their cancer treatment evaluated in clinical trials, which have not been conducted.

## Alternative and Complementary Approaches

Few studies have compared alternative treatments for cancer due to the assortment of different options and the limited information on patterns of utilization and efficacy of alternative treatments for people with cancer.[54] The failure of modern medicine to address common complaints continues to prompt many people to seek support from alternative medicine. The varied practices of alternative medicine are different from *complementary and integrative medicine*, which use a wide range of therapies that complement conventional medicine. As most cancers are related to diet and lifestyle,[55] a new generation of healers is developing alternative and complementary approaches to *preventing* and *treating* cancer.

During the past fifty years, healthcare providers have relied on technology to treat disease and care for patients. A reductionist approach to science and medicine, with a loss of patient-focused care – the "medical model" of health – has progressively led to criticisms that medicine is not holistic. Some claim the prevailing Euro-American version of medicine, based on treating symptoms, has become a model of illness based on *anti*: antibacterial, anti-inflammatory, antidepressant. In *Limits to Medicine*, Ivan Illich – the Croatian-Austrian philosopher, Catholic priest, and critic of contemporary Western institutions – argues the medical treatment of many of life's vicissitudes frequently causes more harm than good and renders many people lifelong patients. Illich was the first to introduce the notion of iatrogenesis, unanticipated complications due to medical treatment or advice. He presents evidence to demonstrate the shocking extent of postoperative side effects and drug-induced illnesses.

James Duke was an American botanist and the author of many publications on botanical medicine, including the *CRC Handbook of Medicinal Herbs*. During the late 1970s, he was chief of the Plant Taxonomy Laboratory, Plant Genetics, and Germplasm Institute of the Agricultural Research Service, US Department of Agriculture. As Duke notes in *The Green Pharmacy*,

> *In 2005, pharmaceuticals killed at least 140,000 people in the United States – that we know of. The number may be even higher. As far back as 2002, the* Journal of the American Medical Association *recognized this problem, calling adverse drug reactions a leading cause of death in the United States. Now compare that with herbs and supplements. Altogether, they caused an estimated 29 deaths in 2005.*

In an analysis of the number of adverse drug reactions in a paper published in the *Journal of the American Medical Association*, researchers calculated 106,000 hospitalized patients die annually from drug reactions while 2,216,000 other hospitalized patients have serious drug reactions.[56] These figures, which would be far higher if the study included nonhospitalized patients, are startling today, and the study was published in 1998.

*The Physicians' Desk Reference*, the leading book for drug information for doctors, consists mostly of warnings, side effects, adverse reactions, and contraindications. One needs only to read advertisements in magazines or listen to commercials on television to appreciate the multiple serious side effects. They spend more time warning people about the many hazards than they do proclaiming the possible benefits – provided you don't react badly. Bernard Rimland, a researcher from San Diego, points out that if these pages were removed from the more-than-3,000–page book, only 150 pages would remain describing benefits of drugs.

Side effects of drugs arise from specificities and redundancies of signaling pathways. If a person touches poison ivy, the body releases histamine, a signal molecule that activates an itchy inflammatory response to ivy's allergen, but only at the site of the itch. That's not so with drugs because most of them act systemically. When a person takes an antihistamine for a rash, the drug is distributed throughout the body, reducing allergic symptoms, but simultaneously causing a side effect of drowsiness. Likewise, women on estrogen-replacement therapy to relieve menopausal symptoms experience side effects that include cardiovascular disease and strokes because estrogen-signaling molecules play an important role in the normal function of blood vessels, the heart, and the brain. They also experience increased risk of breast cancer.

As Andrew Weil notes, "The medical profession has painted itself into a corner by preferring treatments that are so dangerous, so expensive, and so reliant on technology. It has also separated itself from nature. Doctors fail to see that healing is fundamentally a natural process. They are unable to use the power that people give them in the service of healing. We can change that situation. The greening of the pharmacopoeia is an important first step."[57]

As an alternative to our overreliance on drugs and detachment from nature, renowned herbalist Christopher Hobbs suggests that there ought to be a medicinal plant garden in the center of each community.[58] Each community could train young and old alike to care for their own health using vegetable, herbal, and medicinal plants. As few as twenty species of cultivated plants are sufficient to treat most conditions, symptoms, and ailments common in our daily lives. As he suggests, find local plants, learn more about them, grow them, and start to use them.

While this idea may seem novel, it's not. The grounds of Blarney Castle in Ireland prominently display a historical herbal-medicinal garden, and rural villages in Africa, South America, and Southeast Asia still grow herbal gardens. In some Asian and African countries, 80 percent of the population still depends on traditional medicine for primary health care. Traditional Chinese Medicine, Ayurveda, and Tibetan botanical drugs are booming in the West. Developing and commercializing botanical drug mixes is increasing in allopathic medicine. Even in countries where drugs are readily available, botanical mixtures are often prescribed. In Japan, more than 50 percent of allopathic doctors prescribe traditional medicines such as Kampo for their patients. In the UK, up to 48 percent of cancer patients report taking botanical drugs after diagnosis.[59] In the United States the number of visits to providers of complementary alternative medicine in the last decades has exceeded the number of visits to all primary care physicians.[60]

Three herbs derived from Traditional Chinese Medicine have become widely used in the West based on their outstanding performance in scientific studies.[61] Sweet wormwood (*Artemisia annua*), the source of artemisinin, is the preferred antimalarial drug recently approved by United States Food and Drug Administration for use in combination therapies. Thunder god vine (*Tripterygium wilfordii*) is being used for rheumatoid arthritis. Green tea (*Camellia sinensis*) is used as a functional beverage and a component of dietary supplements. Different plants can be used prophylactically or therapeutically. For example: Echinacea (*Echinacea purpurea*) stimulates the immune system; goldenseal (*Hydrastis canadensis*), cleavers (*Galium aparine*), and yarrow (*Achillea millefolium*) reduce inflammation; elderberry (*Sambucus nigra*) and gumweed (*Grindelia* spp.) treat cold symptoms; blackberry (*Rubus villosus*) and marshmallow (*Althea officinalis*) treat sore throat; fennel (*Foeniculum vulgare*) and yellow dock (*Rumex crispus*) encourage digestion; willow bark (*Salix alba*) treats osteoarthritis; cramp bark (*Viburnum opulus*) relaxes muscles; and scullcap (*Scutellaria* spp.) eases muscle tension.

In addition to nurturing the health and well-being of people, partially converting lawns into medicinal, herbal, and vegetable gardens combined with re-establishing locally adapted native plants could add to the overall health of ecosystems. A study by scientists from NASA, in collaboration with researchers in the Mountain West, showed Americans' lawns now cover an area three times larger than any irrigated crop in the United

States – 63,000 square miles of lawn, an area roughly the size of Texas.[62] While lawns can help to mitigate climate change by removing carbon dioxide from the atmosphere, the costs of lawns may outweigh their benefits. Each year, Americans' spend $30 billion on lawn care. We use three million tons of fertilizer – use of nitrogen fertilizer could be cut in half by leaving clippings on lawns to recycle nutrients and build healthy soil. We use more than 30,000 tons of synthetic pesticides on lawns at a cost of more than $2 billion. We burn more than 800 million gallons of gas annually mowing lawns. The amount of gas spilled annually refilling gasoline lawn mowers is 17 million gallons, 1.57 times the amount spilled by the Exxon Valdez off the shores of Alaska. Residential water use outside the home is 30 to 60 percent of total water use. Critically in the arid West, we use between seven billion and nine billion gallons of water daily for suburban irrigation.

## Phytochemical Synergies

Historically, if you wanted to be a doctor or pharmacist in the United States, you had to know botany because most of medicine consisted of giving people preparations of plants as part of the healing process. Today, many drugs are purified extracts of plant origin or synthetic variations of chemicals originally discovered in plants. As Andrew Weil points out, "Today, I'm sorry to say – and I don't mean this with disrespect – it seems to me that the only requirements you really need to be a pharmacist are not to be color-blind and to be able to read and count. All the good stuff that used to go on, like the compounding of natural remedies, is gone."[63]

Synergies created by "compounding" are etiologic in health, as studies with livestock reveal. Animals obtain three advantages when they are allowed to learn to self-medicate using a variety of different medicinal plants.[64] First, they can select the mix and amounts of plants most beneficial for their specific malady. Second, the animals no longer need to be dosed with drugs that some individuals may not need, no longer exacerbating parasite resistance. Third, they can benefit from the synergies provided by phytochemically complex mixtures of plants.

Herbivores can better meet their needs for nutrients and reduce internal parasites when offered a mix of plants with diverse phytochemicals that affect different parasites. As noted previously, goats and sheep in Mediterranean scrublands eat *Pistacia lentiscus,* high in tannins, which adversely affects nematode fertility and prevents infestation with eggs. However, nematode egg excretion in feces is only partly impaired by tannins—terpenes in *P. lentiscus* act in synergy with tannins to further impede the survival of worms. Tannins, terpenes, and alkaloids all have anthelmintic effects.[65] That's likely why populations of Japanese macaques infected by many kinds of parasites eat a range of antiparasitic plant parts.[66]

Combinations of phytochemicals enhance deterrence by increasing toxicity. That's illustrated with alkaloids in tall larkspur. Of the twenty alkaloids in larkspur, only one of them, the diterpenoid alkaloid methyllycaconitine (MLA), is toxic to cattle. However, when MLA is extracted and purified and then given to cattle alone or in combination with the less toxic MDL-type alkaloids in larkspur, toxicity increases with the combination compared to the MLA-type alkaloids alone.[67] The same kind of reaction is illustrated with the tropical rainforest shrub *Piper cenocladum* (black pepper), which has three secondary compounds (amides) with defensive functions.[68] Individual amides are less effective at deterring insect herbivores than a mixture of all three amides. A combination of amides also deterred a broader array of insects, whether they were naive, experienced, generalized, or specialized feeders.

Parasites evolve resistance to a single anthelmintic drugs, and they will continue to do so if people attempt to create the next silver bullet, even with tannins. The rapid evolution of multiresistant strains of parasites worldwide demonstrates that repeated use of the same drugs doesn't work.[69] Animals on over half of the farms in Australia, Brazil, and the United States harbor drug-resistant parasites, which should be an impetus to develop alternative approaches.

Yet, Western pharmaceutical industries and regulatory agencies are skeptical of using complex mixtures of phytochemicals. Plants that contain hundreds to thousands of bioactive compounds encompass multiple molecular mechanisms of action. Identifying molecular targets of all these compounds is seen as a major bottleneck in botanical drug research. In most cases, no feasible mode(s) of action can be revealed with conventional biochemical procedures.[70] That's why industries and agencies are partial to single-ingredient drugs based on synthetic molecules.

Biochemist Jürg Gertsch describes the challenges of synergies for biochemists.[71] He begins with a story about a Mexican physician who claims to cure various cancers with a mix of twelve cytotoxic plants. As he points out, no statistical data exist to prove this physician's intriguing observations, and trying to study the molecular mechanisms of his botanical mixture would be exceedingly difficult. If each plant contained 100 compounds, he would administer more than 1,000 compounds. To discover which of these compounds are bioactive, each compound would have to be analytically identified and determined in blood plasma upon oral administration, an enormous task given the massive number of compounds. In theory, once bioavailable compounds are known, potential synergies and antagonisms between compounds found at concentrations in plasma could be studied in vitro. However, if 1,000 compounds are bioavailable, at the level of synergy, millions of individual experiments would be required to assess all possible combinations and permutations, an overwhelming task. Without solid clinical data, nobody will engage in such a work-intensive and obviously risky project.

With most botanical drugs, multitarget effects are common. For drug discovery and pharmacology, this means the magic bullet may be better thought of as a shotgun. Medicinal chemist Andrew L. Hopkins coined the term *network pharmacology* to stress drug-target networks rather than single drug targets.[72] Integrating network biology and polypharmacology might expand current concepts about drug targets and may help researchers understand the pharmacological action of botanical medicines. To be reliable and compete in the marketplace with conventional pharmaceuticals, polypharmacology will need to find new ways to validate target combinations and optimize multiple structure-activity relationships while maintaining druglike properties. The latter is problematic for developing botanical drugs because they are obtained from plants in nature and thus can't be rigidly engineered.

Companies are beginning to appreciate the need for synergies. The Chinese company PhytoCeutica, Inc., is developing a four-herb Traditional Chinese Medicine (PHY906) with a history of more than 1,500 years of human use.[73] Each of PHY906's four plant species has a distinct pharmacological profile that includes anticancer and antiviral activity, hematological and immunological stimulation, analgesic activity, liver protection, and appetite stimulation. Traditionally, PHY906 has been used to treat diarrhea, abdominal spasms, fever, headache, vomiting, nausea, extreme thirst, and subcardial distention. While this heterogeneity may bewilder Western pharmacologists, Asian practitioners see value in the polyvalent nature of such medicines. The hallmark of traditional phytotherapy is using plant extracts to make complex mixes. Mixing plants to make a more potent medicine is common in phytotherapies, including Traditional Chinese Medicine, Ayurveda, and numerous ethnic pharmacopoeias. Many medicinal plants are now registered and marketed as mixtures under the label of botanical drugs.[74]

## Memories for Medicines

Bodies have remarkable ability to relate flavors of foods with different internal states. Bodies learn to select medicines to rectify different maladies. Assuming John Hoxsey's prize stallion was self-medicating, how did he know which herbs to eat? Perhaps the horse sampled the "profusion of weeds" in the past, but his body found them of no particular use at that time. Can a body remember such experiences and use the knowledge to select aptly when needed?

When I was in kindergarten, I got the mumps, a viral disease that causes painful swelling of salivary glands. There is no particular treatment for mumps, though the symptoms may be relieved by applying ice or heat to the affected area and by taking Tylenol to relieve the pain. Warm saltwater gargles, extra fluids, and soft foods may also help relieve symptoms. These treatments notwithstanding, my body developed an intense craving for chili. That was peculiar because, until that time, I disliked chili. I found the ingredients disgusting. When chili was the evening meal, I went hungry. But during my affair with the mumps, I craved chili and I relished every bite of the phytochemically rich mixture of ingredients – hamburger and beans, onions and tomatoes, chili peppers and cumin spices – that my mother so kindly prepared for me.

How can a food, formerly revolting, suddenly become so delightful? Do cells and organ systems "take note" of various kinds and combinations of phytochemicals in the foods we eat and remember them when necessary? If so, how might cellular memory of primary and secondary compounds become encoded? Bodies develop immunological memories based on encounters with pathogens, the basis for active immunizations that causes a body to generate immunity against a target. This immunity comes from T cells and B cells with their antibodies.

The immune system protects organisms from infection with layered defenses of increasing specificity. Biological, chemical, and physical barriers protect organism from pathogens. Physical barriers prevent bacteria and viruses from entering the body, while biological and chemical barriers destroy pathogens that breach physical barriers. If a pathogen breaches these barriers, the *innate immune system* – the dominant system of host defense in most plants and animals – provides an immediate but nonspecific response. The innate response is triggered when microbes are identified by pattern recognition receptors, which recognize components that are conserved among broad groups of microorganisms. They are also activated when damaged, injured, or stressed cells send alarm signals, many of which are recognized by the same receptors that recognize pathogens. Innate immune defenses are nonspecific, they respond to pathogens in a generic way, and this system does not confer long-lasting immunity against a pathogen. Innate immune systems are found in all plants and animals.

If pathogens evade the innate response, vertebrates have another layer of protection, the *adaptive immune system*, which also has immunological memory, such that each pathogen is remembered by a signature antigen.[75] This system thus generates responses tailored to specific pathogens or pathogen-infected cells. The ability to mount tailored responses is maintained by memory cells. Should a pathogen infect the body more than once, memory cells quickly eliminate the pathogen. Thus, the immune system adapts its response during an infection to improve its recognition of a pathogen. This improved response is then retained in the form of an immunological memory, after the pathogen has been eliminated. That allows the adaptive immune system to mount quicker and stronger attacks when a pathogen is encountered, much like *priming* in plants (as explained in chapter 2). Exposure to a variety of pathogens early in life creates a more robust immune system throughout a person's life.

Newborn infants are especially vulnerable to infection because they have no prior exposure to pathogens. However, mother provides several layers of passive protection. During pregnancy, antibodies are transported from mother to baby directly across the placenta, so human babies have the same range of antigen specificities as their mother.[76] Breast milk or colostrum also contains antibodies that are transferred to the gut of the infant and protect against bacterial infections until the newborn can synthesize its own antibodies.[77] This passive immunity is usually short-term, lasting from a few days up to several months, and the fetus doesn't actually make any memory cells or antibodies. Rather, the fetus gets them from its mother.

These observations raise questions. Cells and organ systems have memories of foods that nourish. Can a body develop passive medicinal memory in utero and early in life, through mother's milk, based on what mother eats? Can a body develop active medicinal memory based on what we eat early in life? If so, how might cells and organ systems accomplish such a feat? If a body develops medicinal memory, that has implications for exposing a body to a wide array of phytochemically rich foods for later use, should the need arise. A body conceived, nurtured in a womb, and raised on junk food has no medicinal memory and is truly at the mercy of the medical and pharmaceutical industries for help with health and disease. Alternatively, a body exposed to a rich array of phytochemicals from conception onward may be armored for life.

# PART III

# Savoring the Artist's Palette

# CHAPTER 8

# Delighting in the Colors

During the years I worked on a ranch in Colorado as a young man, we marked time with events: planting, growing, harvesting, and feeding. The chilly winds that blew down the valley from the high-mountain passes in spring signaled time to sow oats and barley and to plant our vegetable gardens. The warm rays of summer sun melted the snow in the mountains and gave us water to irrigate hay, which we baled, hauled from the fields to the barn, and stacked by hand. The chilly breezes that escorted the first storms of fall were time to put up the last of the meadow hay and haul straw bales of oats and barley, the grain now threshed and stored in the granary. Winter was time to feed cattle, sheep, and goats with hay we'd stacked. It was also a time to savor the vegetables we'd stowed from summertime and the sausages we'd cured in the autumn.

When I look back at those experiences, I recall two overpowering sentiments that remain with me today. I felt then, and still do now, deeply linked with the landscapes where I was born and raised, including those that surrounded the ranch, mountains like Shavano where we ran cattle in summer. The ranch was nestled at 8,600 feet elevation, just east of Mount Shavano, which towers more than 14,000 feet and is named in honor of Ute Chief Shavano. To the south, Mount Ouray, named for another Ute chief, rose prominently to nearly 14,000 feet. Nearby 12,775-foot Mount Chipeta, northwest of Mount Ouray, was named in honor of Ouray's wife.

Throughout those years, I also felt deeply connected with the annual cycles manifest in the seasons, planting during the chilly spring days, irrigating during warm summer days, and harvesting crops in fall. Though I'm not sure how ants "feel" about spring, summer, fall, or winter, I often thought about and felt deeply connected with their seasonal activities as I harvested crops on brisk fall days, just ahead of storms, preparing for winter.

Late-fall preparations also involved butchering hogs. We began by building a small fire under a water-filled caldron, where each hog was scalded just after it was killed. Two people, one on each side of the caldron, held chains, one in each hand, to support the hog's carcass. The hog was scalded on one side until its hair was easily removed, and then turned with the chains onto the other side. After being scalded and scraped, the hog was gutted and taken to the cooler.

The next day, we prepared sausage. We filled the casings – the intestines we'd cleaned immaculately after the hog was butchered – with a unique blend of meat and spices to make soppressata. We also prepared cappicola, traditional Italian pork made from the muscle running along the neck to the fourth or fifth rib of the pork shoulder or neck. And we made hams.

We hung the soppressata, cappicola, and hams in the smokehouse, a cavity hollowed into a hillside decades ago. The sight of the meat hanging from smoke-coated iron rods that crossed the ceiling, along with the smells of the smokehouse, seasoned for decades with the smoke of wood from the apple trees in our orchards, caused us to salivate prodigiously.

As I reflect on these experiences, I realize I was learning traditions, passed down over many generations. Rancher Henry DeLuca and his wife, Rose, had learned how to grow and prepare food for livestock and people from their parents and grandparents and ancestors before them in Italy. Rose made all of her dishes from scratch, and they were rich with flavors from the plants and animals we grew: the sauces made with fresh tomatoes, the many different herbs – including fresh onions, garlic, oregano, parsley, and basil – and the breads made with pumpkin and zucchini. I also remember distinctly that the meals were satisfying, not filling. I didn't have to eat much of the diverse mixtures of ingredients to feel satisfied.

During those years, we never once discussed energy or protein, primary or secondary compounds, phytochemical complexity, vitamins, minerals, or essential fatty acids. Indeed, Henry and Rose wouldn't have known what to make of all that nutritional minutiae. Nor did we ever discuss the latest US Dietary Guidelines for replacing fats with carbohydrates. Henry and Rose never got on that bandwagon. We loved to eat the high-fat sausages we made and the ricotta cheese Rose made fresh with whole milk from the Guernsey cow we milked each morning. We also relished spaghetti and ravioli made from semolina flour and desserts such as pumpkin pie made from our garden-grown pumpkins. We ate a wide variety of wholesome foods, most of which we raised or gathered (mushrooms) and hunted (deer and elk) in the wild.

I've often wondered how all those ingredients, not merely the foods that provided energy and protein but the many herbs and spices that added a richness of flavor, contributed to my liking for the flavors of the dishes and the satisfying nature of the meals. Certainly, the same foods prepared without those herbs would not have been nearly as desirable.

## Nature's Colors in Foliage and Fruits

During summer, seeds we plant in spring transform from packets of potential into seedlings, juveniles, and mature plants that create more packets of potential. Upon ripening, seeds are disguised, some inside plump, red tomatoes, others hidden underneath shucks, still others inside cantaloupes, watermelons, zucchinis, and pumpkins. Their colors change from tans and browns of seeds, to pale green seedlings, to deep green mature plants, to reds, yellows, and oranges of their progeny. Where do plants get their spectacular colors? Why do they have colors at all? The answer to these questions lies in a now-familiar theme: Secondary compounds, along with the many other roles they play in the lives of plants, are the source of their beautiful colors.

**Chlorophyll.** Chlorophyll is the most abundant pigment in the world. It paints Earth green in spring and summer. Chlorophyll is essential for photosynthesis, the process whereby plants convert photons, the small packets of energy that make up light, into the chemical bonds of sugar molecules plants use for energy. Chlorophyll absorbs specific colors (wavelengths) of light and helps to convert them into chemical energy. Photons from different colors of light contain different amounts of energy. The colors in the spectrum – red, orange, yellow, green, blue, indigo, violet – are in ascending order of energy. A photon of red light has less energy than a photon of violet light. Chlorophyll absorbs violet, indigo, blue, orange, and red light, most yellow, but hardly any green light, so green light gets reflected back to our eyes, which is why leaves look green.

The chemical structure of chlorophyll closely resembles hemoglobin, an essential protein located in the bloodstream. Chlorophyll is known for its ability to help increase production of red blood cells, thus preventing anemia, and for blood clotting and wound healing. Chlorophyll is loaded with antioxidants that exert beneficial effects throughout the body, including reducing the risk of cancer by inducing apoptosis, the death of cells that occurs as a normal and controlled part of an organism's growth or development. Chlorophyll strengthens the immune system, helps heal inflammatory disorders such as arthritis and fibromyalgia, and has antiaging properties.

Unlike chlorophyll, other pigments in leaves and flowers absorb light of different colors, so they reflect as red, orange, yellow, blue, and purple light. The pigments in the showy flowers that attract birds, bees, and other insects fall into three classes. Flavonoids (proanthocyanidins) make deep reds, blues, purples, and magentas. Carotenoids (carotene and lycopene) and other flavonoids (flavones and flavonols) create the reds, oranges, and yellows of flowers and fruits. Betalains make colors akin to flavonoids and carotenoids. Deep and varied colors – from purples, blues, and greens to the reds, oranges, and yellows – are an indication of the phytochemical richness in the vegetables and fruits that promote health in humans.

**Flavonoids.** Plants produce more than 6,000 different flavonoids. They are common and not very toxic, so animals can ingest large quantities of them and suffer no ill effects. They are the most common polyphenolic compounds in human cuisines throughout the world. We ingest far more flavonoids from plant foods than from animal foods, and vegetables and fruits are especially rich sources of flavonoids. To name just a few examples, we eat flavonoids in vegetables from broccoli and cabbage to lettuce, onions, parsley, and peppers to sweet potatoes and tomatoes; in fruits from bananas, blueberries, and strawberries to all citrus fruits; in nuts, such as almonds, pecans, and walnuts; and in dark chocolate. We drink flavonoids in red wine and tea.

Flavonoids such as rutin, hesperidin, quercitrin, and tangeretin create yellow pigments in lemons, oranges, and grapefruits. Proanthocyanidins create beiges to shades of black, red, brown, and tan. Anthocyanin pigments in leaves, stems, roots, flowers, and fruits range from red to purple to blue to white and pale yellow depending on soil pH. Anthocyanins are partly responsible for the red and purple colors of olives. Blueberries, cranberries, and bilberries are rich in anthocyanins, as are black and red raspberries, blackberries, blackcurrants, cherries, concord and muscadine grapes, eggplant peel, black rice, and red cabbage.

Flavonoids act as antioxidants that reduce free radical damage to cells and organ systems throughout the body, including neurons in the brain. They also have anti-inflammatory properties. Not surprisingly, since many difficulties in the cardiovascular system involve oxidative stress and inflammation, the antioxidant and anti-inflammatory benefits of flavonoids provide direct support for this organ system. In the bloodstream, flavonoids help protect LDL cholesterol molecules from damage by oxygen free radicals. This LDL protection, in turn, helps to lower risk of atherosclerosis. Flavonoids, including rutin and hesperidin also increase the strength and integrity of the blood vessel walls, lowering the risk of arteriolosclerosis.

**Carotenoids.** Carotenoids absorb blue and indigo light and reflect yellow and orange light. Of the more than 600 types of carotenoids, those most common in the Western diet, and the most studied, are α-carotene, β-carotene, β-cryptoxanthin, lutein, zeaxanthin, and lycopene. Lycopene creates the reds of tomatoes, guava, red grapefruit, papaya, rosehips, and watermelon. Zeaxanthin is the pigment that gives paprika (made from bell peppers), corn, saffron, and goji berries their characteristic colors. Lutein makes yellow pigment in fruits and vegetables, such as kiwi fruit. Carotene creates orange pigment in carrots, yams, mangos, and autumn leaves. It also gives a rich yellow color to the butter from animals on pasture and a yellow tinge to the fat of animals that dine on grasses, forbs, and shrubs. Carotene can color people, too. Eating foods high in β-carotene, for some people three large carrots a day, can give a yellowish tint to skin.

The body breaks down each carotene molecule to produce two molecules of vitamin A. Among other functions, vitamin A promotes healthy eyesight. Vitamin A and carotenoids help prevent premature skin damage and skin cancer. Diets high in carotenoids help prevent damage from ultraviolet light, which can lead to melanoma, aged-looking skin, wrinkles, drying, scaling, and follicular thickening of the skin. Like flavonoids, carotenoids also have antioxidant and anti-inflammatory properties that help us ward off common infections, illnesses, and diseases such as cancer. People who eat diets of fruits and vegetables rich in carotenoids have enhanced immune systems and lower mortality from a number of chronic illnesses.

**Betalains.** Betalains produce colors similar to flavonoids and carotenoids in two subgroups of pigments: red-violet (betacyanin) and yellow to orange (betaxanthin). They create the brilliant colors of the flowers of some cacti and amaranths. They are responsible for the deep red color of beets and the yellow and red in the stems of Swiss chard. The composition of different betalain pigments can vary, giving rise to breeds of beetroot that range from yellow to the familiar deep red. Their intense colors are used commercially as food-coloring agents. Like flavonoids and carotenoids, betalains also have antioxidant properties beneficial for health.

# Passing Colors to Feathers, Fur, and Fish

The fossil record suggests birds emerged during the Jurassic period, 160 million years ago. According to paleontologists, birds are the only group of dinosaurs that survived the Cretaceous-Paleogene extinction, roughly sixty-five million years ago. Nowadays, more than 10,000 species of birds exist worldwide. These "modern dinosaurs" have lightweight skeletons, modified for flight, and their exteriors are covered with feathers of the most amazing combinations of colors.

On this rain-freshened spring morning, I'm admiring the bluebirds, phoebes, finches, redpolls, and flickers landing on the stumps in our backyard. They come in so many patterns, shades, and hues from violet, blue, and cyan to green, yellow, orange, and red. The Artist has created a marvelous range of dyes from just seven colors in the rainbow. The flicker family just arrived. Flickers are large woodpeckers with handsome black-scalloped plumage on their back, a brilliant red strip across the back of their neck, and a breast with big, distinctly round black dots underlain by feathers of russet that blend into gray. When they fly, you see a flash of color in the wings and a bright white flash on the rump. Where do they get their colors?

According to folklore, they got their colors from fire. While flickers are certainly ablaze, we now know their eye-catching colors aren't produced by fire. Rather, birds "steal fire" from plants to make their bodies, and genes interact with their diets to color their feathers. For many birds, the fruits and seeds they eat provide the pigments that paint their plumage.

Three pigments – porphyrins, melanins, and carotenoids – are primarily responsible for the colors of bird feathers. Porphyrins produce reds, browns, pinks, and greens in birds. They are the rarest of the three pigment groups, found in only a handful of bird families.

Different mixtures of porphyrins, melanins, and carotenoids combine to produce wide-ranging and unusual hues and shades. For example, the dull olive green colors of certain forest birds are actually a mixture of pigments from dark brown melanins and yellow carotenoids.

Carotenoids are the source of orange, yellow, and red feather colors in birds. The brilliant pink color of flamingos originates with carotenoids in the crustaceans and algae the birds sieve from the water as they feed. The deep red colors of cooked salmon, red bream, trout, lobster, and shellfish arise from astaxanthin. In a live animal, astaxanthin is linked with a protein to yield a blackish color. When boiled, the protein breaks down to reveal the lobster red of astaxanthin.

A single type of pigment can create plumage of different colors in different species of birds. Carotenoids in the berries and fruits finches eat in summer and autumn create pink shades in the head, neck, and shoulders of male house finches. Carotenoids in the berries a cardinal enjoys eating in summer are laid down in feather follicles that will produce deep red feathers after the next molt. The pigments in one kind of seed might make a cardinal appear pink, while the same pigment might make a house finch appear yellow. If a cardinal or finch has a diet limited to only one type of seed, its feathers become progressively duller with each molt.

Melanin produces black or dark brown coloration in bird feathers. Animals make melanin, but unlike plants, most animals can't make green or blue pigments. Rather, those colors come from structural effects as light is refracted by proteins. A bluebird manufactures melanin and would look almost black, but tiny air sacs in the feathers scatter light and make it appear blue, in a similar way to the sky, which appears blue as gas molecules in the atmosphere scatter light. Melanin is also responsible for the color of hair and fur in mammals. Different types of melanin (eumelanin and pheomelanin) can produce a wide range of colors, from black to sandy to red. Reds and black in fur are also due in part to copper in the diets of mammals.

## Phytochemicals Flavor Meat and Milk

For a brief period during autumn, maple and aspen trees set the hillsides ablaze with greens and yellows, golds and oranges, reds and salmons. If you are a hunter and you wish to find a blue grouse early in the season, you look among the food-rich grasses, forbs, shrubs, and maples. If you wish to find a grouse late in the season, long after the maples have gone to sleep, you look in the conifer trees that provide home and grocery. For blue grouse, not just any tree will do. Grouse are expert at finding trees low in terpenes and high in energy and protein. As discussed, there is a cost involved in eating plant parts that contain too high levels of terpenes. The bill is paid in energy, protein, and other nutrients required to detoxify (liver) and eliminate (kidneys) excess terpenes from their bodies. While grouse aren't likely aware of these nutritional and toxicological niceties, they are adept at finding trees that bear the best needles to eat.

As any hunter knows, the food their quarry has eaten affects the flavor of the meat. The flavor of grouse changes as the birds switch from eating a mix of grasses, forbs, and shrubs in autumn to a diet of Douglas fir needles in winter. Douglas fir gives a hint of terpene flavor to their meat. People often perceive flavors in the meat of wild animals as gamy. Wild animals have hearty, full-bodied flavors due to phytochemicals in their diets. With wild game, and sometimes with livestock, though, gamy flavor reflect how an animal was slaughtered, because stress prior to killing and poor processing create poorly palatable carcasses called *dark cutters*.[1] Dark cutters aside, the robust flavors of wild animals can be distasteful to the uninitiated palate accustomed to eating bland foods, but people who eat wild game don't share the same experience of unpalatability. They learn to eat foods that contain the phytochemicals that create the robust flavors, and those flavors become characteristic of plants and animals in the local cuisine.

Not unlike the flavors of the fruits and vegetables we now eat, the flavors of meat and dairy products have gotten blander during the past fifty years, as Mark Schatzker emphasizes in *The Dorito Effect*. The meat of birds that quickly attain high body weight lacks flavor. And the trend in commercial agriculture for the past several decades has been to speed up production cycles of chickens and turkeys. The hybrid varieties of these birds now commonly raised on poultry farms reach twice the body weight of fifty years ago in only half the time. That leads to diseases of muscle tissue (myopathies) such as white striping and woody breast that also make the appearance, quality, and flavor of the meat bland.[2]

To enhance the flavor of bland foods as Schatzker discusses, Americans now consume more than 600 million pounds of synthetic flavorings a year, according to Euromonitor International. We attempt to make up for lack of flavor by adding synthetic sweeteners, spices, dressings, and sauces to foods. Nevertheless, factory-made flavors lack the phytochemical richness of plant-derived secondary compounds. The flavors we crave and the phytochemicals we need come from plants.

Phytochemicals delight the eyes, arouse the nose, excite the palate, and nourish the cells of our bodies. Odor and taste merge into the flavors we experience when we eat a juicy steak, spicy lasagna, a crisp salad with herbs, or a freshly picked plum, grape, orange, apple, or peach. The richness of colors and flavors link our palates with our health. With each bite of a biochemically rich meal, animals introduce thousands of primary and secondary compounds into their bodies. These compounds interact with one another and with cells and organ systems in complex ways we will never fully understand. Each compound is toxic at high doses but at appropriate doses, has health benefits. Depending on the mix of compounds in foods and the sequence in which foods are eaten in a meal, any phytochemical can be nutritious or toxic.[3]

The evolution of the body's learning to use phytochemicals for nutrition and health has occurred over millennia. Through feedback from cells and organ systems, bodies continually process the diverse array of biochemical interactions and signals generated by thousands of primary and secondary compounds in the plants we eat. As a result, the body of an herbivore knows what to do with the complexity of a day-long meal that includes twenty-five to fifty different species of plants. The body of a human also knows how to process a meal of meat, vegetables, various spices, fruits, and nuts, such as those meals I enjoyed at the DeLuca Ranch.

Despite our efforts to analyze foods to determine which primary and secondary compounds they contain, the fact remains that our bodies, not our brains, are the best judges of what's good to eat. As William Albrecht wrote in *Soil Fertility and Animal Health* many years ago: "Though the cow cannot classify forage crops by variety name or by tonnage yield per acre, she is more expert than any biochemist at assessing their nutritional value." And what's being assessed are messages from soils, expressed in the primary and secondary compounds of plants, selected by herbivores reared in cultures that are evolving in particular landscapes. That, in essence, is what it means to be locally adapted to changing landscapes in which all things are linked and evolving over time.[4]

## Phytochemicals' Role in Terroir and Health

As discussed in chapter 2, the environment where a plant grows – sunlight, nutrients, and water – affects its chemistry, which in turn affects its flavor. Historically, *terroir* referred to how a particular region's climate, soils, and terrain affect the taste of wine. Notions like terroir are now trendy worldwide and they've come to pertain to food as well. Nevertheless, no comprehensive body of research has assessed how the phytochemical richness of different mixtures of forages affects the flavor or health benefits of meat or dairy for people.

The forages livestock eat influence the flavor of milk, which can in turn influence human's preferences for the flavor of milk and cheese. For example, cattle fed low-lipid diets produce milk fat high in cheesy-flavored fatty acids and precursors of blue-cheese–flavored methyl ketones and coconut-peachy-flavored δ-lactones.[5] Cattle fed high-lipid diets make sweet, raspberry-flavored γ-dodecalactone from dietary oleic acid and sweet, raspberry-flavored γ-dodec-cis-6-enolactone from dietary linoleic acid. Botanical diversity of the pasture affects concentrations of phenolics in cheeses such as L'Etivaz and Gruyere.[6] Among other compounds in forages, carotenoids impart a yellow color that affects sensory properties of milk and cheese. Terpenes also influence sensory properties, especially of milk and dairy products derived from natural pastures that include a more diverse mix of plants than managed pastures consisting of grasses only.

When dairy cows graze botanically diverse pastures in Italy, as opposed to being fed a total-mixed ration made of cultivated forages and cereal grains, the flavor and biochemical richness of their milk and cheese are enhanced.[7] People in Italy strongly prefer the flavors of milk and cheese from dairy cows grazing on the botanically diverse swards.[8] Rural peoples in Italy and France select specific cheeses based on the mix of plants in a particular landscape, their palates linked with soil, plant diversity, and herbivore diets.[9] Untrained assessors, who best represent consumers, are often less able than trained assessors to discriminate sensory differences in milk and cheese, which suggests such differences are often subtle. Considerably more research is required to elucidate the many ways that diet affects the flavor of milk and cheese.[10]

As with dairy, people prefer the flavors of foods they are accustomed to eating. Studies show that peoples in Mediterranean and Northern European countries – Greece, Italy, Spain, France, the UK, and Iceland – differ in their preferences for meat. Mediterranean lambs are fattened on grain, while Northern European lambs are finished on pastures. People from Mediterranean countries prefer grain-fattened lamb, while people from Northern Europe prefer grass-finished lamb. Yet another example was demonstrated in a taste-test comparison of the meat from lambs raised with two different feeding protocols. A taste panel from Britain and a taste panel from Spain both sampled the meat of Spanish lambs fed with milk and grains as well as meat from British grass-fed lambs. Both panels found grass-fed lamb had higher odor and flavor intensity. However, the Spanish panel preferred the Spanish milk- and grain-fed lamb, while the British panel preferred the grass-fed lamb.[11]

People in the United States are reared on grain-fed beef. [12] No surprise, then, even American professionals trained to evaluate sensory features of meat often find grass-fed beef less palatable than grain-fed beef.[13] Yet, ratings for grass-fed are inconsistent.[14] Livestock eat different diets during the finishing period and that influences the flavor and biochemical richness of meat and fat to the extent that laboratory analyses can distinguish animals foraging in different landscapes, including feedlots.[15] Among many other phytochemicals, flavonoids, carotenoids, and terpenoids in herbivore diets become part of the flavor of meat and fat. That is illustrated when researchers add garlic, thyme, essential oils from juniper, essential oils from rosemary or clove, or tannins to the diets of lambs and calves – they all improve the flavor and biochemical richness of the meat.[16]

These dietary differences, which are not reflected in the generic label "grass-fed," partially explain why grass-fed beef doesn't have consistent flavor or quality. Consumers need more information about the phytochemical richness of a landscape where animals were raised to make an educated guess about the flavor of meat from those animals.[17]

That nuance is reflected in the comments of Warren Angus Ferris in *Life in the Rocky Mountains*. Ferris was a trapper, cartographer, and diarist. From 1829 to 1835, he traveled to Cache Valley and the Snake River area with the American Fur Company. During that trip, he noted that they stopped along the Bear River, after a very fatiguing and toilsome march of thirty miles. They killed many buffalo and then "feasted luxuriously on the delicate tongues, rich humps, fat roasts, and savoury steaks of this noble and excellent species of game." Bison in poor flesh were the worst diet imaginable, but as they became fat,

> *we grew strong and hearty, and now not one of us but is ready to insist that no other kind of meat can compare with that of the female bison, in good condition. With it we require no seasoning; we boil, roast, or fry it, as we please, and live upon it solely, without bread or vegetables of any kind, and what seems most singular, we never tire of or disrelish it, which would be the case with almost any other meat, after living upon it exclusively for a few days.*

Ferris's comments raise questions about the relationship among phytochemical richness of the diet of the bison, flavor of the meat, satiety, and health outcomes for people who eat the meat of animals fed different diets.

Prospective epidemiological studies – that observe cohorts of people over a long period of time – show positive associations between eating red meat and risk of various diseases.[18] In a sixteen-year follow-up study by the National Cancer Institute of more than 500,000 adults from fifty to seventy-one years of age, people who ate the most red meat had a 26 percent greater risk of dying from cancer, heart and respiratory diseases, stroke, type 2 diabetes, infections, and chronic kidney and liver diseases.[19] Researchers speculate that heme iron in red meat and nitrate/nitrite in cured meat promote oxidative damage and inflammation that cause disease.[20] However, these studies don't distinguish between the effects of eating red meat from livestock fed high-grain rations and livestock foraging on phytochemically rich pastures.[21]

Yet, hunter-gatherers who retain their traditional diets have low levels of heart disease, cancer, diabetes, and osteoporosis, and that is not because they die before these ills develop.[22] The low occurrence of cardiovascular disease and obesity can be explained in part by their higher levels of physical activity compared with people who eat a Western diet.[23] Another explanation for this seemingly contradictory observation may lie in significant details of their diets. The Masai of Africa, for example, eat diets high in red meat and milk, and they add up to twenty-eight herbs to meat-based soups and twelve herbs to milk. These herbs are a rich source of antioxidants that attenuate lipid and protein oxidation, which improve cardiovascular function and prevent cancer.[24] These and other phytochemicals in herbivore diets inhibit cancer by protecting against heme iron in meat and they improve vascular function by thwarting the negative effects of fat on endothelial function.[25]

While most plasma-borne markers of inflammation are not reliably raised after a high-fat meal, they are reduced in many studies when meals included foods such as vegetables rich in anti-inflammatory phytochemicals.[26] Many spices enhance flavor and inhibit protein and lipid oxidation.[27] Adding a polyphenol-rich mixture of antioxidant spices to hamburger during cooking enhances flavor and reduces meat, plasma, and urine malondialdehyde (MDA), a marker for oxidative stress and inflammation.[28]

Many people enjoy a glass of red wine with steak and other red meats, perhaps because the polyphenols in red wine decrease levels of MDA. Plasma levels of MDA rose three-fold after a meal of red meat cutlets, but drinking red wine with the cutlets reduced levels of MDA by 75 percent.[29] When researchers added the polyphenols in red wine to a rat's meat diet, it had the effect of preventing lipid peroxidation in the stomach and absorption of MDA into the plasma.[30]

When herbivores eat phytochemically rich diets, such as the bison Ferris described, antioxidants in their meat and milk may protect us from the oxidation of proteins and lipids that cause low-grade systemic inflammation implicated in cancer and heart disease.[31] Differences in phytochemical richness of herbivore diets may be why people who eat meat of kangaroos that forage on native plants have markedly low inflammatory responses after a meal, much lower than when they eat the meat of Wagyu cattle finished on high-grain diets.[32] In this study, kangaroo meat was presumably sourced from a supermarket (although this was not specified) and would have come from wild-shot animals. Thus, type of diet and type of animal were confounded, and no studies have assessed how different forages herbivores ingest affect the biochemical richness of their meat or fat and how that may be related with humans' inflammatory responses to meat or fat or dairy.

This brings me back to the rich mix of herbs and spices – black pepper, red peppers, garlic, fennel, and marjoram – that we added to the sausages we made on the ranch in Colorado. Those herbs and spices enhanced the flavor of the meat *because* they improved the health-promoting postingestive effects of the meat and fat.[33] That's true even though boiling, microwaving, and grilling decrease the percentage of anti-inflammatory polyunsaturated fatty acids relative to saturated and monounsaturated fatty acids in meat.[34]

## Significance of Terroir

In America, retail sales of grass-fed beef reached $272 million in 2016, up from $17 million in 2012.[35] That amounts to 4 percent of all beef; the market for grass-fed beef grew at 100 percent annually during that time period. With growing interest in consuming grass-fed meat and dairy, people should assess the degree to which grass-fed meat and dairy products are better for health. The lack of research on how herbivore diets affect the flavor and quality of meat and dairy for human consumption reflects the fact that researchers, livestock producers, and consumers are just beginning to appreciate the value of soil health and plant diversity in the diets of herbivores and the implications for the health of soil, plants, herbivores, and humans.

Future studies should evaluate how plant diversity affects the phytochemical richness of herbivore diets, how that influences the flavor and biochemical richness of meat and dairy, and how that affects human health. The need for studies is important given growing interest in grass-fed meat and dairy; concerns about eating red meat and dairy; and the prevalence of animals fed finishing or lactation rations high in cereal grains. Findings from such studies could be a means to achieve two ends. First, they could elucidate interrelationships among flavor, cellular needs, and human health. Second, such studies could help consumers understand how the foods we eat reflect our relationships with landscapes, waterscapes, and airscapes, thus illuminating how palates link soil and plants with herbivores and humans.[36]

When livestock forage on pastures and rangelands, they can be a vital part of regenerative agriculture.[37] Properly managed, foraging by livestock can enhance biodiversity of rangelands, pasturelands, and croplands as the plants they consume nurture life belowground and aboveground while sequestering atmospheric carbon.[38] The wastes that livestock produce nourish the landscapes where they forage. Their meat, eaten in limited amounts, is good for consumers. And this type of livestock production provides a sustainable livelihood for farmers, ranchers, butchers, and various small businesses that raise, process, and sell meat. For the past seventy years, however, most livestock have been raised in a very different way. We remove them from the landscape and confine them by the tens of thousands in feedlots; sustain them with antibiotics that counter illness due to cramped conditions; feed them grain-based rations produced with taxpayer subsidies in ways that can harm the physical and economic health of farmers and consumers; and deny them opportunity to self-select their own feeds. We ignore the fact that this system is not healthy for land, water, air, livestock, or humans.

The ongoing global shift to diets high in processed foods, refined sugars and fats, and red meat is adversely affecting the health of people and the planet.[39] Reductions in atmospheric greenhouse gases could be achieved if erosion-prone cropping systems were replaced by permanent pastures grazed by livestock.[40] Greenhouse gas emissions from domestic ruminants are 11.6 percent (1.58 gigatons carbon annually) of total anthropogenic emissions, while associated cropping and soil emissions contribute 13.7 percent (1.86 gigatons carbon annually), primarily due to soil erosion (1 gigaton carbon annually). Where rainfall is suitable, adding trees to grazing lands through silvopasture is among the best ways to reduce global warming.[41] Of eighty different ways to mitigate climate change, silvopasture ranked ninth and was the most powerful of all agricultural strategies.[42] Thus, appropriately managed livestock foraging on diverse communities of perennial grasses, forbs, shrubs, and trees results in more carbon sequestration than emissions and also results in multiple allied benefits: reduced soil erosion; improved soil ecological function as a result of eliminating damage to soil from tillage, inorganic fertilizers, and biocides; and enhanced biodiversity and wildlife habitat.

Historically, our hunter-gatherer ancestors foraged on plants and animals in the landscapes where they were conceived, born, reared, and died. In turn, those plants and animals foraged on our hunter-gather ancestors after they died. The likings we once developed for foods – and the strong fidelity we felt for the locales where we were conceived, born, and reared – were manifestations of how deeply we were linked with the landscapes we inhabited. Through terroir the cells and organ systems of all creatures that inhabit a landscape become intimately linked.

The sensations Claire Sylvia experienced – after she obtained the heart and lungs of her donor – revealed the disconnect she felt related to her donor's diet and graphically illustrate the profound linkages all creatures have with food and landscapes. To what degree have we lost contact with those life-sustaining energies now that we no longer participate physically and spiritually in the landscapes we inhabit, when the foods we eat are grown anywhere but where we live? Not unlike cardinals and finches fed monotonous diets, to what degree do we become progressively duller as a result of the lack of biochemical richness in the foods we eat?

Likewise, to what degree have we lost any sort of mythological relationships that link us with the mother – Earth – who nurtures our lives? Vivifying mythologies are built upon three pillars, as I explore in greater detail in part five of this book. At the most basic level, they involve the moral and ethical relationships for how to nurture the landscapes that nurture all life. In addition, they include the moral and ethical principles for how to live with integrity and decency in a society – nowadays not just one's own nation, but the larger society of the planet and everybody on it. Ultimately, vivifying mythologies are grounded in spiritual pillars that link societies with those landscapes in awe and with humility in the face of the wonders and mysteries of being. When any one of these pillars begins to crack, civilizations teeter. When they all begin to crumble, as they have throughout the modern world, civilizations collapse.

# CHAPTER 9

# Creating Nourishing Bouquets

The Arkansas River is a major tributary of the Mississippi River. The river begins as snowpack in the Sawatch and Mosquito Mountain Ranges of Colorado. It flows east and southeast as it traverses Colorado, Kansas, Oklahoma, and Arkansas. At 1,469 miles (2,364 kilometers), the Arkansas River is the sixth-longest river in the United States, the second-longest tributary in the Mississippi–Missouri River system, and the forty-fifth–longest river in the world.

The South Arkansas River, known locally as "Little River," is a tributary of the Arkansas River. Little River drains Monarch, Marshall, and Poncha Passes in central Colorado. From the south, Poncha Creek and its tributary Silver Creek flow through valleys between Poncha and Marshall Passes. From the west, Greens, Willow, and Fooses Creeks flow from Mts. Ouray, Chipeta, and Pahlone. The North Fork of the Little River flows through Maysville from glacial lakes – North Fork, Arthur, Island, and Billings Lakes – at 11,000 to 12,500 feet elevation.

As a youth, I loved to fish the Little River from Greens Creek to Maysville, as the river flowed through ranches in the valley bottom. In those days, few people were interested in fishing and none of the ranchers cared if my friends and I fished Little River. Our parents drove us ten miles into the mountains, dropped us off early in the morning, and then picked us up later in the day. I couldn't sleep the night before a fishing outing, every cell in my body filled with anticipation. The mountains from which Little River flows stirred primal sentiments in my soul.

We arrived at the stream early, typically half past six in the morning. The air was fresh with the scent of willows, still wet with dew. The sky was deep blue. We fished upstream toward the lofty granite peaks that served as a backdrop. The cold, crystal clear water made it easy to see our flies as they drifted over the riffles and into the calm holes. The brown trout seemed to appear from nowhere to take a fly. We knew the fisherman's version of Murphy's Law before Murphy's Law had even been written: Taking your eye off the fly is the best way to elicit a strike. As we knew so well, that was also the best way to miss hooking a fish, which had already tasted and spit out the artificial fly. Three flies – Rio Grande King, Plain Coachman, and Royal Coachman – worked well, but we never knew which would work best on any particular day. To increase our chances of inspiring a fish to rise, we offered fish two flies attached to our leaders. On some days we caught many fish, while on others we caught none.

We fished all the tributaries of the Arkansas River. We learned which flies worked best in the different streams and glacial lakes from which the streams flowed. We learned how finicky fish could be and how fortunate we were to be there when they were interested in what we had to offer. Like the fish, we also came to appreciate the hatches of different insects seasonally. You see, fish can feed only on what each stream has to offer and that varies seasonally.

I like to draw a parallel between the cells in our bodies and the fish feeding in brooks. Blood flows from the heart through our arteries, which branch and narrow into arterioles, and then branch further into capillaries, the "brooks" where nutrients and waste products are exchanged. Capillaries occur as interwoven networks that supply nutrients to cells. The capillaries then join and widen to become venules, which further widen and converge to become veins, which return blood to the heart through the great veins to be oxygenated by our lungs.

Like fish in a brook, cells can forage only on nutrients flowing in the capillary beds in their vicinity. The more metabolically active a cell, tissue, or organ, the more capillaries are required to supply nutrients and carry away waste products. Cells forage selectively through the cell membrane, which separates the interior of the cell from the outside environment. The membrane controls movement of nutrients and waste products into and out of the cell. Primary compounds such as glucose or amino acids enter a cell through transmembrane protein channels and transporters. Similar selective foraging occurs for secondary compounds.

Cells can also exclude compounds. With insulin resistance, for example, the membrane excludes glucose from entering the cell. Insulin resistance is a cellular antioxidant defense mechanism, in response to excess energy, and it can be reversed rapidly.[1] Indeed, when necessary, insulin resistance can prevent muscle cells from using glucose, thereby conserving glucose for red blood cells and the brain. Insulin resistance is also strengthened during pregnancy, when demands for glucose are high for the growing body and brain of the fetus.[2]

A person with type 2 diabetes can increase their sensitivity to insulin through a combination of diet, regular exercise, and weight loss. Through diet, people can reduce the amount of glucose in the body. Through mental and physical exercise, people can increase the amount of glucose used by organ systems throughout the body. Through weight loss, people can increase their need for more energy from glucose. That combination of behaviors can even enable some individuals to return to their prediabetic state. In the longer term, diet also has the potential to change the ratio of polyunsaturated to saturated phospholipids in cell membranes, thus changing cell membrane fluidity. The percentage of polyunsaturated phospholipids in cell membranes is strongly inversely correlated with insulin resistance.[3] Increasing cell membrane fluidity by increasing polyunsaturated fatty acids may enhance the number of insulin receptors, increase the affinity of insulin to its receptors, and reduce insulin resistance.[4]

The ability to forage selectively enables fish living in streams to meet their bodies' overall nutritional needs and enables cells foraging in capillaries to meet the needs of the organ systems of which they are a part. Unlike fish foraging in a stream, however, cells and organ systems have the ability to manipulate the foraging behaviors of their host. Flavor-feedback relationships are the basis of this influence. And the influence of cells and organ systems, including the microbiome, on food preferences may help to answer the question: Why is it that large herbivores and humans prefer to eat a variety of foods within and among meals?[5]

## Food Combinations and Meal Courses in Herbivores

Three explanations account for why herbivores eat a variety of foods and forage in different locations. One emphasizes food flavors, another nutrients, and another secondary compounds. Each of these explanations is consistent with the notion that a diverse mix of phytochemically rich foods available in time and space is critical: No single food can meet the needs of cells and organ systems and simultaneously prevent toxicity.

**Flavor-specific satiety.** In studies where sheep or cattle are offered flavored food – such as maple- or coconut-flavored grain or straw – for two hours on one day, they prefer food with a different flavor the next day.[6] Studies such as these confirm the explanation that livestock satiate on flavor. Of even more importance is that the *degree* to which their preference declines depends on how adequate the food is relative to their needs, such as a need for energy or protein. The more adequate the food, the more likely they are to eat more of the same-flavored food in an ensuing meal. The less adequate the food, due to an excess or a deficit of primary or secondary compounds, the less likely they are to eat much of the same-flavored food in the ensuing meal.

**Nutrient-specific satiety.** As previously discussed, herbivores must eat a variety of foods to meet their needs for energy, protein, minerals, and vitamins, as no one plant species contains all these nutrients in amounts animals require.[7] Herbivores can easily do that when habitats are phytochemically rich. To diversify their diets when grazing in habitats that are less rich botanically, cattle alternate among locations that contain different species of plants.[8]

**Secondary compound–specific satiety.** This third explanation is already familiar, because it figured in discussions in chapters 2 and 4. Secondary compounds can be toxic, but because they are ubiquitous in plants, animals must ingest secondary compounds in order to meet their primary need for energy and protein.[9] The presence of secondary compounds limits animals' intake of any one kind of plant (to avoid toxicity) and thus causes animals to eat a variety of plants and to forage in a variety of places.[10]

**Satiety hypothesis.** Each of the previous explanations highlights part of a functionally integrated response by organ systems throughout the body. The *satiety hypothesis* unites these notions by ascribing changes in preference within and among meals to transient aversions that arise as primary and secondary compounds interact to cause satiation and satiety.[11] Eating any mix of foods decreases liking for those foods, and the degree to which liking declines is stronger and more persistent when the mix of foods is deficient or excessive in phytochemicals relative to needs. These flavor-feedback relationships cause herbivores to eat a variety of foods that have complementary phytochemical profiles, both within and among meals.

My first experience of these phenomena occurred watching sheep eat a smorgasbord of crested wheatgrass and ten species of shrubs on rangelands in central Utah.[12] During winter, shrubs made up more than 50 percent of their diet, which is not surprising because shrubs are especially valuable as sources of protein and secondary compounds. Sheep also pawed through the snow to eat the energy-rich, but protein-poor, dry grass. Their pattern of use of shrubs was intriguing. Each morning when they left the pen, the sheep walked straight to the patch of sagebrush, which was a high-energy and high-protein breakfast that made up one-third of their diet. During the day, they ate the dry stems and green basal leaves of crested wheatgrass along with the other shrubs. For dinner, they ate four-wing saltbush, a source of secondary compounds, protein, and more slowly digestible energy to sustain them through the night.

Sheep in the United Kingdom also eat meals in courses. They prefer to eat white clover in the morning and grass in the afternoon. Why? Hungry sheep initially prefer clover because it is higher in energy and protein than grass. As they eat the clover, however, they acquire a mild, transient aversion, as they satiate on readily fermentable carbohydrates and proteins as well as cyanogenic glycosides.[13] The mild aversion causes sheep to seek the more slowly digested grass, which can sustain them through the night. During the afternoon and evening, as they digest and detoxify the clover, the aversion subsides. By morning, they're ready for more clover.

These complementarities also involve specific meal courses, as illustrated by scientists at the US Sheep Experiment Station in Dubois, Idaho. By retaining sheep that ate large amounts of sagebrush, these researchers created a flock of sagebrush eaters. While they initially thought differences in form and function enabled sheep to cope with terpenes in sagebrush, the scientists found that sheep with a high preference for sagebrush ate much more bitterbrush than sheep with a lower preference for sagebrush: Bitterbrush as an appetizer helps the sagebrush go down.[14] In previous chapters, I've mentioned several examples of the complementarity of tannins and alkaloids. In this case the tannins in bitterbrush bind with the terpenes in sagebrush.[15]

These relationships are also illustrated in studies where sheep are first fed alfalfa (high saponins) or birdsfoot trefoil (high tannins) as an appetizer for thirty minutes, followed by a meal of either endophyte-infected tall fescue (alkaloids) or reed canarygrass (alkaloids). Sheep eat more and better digest the dry matter, energy, and protein in the meal of fescue or canarygrass when they are first provided with an appetizer of alfalfa or trefoil.[16] Like sheep, cattle forage actively for thirty minutes when fed an appetizer of alfalfa or trefoil and then continue to forage eagerly on fescue or canarygrass during the ensuing meal. Conversely, when the sequence is reversed, they eat little fescue or canarygrass.[17]

These studies demonstrate the importance of food combinations, and the sequences in which foods are eaten, in food selection and nutrition.

Little wonder, shepherds in France value appetizers and desserts as ways to stimulate food intake, animal production, and health.

## Guiding Naive Livestock to Better Food Choices

In the United States, we've come to rely on fences and grazing systems to influence the foraging behavior of livestock and in turn the health of soil, plants, and herbivores. But that wasn't always the case. In 1918, Moroni A. Smith wrote a book titled *Herding and Handling Sheep on the Open Range in the U.S.A.* in which he describes shepherds' collective knowledge herding sheep on open ranges in the West. He talks about the importance of unlimited patience, kindness, and gentleness when moving sheep; getting the right start in the morning so the sheep forage along the path the shepherd desires; the correct way to turn and place sheep during the day and how to encourage sheep to make the moves the herder wants them to make. In essence, Smith is describing how to use low-stress techniques to move and place sheep on rangelands, blending herding and grazing as people had done for eons before we turned fences into babysitters. He also described the mutual learning among herders and sheep.

During the last century, people came to rely on fences as "livestock sitters" to manage relationships among plants and herbivores. More recently, people have turned to management-intensive and short-duration grazing, mostly using electric fences, to intensify management to benefit soil and plant resources and enhance livestock performance.[18] Rather than relying on fences as livestock sitters, farmers and ranchers can create complementary grazing resources through close herding, which is a nuanced way to rekindle our relationships with livestock and landscapes – people learning from animals, animals learning from people. By far the greatest sophistication in managing grazing can be achieved through the relationship between a herder, a herding dog, a flock, and a landscape of diverse forage species, some palatable and others unpalatable.[19] An experienced herder and dog can influence where and how long a flock enjoys various courses in meals that meld phytochemistry, nutrition, behavior, and landscape ecology.

For herders to be effective, the animals in their care must know about grazing conditions and forages. Presently in France, many herders are entrusted with flocks of sheep familiar with simple plant mixes on pastures, but that are naive to grazing on rangelands. Naive sheep spend most of their time searching for plants that resemble foods they know on pastures.

Training livestock involves three steps. On arrival at the grazing allotment, a herder places animals in a small fenced pasture – with familiar, palatable forages – for a few days. This allows the herder to observe how animals behave in the new environment. The herder then mixes naive with experienced animals and herds the group in a specific "schooling area," mostly composed of somewhat familiar plants, but surrounded by mixed vegetation patches with edible forages as yet unknown to naive animals. When naive animals see experienced animals eating other forages, they become curious to smell and taste the novel plants. Progressively, the herder leads animals on larger daily grazing circuits with longer phases on familiar forages, sequenced with shorter ones on patches with novel forages. This step takes place over weeks.

Skilled herders in France design grazing circuits – various courses for a meal – to increase appetite and intake, to create synergies among meal phases, and to increase intake of abundant but less palatable forages. They partition landscapes into grazing sectors that are carefully sequenced within daily grazing circuits. Meals are based on complementary blends of *terrain* and *plant communities* within and among sectors, *not on particular plants*. By exposing animals to a variety of plant communities during the day, herders *enable individuality* as each animal can pick and choose from plants that vary in primary and secondary compounds.

Herders identify and ration various sectors into phases of a meal: appetite stimulator or moderator, first course, booster, second course, and dessert sectors. According to herders, animals develop a "temporary palatability scoring" as they judge if, in a comparative way, the foods in an area are satisfactory. Herders can successfully modulate palatability scoring by organizing access to sectors that enable minor foraging transitions over several days.

Herders avoid two situations. First, herders want to prevent the herd from having a much better foraging experience on one day than on others. Thus, herders avoid offering a desirable, but rare forage, because that can lead to frustration and reduced food intake as the animals search for the rare plant, ignoring other forages easily available to them.[20] Second, they avoid offering the same foods continually because that leads to "weariness." Livestock become averse to foods and locations when they are persistently offered the same blends of locations and plant communities. To avoid frustration and boredom, herders make use of different vegetation patches predictably during a half-day or day. Herders save highly appreciated dessert forage(s) for the end of a circuit to prevent animals from searching for them during the day.

In so doing, herding unites two areas of interest in the United States – low-stress techniques for moving and settling animals, and management-intensive approaches for grazing. People are learning to move and place livestock in landscapes in ways that minimize stress on livestock and people. They are also using management-intensive grazing – moving livestock at least once if not several times each day. That is creating a mind frame that can accept herding. People can use low-stress techniques to manage grazing intensively by moving livestock throughout the landscape – from meal to meal during the day – in ways that stimulate appetite to improve nutrition, health, and production of animals. Who better to enhance the health of soil, plants, and herbivores than experienced shepherds – ecological doctors – practicing herding? And for humans, who better than ecological doctors, who encourage their flocks to eat biochemically rich foods from conception throughout life, to enhance their health?

## Shepherding Practices for Humans

Not unlike sheep who are naive to grazing on rangelands with diverse arrays of plants, many people nowadays are naive about how to eat in ways that engender health. Like French herders who tend those sheep, people learn to organize meals by courses to stimulate appetite and intake. Offering a variety of foods in a meal increases intake, and the more foods differ in their flavor and nutritive characteristics, the greater the boost.[21] For example, most of us know a tempting dessert can rekindle appetite even after a satisfying appetizer and main course.[22]

To manage obesity, though, scientists are exploring ways to reduce food intake by increasing satiation and satiety.[23] Unlike with research on herbivores, their emphasis hasn't been on how the mix of primary and secondary compounds within a meal affects satiation or satiety.[24] Rather, they are investigating how foods differ in their abilities to maintain satisfying sensations of fullness following a meal, even when they contain the *same amount* of energy.

In one study, researchers fed equal-calorie portions of thirty-eight foods to groups of eleven to thirteen people and then measured changes in satiety for the next two hours. The various foods produced a seven-fold range of satiety responses.[25] The most filling foods were high in protein, carbohydrates, fats, fiber, or water.[26] They included steak, fish, potatoes, porridge, brown pasta, baked beans, apples, and oranges. We know carbohydrate-rich foods that are digested at a slow rate prolong activation of tension receptors in the gut, which maintains feelings of fullness. Unrefined foods high in fiber also satiate more than refined foods. Thus, whole-grain breads satiate more than white bread; brown pasta satiates more than white pasta; and porridge and high-fiber breakfast cereals satiate more than other cereals. Foods such as beans and lentils that delay absorption of food from the gut also have a satiating effect.

Energy density plays a multifaceted role in satiety. When the energy and protein contents of foods are varied, but their energy density (kcal/g of food) is held constant, the effects on satiety are similar for fat, carbohydrate, and protein.[27] When people eat meals that are less energy-dense, they eat more food – about a pound per day – but they ingest fewer calories and lose more weight.[28] Humans have the capability to dilute energy density – calories per bite or per portion – of dishes. There are nearly endless examples of how to reduce the energy density of what we eat. For example, adding water to foods adds weight and volume but no calories, and thus it dilutes the energy density. We can add vegetables to a casserole, substitute a small amount of olive oil for a cream sauce with pasta, use whole-wheat as opposed to white flour, and so on.

When the energy density of the main dishes eaten during a day is cut by adding puréed vegetables, both adults and preschool children consume fewer calories.[29] In year-long trials, people advised to reduce energy density by eating more fruits and vegetables and reducing intake of fat, lost more weight than those advised merely to reduce intake of fat.[30] Both energy content and portion size can be reduced without people noticing. People report similar ratings of hunger and fullness, even when energy intake was reduced by 25 percent over two days.

Some of the most interesting "shepherding practices" for humans concern how "grazing circuits" can allay diabetes. Eating a first course of a low-energy food – such as soup, salad, or fruit – can decrease energy intake during a meal and lower glucose and insulin levels after a meal.[31] Conversely, one study showed that glucose and insulin are much higher when the food order is: carbohydrate (ciabatta bread and orange juice) followed fifteen minutes later by protein (skinless grilled chicken breast) and vegetables (lettuce and tomato salad with low-fat Italian vinaigrette and steamed broccoli with butter) as opposed to when the food order is reversed. [32] The effect on glucose levels – which were reduced by 73 percent when vegetables and protein were eaten before carbohydrate – is similar to pharmacological drugs that target glucose. Another study showed that eating a high-energy breakfast and a low-energy dinner can decrease hyperglycemia throughout the 24-hour day and increase GLP-1, a hormone that increases insulin sensitivity and satiety and reduces intake.[33]

Adding some fruits and spices to a meal can also reduce hyperglycemia and insulin resistance. Fenugreek and gooseberry (*Emblica officinalis*) consistently lower fasting glucose levels in diabetic patients, as do bitter melon (*Momordica charantia*), blueberries, cinnamon, and ginseng.[34] Several, though not all, studies have demonstrated that vinegar added to salads and dishes can also help reduce hyperglycemia, hyperinsulinemia, hyperlipidemia, and obesity.[35]

Food combinations also affect glucose levels. Blood glucose levels rise when we eat carbohydrates. How high they rise and how long they stay high depends on the quality of the carbohydrates – the glycemic index (the GI) – as well as the quantity.[36] Glycemic load (GL) combines both the quantity and quality of carbohydrates, so it is a better way to compare blood glucose values of different *kinds* and *amounts* of foods we eat. Mixing high- and low-glycemic carbohydrates moderates GI and GL values, though that doesn't change their hierarchical relationships: Eating peanut butter on a piece of white bread lowers the GI of that slice of bread, but the GI of white bread is still higher than that of whole-wheat bread.[37] Adding butter or oil to foods lowers the GI/GL of a meal, though the GI/GL ranking of the individual foods in the meal doesn't change. Eating foods high in fat or protein along with high-GI/GL foods lowers the overall GI/GL of a meal. The fiber, fat, and protein in peanut butter blunt the impact of the carbohydrates by slowing digestion and absorption, lowering GI/GL and insulin response.

Diets higher in protein reduce GI/GL and help people maintain weight loss.[38] Higher GI/GL starches, refined grains, and sugars satiate less, and they activate regions in the brain associated with reward and craving, a combination that causes overeating and weight gain.[39] Conversely, foods high in protein reduce weight gain, as do whole grains, fruits, and vegetables.[40] Lowering the fat content of dairy or other foods *increases* intake of carbohydrate in adults. That's why weight gain usually doesn't differ for low- versus high-fat versions of foods.[41] These findings for adults are probably similar for children, as kids who consume low-fat milk gain more weight than those who consume whole milk.[42]

Some combinations of foods support overall health better than others, as illustrated in one study that identified two dietary patterns. The prospective cohort, followed for eight years, had nearly 45,000 men aged forty to seventy-five years who were free from cardiovascular disease when the study began.[43] During the study, 1,089 men developed coronary heart disease. The study designers defined different types of diets followed by the cohort. People who ate a "Western" diet had relatively higher intake of red meat, processed meat, French fries, refined grains, high-fat dairy products, sweets, and desserts. People who ate a "prudent" diet had relatively higher intakes of vegetables, fruit, legumes, whole grains, fish, and poultry. After adjusting for age and coronary heart disease, researchers found that the prudent diet reduced risk by 30 percent while the Western diet increased risk by 64 percent.

A Mediterranean-type diet – which like the prudent diet described above includes high intake of fruits and vegetables, legumes, and cereals high in fiber – reduces the rate of recurrence of myocardial infarction by 28 to 53 percent after the first infarction. Simply adding nuts to your diet is associated with reduced risk of cardiovascular disease, cancer, and all-cause mortality.[44]

## Shepherding Practices for Preventing Cancer

Each of us will experience some ten thousand trillion cell divisions during our life, some of which will result in cell mutations that can lead to cancer. As we age, the likelihood of cancer-promoting mutations increases. Damage to healthy cells can lead to cancer through three stages. During *initiation*, damage to DNA in a parent cell is passed on to daughter cells and their progeny. During *promotion*, damaged cells can begin to grow, but how they grow depends on whether or not the kinds and combinations of foods we eat encourage or suppress growth of cancer cells. During progression, cancer cells grow without limits and may wander from the site where they originated to invade neighboring locations or distant tissues. When cancer cells assume this lethal property, they are malignant. When they move from their location of origin and meander to other locations throughout the body, they are metastasizing.

Over thirty years ago, eminent British epidemiologists Sir Richard Doll and Sir Richard Peto compiled a comprehensive review on avoidable risks of cancer in the United States. In a 120-page review, they concluded that man-made chemicals, pollution, occupational exposure, and food additives play only a minor role in cell mutations and cancer.[45] Diet plays a primary role. Diet doesn't change the *nature* of cancer cells – the mutations that lead to cancer – it changes the *nurturing* of those cells. Cells that become cancerous and begin to grow prolifically do so by gaining selective advantages over normal cells. One way they do that is by expressing more insulin receptors, which enables them to take up more glucose to fuel rapid growth and propagation.[46] Diets high in refined carbohydrates provide a readily available source of glucose. Cells also express more IGF (insulin-like growth factor) receptors, which provides cells with strong signals to multiply and inhibits the programmed cell death that ordinarily prevents damaged cells from flourishing.[47] We can imagine cancer cells providing feedback to the palate through signals that enhance liking for the refined carbohydrates that provide a readily available source of glucose; the larger the cancer, the stronger the signals.

I previously discussed the significance of herbivores nibbling on fifty or more plants daily to enable their health prophylactically. That occurs when wild or domestic herbivores forage on phytochemically rich landscapes, is less common when domestic herbivores forage on monoculture pastures, is close to zero for herbivores in feedlots, and is increasingly rare for people who forage in modern food outlets. When we prevent humans and the animals in our care from eating phytochemically rich diets, we pay the price in ill health.

Researchers are studying the ability of mixtures of phytochemicals to prevent cancer, due to concerns about the ability of cancer cells to doggedly adapt to chemotherapy, the side effects of chemotherapy, and the demand for more affordable therapies.[48] The mechanisms of action of phytochemicals differ from chemotherapy. Anti-inflammatory compounds such as polyphenols, isothiocyanates, and epicatechins as well as micronutrients such as folic acid and selenium affect gene expression. Nutriepigenomic studies are identifying phytochemicals that reverse epimutations and prevent cancer. That research is promising because epigenetic changes are more easily reversible than genetic mutations.[49] Phytochemicals also exhibit low toxicity, so adverse side effects are less of a concern. And they are also readily available at low cost.

Diet-induced changes in DNA-methylation (epimutations) can counter all of the hallmarks of a cancer cell, including limitless replicative potential, self-sufficiency in growth signals, insensitivity to growth-inhibitory signals, evasion of programmed cell death, sustained angiogenesis, and tissue invasion and metastasis. Tumor cells acquire these properties from cumulative epigenetic changes in multiple genes and associated cell signaling pathways, most of which are linked to inflammation. The preventive and protective activities of phytochemicals lie in their combined effects on cellular defenses including detoxification and antioxidant enzyme systems as well as anti-inflammatory and antimetastasis responses.[50] Biochemically rich combinations of foods are far more effective in protecting against and treating cancer than isolated compounds. Drugs that attempt to target a single gene product are unlikely to be useful in either preventing or treating cancer. Rather, people need safe and effective "drugs" that act throughout life on cell networks, rather than on single targets at specific points in time.

Cancers often have latencies of twenty years or more when they are undetectable. By the time they are clinically perceptible, the body may have changed in ways that are beyond repair. Thus, the timing of various dietary exposures – including preconception, pregnancy, lactation, neonatal life, early life, puberty, and pre- and post-menopause – may be critical in health, as epigenetic plasticity changes continually from conception to death.[51] If intervention begins only after critical events have taken place, opportunities for modulating effects will likely be missed. The best way to do that is to eat a biochemically rich diet of wholesome foods throughout life.

Human health is consistently enhanced as the biochemical richness of our diets increases. Thus, we can choose to take a particular fatty acid supplement such as EPA or DHA, or we can take a complex mixture of omega-3 fatty acids. Much better still, we can eat a portion of oily fish that contains hundreds of compounds in addition to omega-3s or a full meal of oily fish, vegetables, and fruits that contains tens of thousands of compounds.[52] The potential value of this approach has been demonstrated with trials in breast, prostate, and colon cancer.[53]

The old saw of "an apple a day" is true, especially if you eat the skin as well as the flesh. Eating the skin and flesh nearly doubles the antioxidant capacity of apples compared to eating the flesh only.[54] Phenolics and flavonoids are highest in the skin, followed by the skin plus flesh, followed by flesh. Triterpenoids in apple skins have potent antiproliferative activities against liver, colon, and breast cancer cells and may be partially responsible for the anticancer activities of whole apples. The ability to inhibit propagation in a colon cancer cell line was greater (43 percent) using apple extract with skin compared to apple flesh without the skin (29 percent). The liver cancer cell line was inhibited more by skin (57 percent) than flesh (40 percent).

Studies of rats illustrate how synergies compound. When rats eat five to seven servings of tomatoes a week, their risk of prostate cancer is reduced by 30 to 40 percent.[55] Simply providing rats with lycopene (derived from tomatoes) doesn't reduce prostate tumor weights, but eating whole tomatoes reduces tumor weights by 34 percent. Tomato oleoresin has up to five times the ability to inhibit LDL oxidation compared with pure lycopene because tomato oleoresin also contains vitamin E, flavonoids, and phenolics.[56] When rats eat broccoli, prostate tumor weights are reduced by 42 percent. The beneficial effects on colon and liver cancer are also greater when people eat fresh broccoli with a full complement of glucosinolates – pungent compounds in mustard, cabbage, and horseradish – than when they take a supplement that contains only one glucosinolate (glucoraphanin) processed into the anticancer compound sulphoraphane.[57] When rats eat broccoli plus tomatoes, prostate tumor weights are reduced by 52 percent.

Many popular herbs and spices – coriander, dill, oregano, parsley, rosemary, sage, thyme, cinnamon, cloves, ginger, nutmeg, and turmeric – are rich in polyphenols that can thwart both cancer and cardiovascular disease.[58] The benefits of adding herbs and spices to enhance the flavor of foods is illustrated with marjoram. Sprinkled to taste on salad, marjoram increases the antioxidant capacity of a salad by 200 percent.[59] And fortunately for us humans, neither cooking nor digestion diminishes the antioxidant or anti-inflammatory activities of herbs or spices.[60]

The power of diet to prevent cancer is illustrated by observations that women who eat more high-fiber foods rich in phytochemicals during adolescence and young adulthood have less risk of breast cancer than those who eat less fiber.[61] These findings are based on over 90,000 women who participated in the Nurses' Health Study II, among the largest studies funded by the NIH to assess risk factors for major chronic diseases in women. The results suggest that both small shifts in diet by adults, as well as exposure to a wholesome diet beginning early in life, are associated with a reduced risk of getting cancer. Breast cancer risk was 12 to 19 percent lower for women who ate more fiber in early adulthood. High intake of fiber during adolescence was associated with a 16 percent lower risk of overall breast cancer and 24 percent lower risk of breast cancer before menopause. For each additional 10 grams of fiber intake daily – for example, about one apple and two slices of whole wheat bread, or about half a cup each of cooked kidney beans and cooked cauliflower or squash – during early adulthood, breast cancer risk dropped by 13 percent. The greatest benefit came from fruit and vegetable fiber. All-cause mortality is reduced when women eat fiber from whole grains, but the effects are lost with refined grains.[62]

Studies such as these are why the US Dietary Guidelines advise eating five to ten servings of fruits and vegetables daily to help prevent diseases.[63] Eating combinations of dark green and yellow- or orange-fleshed vegetables and fruits, legumes, nuts, seeds, and whole grains containing antioxidant phenolics, fiber, and numerous other phytochemicals enhances antioxidant capacity compared to consuming any one fruit or vegetable alone.

## Redesigning Grazing Circuits for Humans

Indigenous Americans learned to balance their consumption of maize (corn) with beans and other sources of protein, such as amaranth and chia as well as meat and fish, to acquire the complete range of amino acids bodies require for protein synthesis. These food combinations are satisfying (satiating) because they enable people to meet their needs for essential nutrients.

The traditional method of preparing maize, developed by indigenous Americans who domesticated the plant, required soaking maize in alkali water made with ashes and lime. This treatment makes niacin available, reducing the likelihood of developing pellagra. In addition, older varieties of corn were inadequate in the essential amino acids tryptophan and lysine. Treatment with lime also increases the availability of tryptophan and lysine. When corn cultivation was adopted worldwide, this method of preparation was not universally implemented because the benefits of the treatment weren't understood. Pellagra became common in some areas when corn eaten without this traditional preparation became a staple.

As humans learn to eat combinations of foods that satisfy their nutritional needs and promote health and that knowledge becomes tradition, individuals within the community no longer consider which foods they eat or why they eat them. With time, people little understand or appreciate the biological or cultural origins of their diets. Nor do they realize when those norms change, as they have during the past century, in ways that are harmful. That's especially true when peoples become detached from the acts of growing and harvesting foods. We've lost knowledge of how to combine wholesome foods into meals that nourish and satiate ourselves. By raising our level of awareness of the knowledge we've lost, we can redesign "grazing circuits" that better enable the health of herbivores and humans and the landscapes we inhabit.

# CHAPTER 10

# Painting Your Canvas

During the nineteenth century, explorers of the Canadian arctic who were forced to live only on the lean meat of wild rabbits experienced a condition called "rabbit starvation." No matter how much they gorged on rabbit, they were unremittingly hungry. After just seven days, explorers were eating three to four times more rabbit than at the beginning of the week, and by day ten the swelling of their distended bellies was visible through layers of clothing. When they ate only lean meat, they became ravenous for fat. They were attempting to eat enough fat, present only in low amounts in rabbit, by overeating rabbit. Given a chance, they'd eat a large quantity of pure fat, even oily fat, without nausea. Without fat, they died in a few weeks.

In *People of the Deer*, Farley Mowat describes his experience of fat deprivation during a canoe trip when he and his friend Franz were out of supplies, except for a little tea and lard. They were living on deer that were thin due to long summer months of torture by flies. Mowat began to experience malaise – diarrhea, a sick lassitude, and increasing loss of will to work.

> *For the first few days I made out very well on three meals of lean meat a day, but before the end of the week I was smitten with an illness which for want of a better name I called mal de caribou. Then Franz turned physician. One evening he took our half a pound of precious lard, melted it in a frying pan, and when it was lukewarm and not yet congealed, he ordered me to drink it. Strangely, I was greedy for it, though the thought of tepid lard nauseates me now. I drank a lot of it, then went to bed; and by morning I was completely recovered. This sounds like a shock cure, but in fact I was suffering from a deficiency of fat and did not realize it.*

I experienced an aversive reaction similar to Mowat's when eating at a restaurant that specializes in wild game meats. As an appetizer, we ordered six large bones from bison. We scooped marrow from the bones and spread it onto small pieces of toast. But my body wasn't greedy for marrow, and the more I ate, the less I liked it. That night and the next day, I was satiated to the point of mild nausea. Sue and I eat a diet high in fat, and I attribute my lack of interest in the high-fat marrow to my lack of need for the fat and any of the nutrients the marrow had to offer. Yet, I have no doubt my body would crave marrow if the need arose.

We *experience* life through gradations in physical and emotional *sensations*. Our bodies translate biochemical and physiological interactions among cells and primary and secondary compounds into sensations through molecules of emotion. *Palatability* mediated by *metabolic feedbacks* is a continuum from *dislike* (mix of foods deficient relative to needs) to *like* (mix of foods adequate to meet needs) to *dislike* (mix of foods excessive relative to needs). A body craves what it needs and the greater the need, the greater the craving. As Mowat wrote, "I was greedy for it." However, a body becomes averse to an excess of what it doesn't need, and the greater the excess, the greater the aversion: "The thought of tepid lard nauseates me now."

*Hunger* is my experience of the needs of cells and organ systems for primary and secondary compounds. *Satiety* and *malaise* are my experiences of the *benefits* and *costs*, respectively, of mixing primary and secondary compounds in various combinations in the meals I eat. Ingesting *appropriate* blends of primary and secondary compounds results in benefits, experienced as *satiety* and a *liking* for the flavors of the foods. Ingesting *excess* primary or secondary compounds imposes physiological costs, experienced as *malaise* and a *dislike* for the flavors of the foods. These responses change as cells and organ systems in my body gently guide my choices to meet their collective needs.

## Eating Mindfully

If we are attentive, we can become aware of the relationship between the palate and feedback from the body. That's the point made by Mireille Guiliano in *French Women Don't Get Fat* and by Charles Eisenstein in *The Yoga of Eating*. They discuss the importance of simply becoming aware of the signals that emanate from the body through to the palate.

As one example, Eisenstein reviews evidence for the importance of fat in our bodies and the lack of scientific evidence that fat and cholesterol cause atherosclerosis. As he then writes, from the point of view of the yoga of eating (the wisdom body), all of that is a moot point. The body – not nutrition scientists and medical doctors – is the final authority in our food choices.

Eisenstein describes an experience eating coconut oil. "I'd read that contrary to conventional views," he writes, "coconut oil is very healthful for its plethora of medium-chain fatty acids. I purchased some high-quality coconut oil to try it out, and sure enough, my body responded gratefully – at first. Yet before long, eating moderate quantities every day, the very thought of the stuff disgusted me. My body was telling me I'd had enough. I've experienced the same repugnance when I've eaten large quantities of other fats, too, particularly nuts."

Eisenstein's experience is consistent with studies that show that eating mindfully – consuming food in response to physical cues of hunger and fullness – is as effective as adhering to nutrition-based guidelines in reducing blood sugar levels and weight in adults with type 2 diabetes.[1] In one study of the effectiveness of these two interventions – which were titled Smart Choices and Mindful Meditation – participants lost the same amount of weight.

The Smart Choices treatment group followed an established diabetes self-management education program with a strong emphasis on nutrition information. The Mindful Meditation group was trained in mindful meditation and a mindful approach to food selection and eating. Both interventions, which involved weekly group meetings, also recommended physical activity.

The Smart Choices program focused on factors that lead to complications with diabetes, including heart disease, kidney and nerve damage, eye problems and stroke, the importance of blood sugar control, and appropriate food choices when blood sugar levels spike. Every session had a discussion on topics such as calorie-intake goals, percent of fats and carbohydrates in an ideal diet, and portion control. Sessions included a fifteen- to twenty-minute walk to further emphasize the recommendation for physical activity. The program also focused on problem solving regarding choosing healthy foods in high-risk situations, such as the holidays.

In contrast, trainers in the mindfulness program encouraged participants to cultivate *inner wisdom*, or mindful awareness of eating, and *outer wisdom*, which is personal knowledge of optimal nutrition choices for people with diabetes. The training encouraged people to be aware, stay in the moment, and live and eat in response to hunger instead of habits and unconscious eating. Each session included guided meditation oriented toward people's experiences and emotions associated with food. Participants received CDs to help with home meditation practice. They were encouraged to tune out the many environmental cues that cause us to overeat and tune into our normal physiological signals to eat. Being mindful means stopping long enough to become aware of these physiological cues. The mindful intervention also included information about the relationships among calories consumed, carbohydrate and fat intake, weight regulation, and high blood sugar. Overall, the participants in both programs lowered their long-term blood sugar levels significantly and lost an average of 3.5 to 6 pounds after three months.

Being mindful of their body's nutritional wisdom may be why French women aren't obese, despite all the energy-rich appetizers, main courses, and delectable dessert pastries they eat. They know when to stop eating. They pay attention to internal cues, such as whether they feel full, and they are also mindful of external cues, such as portion sizes, proximity to food, and social settings, which can lead us to overeat. We are mostly unaware of our eating habits.

We can cultivate mindfulness by eating a variety of wholesome food and being aware of the sensations, including the changes in liking for the flavors of those foods, we experience during a meal. I find the wholesome foods I eat *taste much better* and I am *more aware* of feedback from my body regarding my needs when I eat less food and when I eat many foods in moderation. My food choices range from the meat of wild and domesticated animals and full-fat dairy products to vegetables, fruits, nuts, and small amounts of foods made from some refined carbohydrates. The feedback from my body is also clearer when I avoid fortified, enriched, and processed foods and nutrient supplements. I also believe that enriching, fortifying, and processing foods, along with winning "expert" endorsement for supplements, are stealthy ways food and supplement companies encourage overconsumption to benefit their bottom line.

My point isn't to recommend that you follow my choices of what to eat and not eat. The point is I've taken the time to figure out what works for me. That may or may not be what works for you. Research studies and clinical practice have both shown that the best way to improve your ability to stick to a diet and lose weight is to review a broad spectrum of diet options designed to best match your food preferences, lifestyle, and disease-risk profile.

## Eating Mindlessly

Herders in France design grazing circuits specifically to stimulate the appetites of sheep by offering them different meal courses throughout the day. It is a mindful approach to meal planning. Many humans do not take such an approach, and they end up overeating because they are unaware of the many covert influences that cause us to eat mindlessly. For example, people eat more of the food that's placed right on the dining table than the food that's set on a counter away from the table. With this in mind, we are better off leaving salad and vegetables on the table throughout a meal, but placing energy-dense foods on the counter after portions are served.

Much as movements and proximity to food increases food intake by livestock in grazing circuits, distance to food influences how much people eat. In one study of this phenomenon, secretaries who won an award were given all the candy they could eat during a one-month period.[2] The candy was placed either on their desk or six feet away from their desk. A "typical" secretary ate nine pieces of chocolate each day – about 225 calories – if the candy was on the desk, but only four pieces – about 125 calories – if the candy was six feet away. Over the course of a year, that translates into 11 to 12 pounds of weight they would gain by having a bowl of candy on the desk as opposed to six feet away. Also, the secretaries ate two more pieces of chocolate each day when the candy was placed in a clear bowl rather than in an opaque bowl.[3]

We know sheep and cattle eat more when they are offered a variety of forages. Planting forages in patches minimizes the time animals spend searching for different foods and stimulates appetite, which improves health, nutrition, and production. When sheep and cows forage in an area where grass and clover have been planted in patches, the sheep consume 25 percent (over half a pound per sheep per day) more and dairy cows produce 11 percent (over five pounds per cow per day) more milk.[4] Like herbivores, humans also forage more efficiently when foods are available in "patches," such as an array of different containers in a smorgasbord. Unfortunately, that's a recipe for disaster because, rather than making mindful choices, we default to scooping something out of every container and piling it on our plates.

In social species such as goats, sheep, and cattle, companions strongly influence where their friends eat, the kinds of foods they eat, and how much they eat. This occurs even when those companions have been trained to avoid eating a particular food by pairing it with nauseating postingestive consequences of lithium chloride. When the trained animals are eating alongside untrained buddies, the trained animals see their companions eating the food they were trained to avoid, and they can't resist sampling it themselves. For example, in a three-year study, cattle taught to avoid eating larkspur avoided larkspur throughout the summer for three years – they abstained until they were mixed with buddies who ate larkspur. Gradually during the ensuing month, the cattle that had eschewed larkspur increased their larkspur intake.[5]

People are at least as social as livestock when it comes to foraging.[6] For example, when we eat next to a person who is eating quickly, we tend to eat more quickly and ingest more calories than when we eat alone. When we eat with someone who eats more slowly than we do, we ingest fewer calories. When a woman follows another woman through a buffet, the follower is more likely to take portions that mimic the serving size taken by the woman in front of her. People also eat more when they are engaged in conversations with companions.

Visual cues influenced the amount of food fifty students ate when they were seated at twenty-one tables, randomly assigned either to be bused or unbused, at an all-you-can-eat sports bar. Students at unbused tables saw how much they had eaten – in this case, leftover chicken-wing bones – and ate 27 percent less than people who had no such environmental cues.[7] The difference between "bused" and "unbused" groups was greater for men than for women.

When people were asked when they would stop eating a bowl of tomato soup, 81 percent said they'd stop eating when the bowl was half full or empty.[8] Only 19 percent said they'd stop eating when they were no longer hungry. People who unknowingly eat from a constantly refilling bowl eat 73 percent more soup than people who eat from a conventional bowl. When asked if they were full, they replied: "How can I be full? I have half a bowl of soup left."

Studies of portion size illustrate how we learn to overeat. When three- or five-year-old children were offered either medium- or large-size servings of macaroni and cheese, the three-year-olds ate the same amount regardless of what they were offered.[9] They ate until they were full and then they stopped. The five-year olds, on the other hand, ate 26 percent more when they were given the bigger servings. And that's what happens with adults. We let what's offered and the size of the serving influence how much we eat, little aware that most of us were trained as children to "clean our plates." In another study, people who ate dinner and then went to the theater – presumably satiated from the meal – were given five-day-old stale popcorn in buckets of two sizes.[10] The larger the bucket, the more popcorn they ate while watching the movie, paying little attention to how much popcorn they ingested.

## Dieting and Exercising

Notwithstanding the many books written on the topic, diets don't work. Very few people experience sustained weight loss while dieting. Most people quickly regain the five to ten pounds they lose once they quit the diet. Sticking to a diet is key, and most people can't stick to a diet.[11] Indeed, this practice is an excellent way to gain weight. The small amount of weight loss observed from exercise is due to low amounts of energy expended along with an increase in energy intake.[12] People are better off not dieting. The vicious cycles of losing and gaining weight are linked to cardiovascular disease, stroke, diabetes, and suppressed immune function.

To assess how diet influences weight, a year-long study called the A to Z Weight Loss Study assigned more than 300 free-living, premenopausal overweight women to one of four diets.[13] The regimens varied from low carbohydrate–high fat to low fat–high carbohydrate: Atkins (extremely low carbohydrate), Zone (low carbohydrate, high protein), USDA/Food LEARN (high carbohydrate, moderate to low fat), or Ornish (very low fat). Within each group, some women lost as much as fifty-five pounds while others gained as much as eleven pounds – a range of sixty-six pounds – that reflects the individuality we've discussed throughout this book.

The DIETFITS study sponsored by Stanford University showed this same kind of variation. Researchers assigned 600 healthy, free-living adults randomly to either a healthy low-fat or a healthy low-carbohydrate diet. People in both groups were told to eat as little fat, white flour, and sugar and as many vegetables as possible, but they weren't told to cut calories.[14] By avoiding energy-dense processed foods, each group cut 500 calories a day from their diets. After a year, each group lost an average of thirteen pounds, but individuals within each group varied markedly along a continuum from losing as much as sixty pounds to gaining as much as twenty pounds. In the context of these two common weight loss diet approaches, neither of the two hypothesized predisposing factors (genotype pattern or insulin secretion) was helpful in identifying which diet was better for different individuals.

In 2013, the Obesity Society, the American Heart Association, and the American College of Cardiology issued a report on ways to reduce obesity.[15] They concluded no two people are alike, and thus, one diet will never work for all people. As the lead author, researcher Michael Jensen, stated in an interview published in *Nutrition Action Healthletter* (April 2015),

> *We started off with different, almost religious, beliefs as to what diet would be better, and we wound up being agnostic. We looked for studies that had results for at least a year, and preferably two years, because for most of us who treat patients with obesity, it's not*

> *what they did in six months that matters, it's what they did in two years that really drives our decision making. For any given person, it's really more a matter of what can they stick with, rather than whether there's a specific diet that we know is best for them.*

As with diet, some people who exercise lose weight while others gain weight. Researchers in the UK had thirty-five overweight or obese men and women exercise – on a stationary bike, treadmill, stepping machine, or rowing machine – enough to burn 500 calories a session, five times a week for twelve weeks.[16] At the extremes, some people lost thirty-two pounds, others gained four pounds, and the "average" person lost eight pounds. Compared with those who lost more weight, those who lost less weight were hungrier and ate more each day near the end of the study.

In a rigorous study of physical activity and weight loss, scientists recruited 200 overweight, sedentary adults and put them on an aggressive exercise program.[17] To isolate the effects of exercise on weight, people were told not to change their diets. Participants were monitored to ensure they exercised five to six hours a week, more than double the 2.5 hours per week of exercise recommended in federal guidelines. After a year, the men lost an average of 3.5 pounds, the women 2.5 pounds. Nearly everyone was still overweight or obese.

When I left Colorado in September of 1975 to attend graduate school at Utah State University, I weighed 140 pounds, and I was in top physical condition due to long hours of hard physical work each day. By June of 1976, I weighed 170 pounds, and I was in poorer physical condition due to a drastic decline in the amount of my physical activity. My diet didn't change, but rather than working vigorously ten hours a day, I was studying intently ten hours a day.

Most people nowadays are sitting at a desk rather than doing strenuous physical labor. These marked declines in work-related physical activity have contributed to obesity during the past five decades.[18] Indeed, some people argue sugar consumption now appears to be deleterious only because levels of physical activity are too low, relative to sugar consumption and glucose use, to support metabolic health.[19] They contend that until the pathologies due to physical inactivity are corrected, metabolic health of populations will continue to decline.

Exercise little influences weight loss because so few calories are burned in workouts of only thirty to ninety minutes a day. People grossly overestimate how many calories they burn while exercising.[20] You have to walk three miles to offset one 12-ounce Coca-Cola, which has roughly 10 teaspoons of sugar and 140 calories. Most people lose more weight by cutting calories than by exercising. Dieting and exercising combined provides the best results.[21]

Dieting and exercising interact to influence *weight* and *fat* loss. Israeli researchers randomly assigned 278 sedentary adults (mostly men) who were obese or had high triglycerides to one of two diets equal in calories – either low-fat or low-carbohydrate Mediterranean – for eighteen months.[22] For the last twelve months, half of the people were assigned to an exercise routine that involved forty-five minutes of aerobics and fifteen minutes of strength training three days a week. People in the low-carbohydrate group were told to eat more vegetables, beans, poultry, and fish rather than beef or lamb. They were also given an ounce of walnuts to eat each day. The low-fat group was told to eat vegetables, fruits, whole grains, and beans and to cut back on sweets and high-fat snacks. Each group was served either the low-fat or low-carbohydrate meal for lunch – the main meal of the day – at work. After eighteen months, both groups lost roughly six pounds. However, people who exercised lost more visceral (deep belly) fat, regardless of diet. Waist size, liver fat, and triglycerides decreased more in the low-carbohydrate group, whether they exercised or not.

Exercising and fasting interact to improve body composition and health. Young adult men who did resistance training while following a regimen of eight weeks of *time-restricted fasting*, eating only during an eight-hour period each day, lost fat mass while retaining lean mass and improving muscle endurance and maximal strength.[23] This eating and exercise regimen also improved basal metabolism and markers for inflammation and cardiovascular risk factors. The hard physical work we once did, in combination with the wholesome foods we once ate, surely would have contributed to the health and well-being of our ancestors. Diet, exercise, and fasting are just as important today.

## Fasting Trumps Dieting

Until retirement, Sue and I ate three meals a day, usually between 6 and 7 a.m., at noon, and between 6 and 7 p.m. Since retirement, we eat only two meals a day typically, though not religiously within the span of a six- to eight-hour period. Sue changed our patterns of eating to attenuate her acid reflux, which bothered her if we ate too late. The mid-afternoon meal also allowed us to take long walks in the evening. At the time, we didn't have a clue that we were practicing *time-restricted fasting*. Another variant on this theme is *intermittent fasting*, in which people eat little or no food for extended periods of 16 to 48 hours, with intervening periods of normal food intake, on a recurring basis.

Time-restricted fasting is one way to burn fat without dieting. Most bodies require roughly six to eight hours after a meal to metabolize circulating glucose and glycogen stores in the liver. Once that's done, a body shifts from using carbohydrates to using fat as the source of energy. In humans, both daily calorie restriction and alternate-day fasting produce similar beneficial effects on weight loss and weight maintenance.[24]

Fasting improves health in laboratory mice and rats and humans.[25] In studies of obesity, rats and mice in control groups are typically sedentary, which, along with ad libitum feeding, renders them like human couch potatoes.[26] Fasting reduces the copious and constant stream of calories that adversely affects health. In so doing, fasting improves metabolic function, because abstaining from food causes cellular changes that promote longevity.

Fasting increases circulating levels of human growth hormone, commonly called the "fitness hormone," which improves health and longevity. By increasing rates of metabolism, fasting promotes muscle growth and enhances fat loss. Like people, rats and mice maintained on fasting diets show improved physiological indicators of health, including sensitivity to satiety hormones such as insulin and leptin; elevated levels of ketones, which indicate a body is using fat for energy; reduced resting heart rate and blood pressure; and increased resistance of the heart and brain to stress. Fasting can thus delay the onset and slow the progression of neuronal dysfunction and degeneration in Alzheimer's, Parkinson's, and Huntington's diseases.

Rats and mice fed a high-fat diet ad libitum develop hyperinsulinemia, obesity, and inflammation, all of which are prevented by restricting eating to eight hours a day.[27] The antidiabetic effect is not due to caloric restriction, because mice fed just eight hours a day eat the same amount of food as control mice. The metabolic shift to use of ketones and responses of the brain and autonomic nervous system to food deprivation play key roles in the fitness-promoting effects of fasting. Fasting increases alertness and arousal as well as mental acuity.[28]

While overweight and obesity promote inflammation, intermittent fasting suppresses inflammation. Obese women who changed from eating multiple daily meals to alternate-day energy restriction significantly reduced levels of inflammatory markers.[29] High-fat and high-carbohydrate meals both induce inflammatory stress, and that inflammatory stress is much greater and more prolonged in obese than nonobese people.[30]

The best way to lessen inflammation is to lose excess weight, and for many people the way to lose weight and metabolize energy is to eat a diet low in refined carbohydrates.[31] That was illustrated when forty overweight men and women ate either a low-carbohydrate or a low-fat diet for twelve weeks. People on the low-carbohydrate diet lost weight, and six inflammatory markers decreased more for the low-carbohydrate group than for the low-fat group.[32]

Compared with those fed ad libitum, the life spans of organisms can be extended by dietary energy restriction.[33] The antiaging effect of intermittent fasting occurs in species from yeast, nematodes, and worms to mice and monkeys and no doubt to humans as well. Rats fasted on alternate days beginning when they are young live nearly twice as long as rats fed ad libitum. Even when alternate-day fasting is delayed until middle age, rats live 30 to 40 percent longer than rats fed ad libitum, and this life extension is further extended with regular exercise.

That also appears to be true for humans, as illustrated when scientists tested the effects of calorie restriction on fifty-three healthy men and women from twenty-one to fifty years of age.[34] For two years, one-third of the participants ate their normal diet, while the rest reduced their calorie intake by 15 percent. Those who consumed fewer calories lost about twenty pounds, and their metabolic rate, which governs the amount of energy the body needs to sustain normal daily functions, dropped by 10 percent. Oxidative stress was also reduced, supporting two long-standing theories of how aging occurs.

## Happiness and the Obesity Paradox

As a child growing up in the small town of Salida, Colorado, in the 1950s and 1960s, I was in constant contact with nature, and I lived in an extended family that included not only my immediate family but also the community at large. I lived with my parents and siblings, and I often saw my grandmother and great aunts, as well as aunts, uncles, and cousins. Beyond my immediate family, my family knew all the other families in the community. So, when I told someone, "I'm Fred Provenza," they recognized my family name and it had meaning to them – for better or worse. I felt the love and nurturing of my family, extended family, and community.

That's all changed. Most of the folks I knew in Salida are either buried in Fairview Cemetery or, like me, they left our hometown long ago. We were told as youngsters to get an education and leave, as there was no future in Salida. The few eighty-year-old natives who are still alive have become strangers in what was once their hometown. As a nation, we all moved helter-skelter. Slowly and incrementally, the loss of love and nurturing provided by extended families and close-knit communities has adversely affected the health of the nation.

This change has been documented in situations like that described by physician Lissa Rankin in *Mind over Medicine* for the people of Roseto, Pennsylvania. These Italian immigrants, originally from Roseto Valfortore in southern Italy, moved to the United States in 1882, seeking a better life. During the day, the men of Roseto worked in the stone quarry while the women worked in the blouse factory. In the evening, people strolled the village's main street, visiting with neighbors and friends. As church bells rang, women gathered in communal kitchens to prepare classic Italian feasts of pasta, sausage, and meatballs fried in lard. The people of Roseto were one big extended family who knew and cared for one another.

Based on today's nutrition recommendations, the people of Roseto ate poorly. They routinely ate foods like spaghetti that are high in refined carbohydrates. They ate pizza loaded with homemade high-fat sausage, pepperoni, salami, and eggs. They cooked with lard. Indeed, 41 percent of their calories came from fat. Moreover, the people of Roseto were not physically fit. Most were sedentary and obese. To add insult to injury, they also smoked and drank.

Yet, compared with people from surrounding towns and the national average, the rate of heart attacks in Roseto was half the national average, nearly zero for men under sixty-five. Their death rates from all causes were 30 to 35 percent lower than average. Their health couldn't be attributed to genes, to where they lived, or to the type of medical care they received. Rather, a supportive, tight-knit community was a better predictor of their health than was their "poor" diet or use of tobacco and alcohol. No doubt, their levels of stress were very low. In a six-year prospective, population-based study of older adults in Italy, individuals with the lowest levels of the stress hormone cortisol, with or without preexisting cardiovascular disease, had a five-fold lower risk of cardiovascular mortality.[35] Epidemiological studies show that chronically elevated levels of cortisol are associated with increased adiposity, insulin resistance, and cardiovascular disease.[36]

Ironically, being overweight is correlated with a lower risk of death for older people. The effect has been studied most with cardiovascular diseases, but a protective effect also has been shown for people who have suffered a stroke or those with diabetes.[37] People over age sixty-five – with a body mass index (BMI) between 25 and 29.9 – had the lowest all-cause mortality rate of any of the BMI categories, lower than people with a BMI considered ideal (18.5 to 25), as opposed to overweight (25 to 30) or obese (over 30). A 2014 meta-analysis found the lowest mortality rate occurs in people over age 65 with a BMI between 24 and 30.9. But there are risks with being overweight – from diabetes to blood pressure to heart disease – and the science of the obesity paradox is uncertain enough that physicians don't generally think normal weight people should gain weight. The obesity paradox seems most apparent in people with low fitness. For people of average fitness or above, staying slim correlates with better health.

Sadly, as the dreams of the older generation were realized – a better life for their children – Roseto died. Extended families dispersed. Nightly celebrations gave way to the isolation that now characterizes life for most in the United States. Neighbors no longer stopped by for casual visits. Evening meals where adults sang and children played became nights in front of the television. These changes were reflected in the poor health of the community. Heart attack rates doubled. The incidence of high blood pressure tripled. The number of strokes increased. In the end, the number of fatal heart attacks in Roseto increased to the national average.

According to the United Nation's *World Happiness Report* 2018, people who live in some of the coldest and darkest parts of the world, seemingly the dreariest parts of the planet, are the happiest people on Earth – Finland, followed by Norway, Denmark, and Iceland. Top-tier nations have high values for factors that enable well-being: income, life expectancy, social support, freedom, trust, and generosity. In 2018, the United States dropped four places to eighteenth. Why does the United States, a nation founded on the promise of individual liberty and the pursuit of happiness, rank in the second tier and why is our ranking continuing to drop?

Much of our happiness, as anthropologists and social psychologists point out, comes from our relationships with others, our shared sense of community and purpose. On these measures, America is in decline. We have little faith in institutions, from corporations and government to the media. Work – the endless pursuit of money, security, and things – dominates the lives of most people. We have little time for family or friends, the activities that create community and meaning. Loneliness is pervasive, along with consoling addictions to painkillers, unhealthy foods, and technology. Rather than engage with one another and the awe-inspiring natural environments we inhabit during our moment on Earth, we live in virtual worlds. The value a culture places on time spent in nature and in enjoyable, meaningful contact with family and friends separates top-tier countries from lower-tier countries. The Danes have a word for such harmonious relationships: *hygge*. As William Falk, editor-in-chief of *The Week*, puts it, "Richness comes from human connection. GDP matters less than hygge."[38]

A supportive family and community, a romantic relationship, and spirituality support happiness and longevity. And they do so by affecting us physiologically through stress, or a lack of it. When we experience chronic stress – fight or flight responses – the stress hormone cortisol suppresses immune responses to reduce inflammation that would accompany "wounds" that a "predator" might inflict. Those chronic responses cause adverse effects on our gut, manifest in "irritable gut" syndromes. Given time, a body can no longer relax or repair itself. Organs like the heart are damaged and the cancer cells, which we make naturally every day, are enabled to proliferate. Conversely, when we create nurturing environments, our immune system operates, and the body repairs itself naturally, preventing illness and treating diseases. That's when a diet rich in nutritious foods can usefully provide a body with what it needs to function best.

Indeed, despite their high prevalence of risk factors such as obesity, smoking, and sedentary lifestyles, first-generation Greek immigrants to Australia continue to display more than 35 percent lower mortality from cardiovascular disease and overall mortality after living at least thirty years in Australia compared with Australian-born controls – the so-called morbidity mortality paradox or Greek-migrant paradox.[39] Despite changes in food selection that contributed to significant weight gain, Greek immigrants continued to eat large amounts of leafy vegetables, herbs, and spices added to dishes prepared according to Greek cuisine. In addition, Greek immigrants returned to traditional Greek cuisine as they age, which helps to explain why they have more than twice the circulating concentrations of antioxidants compared with their Australian counterparts. Findings like these for immigrants from Greece to Australia and Italy to America illustrate the influence of diet, lifestyle, and healthy relationships in well-being.

## Two Spiritual Dangers

In autumn here at our mountain home, the aspens are so bright yellow they glow at night, even without moonlight. The skiff of snow beneath the aspens in our woodlot and on the high peaks means winter isn't far off. The high peaks will soon be off-limits, so Sue and I decided to savor a trek into the Mount Massive Wilderness Area. The brightly colored carpet of *Vaccinium* (blueberries) leaves that blanket the peaks are visible from far away. Their once-green leaves are now bright yellow, orange, and red. The berry-laden coyote scat along the trail tells us the blueberries are ripe. The deep bluish purple berries are sweet, but their rich flavor goes far beyond sweet, in ways my palate appreciates but my words are unable to capture.

Nor can my words capture the serenity I feel as my being merges with the vastness of the surroundings that embrace and comfort me. When I'm in the high country, the cultural bonds that can easily muddle my physical and spiritual being, with all the daily "doing," dissolve into the illusions they are. Here, I enter a state of authenticity, as the chatter created by *doing* dissolves into *being*. This is a dimension of heaven. I feel the Artist in every quark in the galaxy. I understand why monks throw off civilized life to dwell in seclusion in places like this. In a word, this is a meditation. I can sit for hours, quietly absorbing the beauty that surrounds me. I often see rabbits sitting quietly for long periods in our woodlot. One sat on our deck an entire day without moving. I wonder if they, too, are in a meditation, if they, too, feel the at-one-ness.

In the evening, we feel the seven miles we walked. Our once-young bodies no longer have the strength and endurance they had. Nor do they have the resilience. We savor a simple meal as darkness falls on the woodlot. While we still eat all the variety of foods we've eaten throughout our lives, the amounts and proportions of what we eat have changed. We no longer need the amounts of energy or protein we once needed. People say we lose our senses of taste and smell (retronasal olfaction) as we age. That may be so, but the expression misses a larger point. Our retronasal tastes change as we age, too. They change as a function of what we need.

Palatability is more than a matter of taste – and health is more than a matter of food. Today, my eyes dined on plant-clothed shoulders of granite that caressed glacial lakes and streams enveloped in a deep, fall blue sky. My eyes and ears partook of streams as they splashed down grassy slopes and riffles as they danced across lakes and splashed into the shore. Every cell in my body feasted as the scent of terpenes from pines flowed from my nose through my lungs into my bloodstream on their way to capillaries that nourish my cells.

Spending time "ingesting nature" – forest therapy – enhances immune function.[40] Studies show time in nature increases natural killer (NK) cells that help us fight infections and cancers. A natural killer cell is a kind of white blood cell that sends self-destruct messages to tumors and virus-infected cells. Stress, aging, and pesticides can reduce NK cells.

Qing Li, a medical doctor at Tokyo's Nippon Medical School and author of *Forest Bathing*, is one of the world's foremost experts in forest medicine. Years ago, Li and his colleagues suspected that trees nourish our health. Specifically, they wondered if NK cells are affected by aromatic volatiles called phytoncides – the pinenes, limonenes, and other aerosols emitted by evergreens and many other trees and shrubs. Scientists have identified 50 to 100 of these phytoncides in the Japanese countryside and virtually none in city air that's not directly above a park.

To test the phytoncide hypothesis, they sequestered twelve people in hotel rooms. In some rooms, they rigged a humidifier to vaporize stem oil from common Hinoki cypress trees; other rooms had only a humidifier. The cypress smellers had a 20 percent increase in NK cells during their three-night stay and reported feeling less fatigued. The control group saw almost no changes. For urban dwellers, NK activity is enhanced by 56 percent on the second day of forest therapy. They maintained 23 percent for one month after they returned to urban life.

The researchers also brought a group of middle-age Tokyo businessmen into the woods. For three days, the men hiked in the morning and again in the afternoon. By the end, blood tests showed their NK cells had increased 40 percent. A month later, their NK count was still 15 percent higher. By contrast, NK levels didn't change with urban walks. Because most people can't spend three days a week walking in the woods, Li's research group was curious to learn whether a one-day trip to a suburban park would have a similar effect. And in fact, the one-day outing boosted levels of both NK cells and anticancer proteins for at least seven days afterward.

Forest therapy also relieves stress. Compared with an urban control group, people in Japan who sit quietly in natural surroundings have less stress; they are calmer; they have lower blood pressure and heart rate.[41] In addition, their parasympathetic nervous activity is enhanced by 55 percent, indicating a relaxed state. Walking in the woods has a similar effect.

Unlike our ancestors, most *Homo sapiens* are detached from the rhythms of nature. In *The Shallows*, author Nicholas Carr cites the statistic that the average American spends at least eight hours a day looking at an electronic screen. Then we "relax" by watching television. Kids between eight and eighteen years of age spend more than seven hours a day looking at screens, often using more than one media platform at a time. We become stressed and short-tempered with all that sitting and watching. In *Your Brain on Nature*, physician Eva Selhub and biophilosopher Alan Logan assert that, since the age of the internet, North Americans have become more aggressive, more narcissistic, more distracted, more depressed, and less cognitively nimble.

Psychologists Rachel and Stephen Kaplan noted that, beginning in the 1970s, distress frequently became related to mental fatigue.[42] Modern life demands sustained attention to tasks that are both important and mundane – for instance, checking email, working a desk job, finding a parking spot. Our brains need rest from all the tedium of our lives. That's what happens when we look at a flower, watch a sunrise, or listen to the rain. That's why they recommend nonathletic approaches to being in nature. Pursuing an outdoor sport isn't necessarily the same as being in nature. Of course, we get some mental and physical benefits from working out in the open air, but to get the most out of natural surroundings, we need to be present, not distracted by all the trappings of modern civilization. When we spend time outside but distract ourselves – for example, by listening to an iPod while going running – we may end up more irritable and impatient later, less able to stay on task, focus, and plan than our nature-engaged peers.

*Homo sapiens* evolved as a species that spent all of its time in nature. In contemporary times, most people still achieve feelings of comfort when their rhythms are synchronized with those of nature. In *The Nature Fix*, journalist Florence Williams reviews the science that underlies those rhythms and the value for our health and well-being. Nature, she notes, is restorative for individuals and cultures. Nowadays, we don't spend enough quality time in nature to experience those healing effects. Merely exercising outdoors isn't enough to reap the benefits. We must become untethered from modern technology – unstrap the heart monitor and unplug the headset. Leave our phones at home. Don't clock our wind sprints. Experience being in nature.

It is a measure of the degree to which we've become hitched to technology and untethered from nature that we now need scientists to remind us that wildflowers and birds, mountain streams and lakes are good for our well-being. But then again, if the benefits of being in nature are so natural, why don't more of us do it? We don't because we've so lost contact with nature that although we feel the need to search, we don't realize what we've lost or where to find it. During a time when we could use a nature fix more than ever, nature-based recreation has declined markedly in the United States during the past four decades.[43]

Nor do we need scientists to tell us how nature nourishes our bodies. We simply need to experience that. Our ancestors' palates were linked with the landscapes they inhabited through the hunting of animals and the gathering and growing of plants. While most people no longer hunt or gather, and few people work in agriculture, we can still eat wholesome foods, grown on fertile soils. We can also grow gardens, a modest act that can profoundly affect health. As with a walk in the woods, tending plants in a garden provides the alone time we human beings need to connect with our inner selves through meaningful contact with the natural world.

Will having more data about how intimate contact with nature improves the well-being of our minds and bodies lure us into the woods or the garden? That's not likely because we can't appreciate something we've never deeply experienced, and from the womb to the tomb, we are ensnared in the traps of technology under the guise of contentment. When none of that nourishes our bodies and souls, as we are finding, we turn to opioids to ease our pain and anxieties.

Our decisions about nature – whether or not to nurture the landscapes, watersheds, and airscapes that sustain human communities – are based upon our relationship with nature. As most people never experience nature and thus have insignificant understanding of the services nature provides, we the people are unlikely to choose a viable future for ourselves or the plants and animals who cohabit this planet with us. As the keystone species on the planet, our detachment from nature is an ominous sign for their survival – and ours.[44]

Though we still depend on nature for our bounty, we are in the grips of technology. We erect houses for shelter and we live in cities. Nowadays, over 98 percent of the population of the United States no longer lives on farms or ranches. Most people don't grow or gather plants, hunt and kill wild animals, or nurture and slaughter domestic animals for their sustenance. As a result, we've distanced ourselves from these sacred acts. Unsuspectingly, we've lost our linkages with the landscapes that sustain our physical and spiritual health.

As Aldo Leopold cautioned in *A Sand County Almanac*:

> *There are two spiritual dangers in not owning a farm. One is the danger of supposing that breakfast comes from the grocery, and the other that heat comes from the furnace. To avoid the first danger, one should plant a garden, preferably where there is no grocer to confuse the issue. To avoid the second, he should lay a split of good oak on the andirons, preferable where there is no furnace, and let it warm his shins while a February blizzard tosses the trees outside. If one has cut, split, hauled, and piled his own good oak, and let his mind work the while, he will remember much about where the heat comes from, and with a wealth of detail denied to those who spend the weekend in town astride a radiator.*

# CHAPTER 11

# Linking Palates with Landscapes

Most humans mark time with calendars. Bluebirds mark time by changes in seasons. Sue and I mark time by the comings and goings of bluebirds. Each year when they arrive in our neck of the woods, we know winter will be ending soon. The birds spend spring tending their eggs and offspring. During summer, bluebird families join us each morning for breakfast. In October, those families leave for warmer climes. Though we don't know for sure about our bluebirds, ornithologists who band birds tell me the same individuals return to the same locations each year.

Not unlike humans, mountain bluebirds are linked at the heart—through molecules of emotion—with the landscapes where they were conceived, born, and reared. Those bonds develop as cells and organ systems create linkages with the biophysical environments we all inhabit. Over generations, these relationships create cultures that learn how to navigate their local environments. It's the home-field advantage often touted in sports and absolutely essential in life.[1] Creatures learn where and where not to go, what and what not to eat, and what's a predator and what's not. Our ancestors sorted through beneficial and harmful plants, learned to prepare and mix foods in ways that decreased toxicity and increased health, used local materials for clothes and homes, and created mythologies that linked physical with spiritual realms.

I've traveled to many places on this planet, and I've savored them all—from Iceland and Ireland, to the Alps of France and Switzerland, to the plains of Africa, to the eucalyptus forests of Australia. But I never felt as comfortable in unfamiliar terrain as in the haunts where I grew up. I know my home country in a way "a stranger in strange lands" can never know a new place. The place is in me. I developed an intimate relationship with my home terrain. I know where and when various plants will emerge during spring, summer, and autumn. I know where and when to find edible and delectable foods such as boletus and cauliflower mushrooms. I know which dry flies will work best for catching fish at which times in which glacial lakes and streams. I know where to find grouse, ptarmigan, hares, mule deer, elk, and mountain goats.

If home is where the heart dwells, my home will always be in the high country of Colorado where I grew up. I gaze in awe upon the beauty of mountain slopes covered in bouquets of alpine plants, a kaleidoscope of greens, blues, purples, yellows, oranges, pinks, salmons, and reds. For me, nothing can equal the beauty or serenity of a glacial lake encircled by splendid peaks, jutting into deep blue sky. Creatures who live in these haunts have always captivated me, from brook and cutthroat trout, to furry picas and marmots, to feathered ptarmigan and blue grouse, to mule deer, elk, and mountain goats.

I once knew a psychologist who raved about his love for Tintic Valley in central Utah. I had worked there for many years as a technician, studying soils, plants, livestock, and wildlife, but never found the valley the least bit appealing. For me it was a hot, dry, dusty, windy place. So I was astonished and curious about his deep attachment to Tintic Valley. When I finally asked where he was born and raised, I wasn't surprised to learn he came from Tintic Valley.

## How Palates Link Wild Animals with Landscapes

While most humans record events in logbooks, rabbits record events internally. The plants they eat seasonally are logged in their meat and fat, which serves as a transcript for their offspring. Decades of research by Dr. Vilmos Altbäcker and his colleagues at Kaposvar University illustrate how that happens. They began by showing rabbits that live in one part of Kiskunság National Park prefer juniper, while rabbits that live in another part of the park shun juniper. Studies with transplanted junipers show the difference is not due to the plants, but a matter of the local rabbits' culinary traditions. Kittens with experience eat much more juniper than control litters that have never eaten juniper.[2] When Altbäcker supplemented the diet of mothers in the lab with juniper berries or thyme leaves, both of which are eaten by wild rabbits, offspring later preferred the aromatic plant that their mother had eaten, and they digested significantly larger amounts compared with the control litters.[3] Thus, the diets of rabbits are a result of social transmission of food preferences.[4] Offspring remember and prefer their mother's diet for at least six months after weaning, even if they have no access to those foods in the meantime.[5]

Given these results, the question arose whether the social transmission of food preferences, if targeted cleverly, could provide an opportunity to control the spread of several invasive plant species. Persuading animals to acquire a taste for a previously shunned plant is not unprecedented. Ranchers and farmers train their livestock to eat weeds, and offspring will even pick up the habit from the example of their elders.[6] Selective grazing by rabbits can significantly alter vegetation composition, and thus a group of biologists, led by Altbäcker, are now seeking to determine whether wild rabbits can be trained to eat and thus control milkweed.[7]

Cultural transmission is harder to initiate with wild as opposed to domesticated herbivores for several reasons. Young rabbits leave the nest as soon as they are weaned, providing little chance to learn from the mother. But directly observing what the mother eats isn't the only way offspring learn what to eat. They have three other sources of information about the mother's diet.[8] Amniotic fluid provides clues to the mother's diet during pregnancy, and the mother's milk provides clues to her diet during lactation. Fecal pellets, deposited in the nest by the mother during nursing, contain remnants of plants eaten by the mother, and baby rabbits sniff and ingest these maternal pellets.

Milkweed (*Asclepias syriaca*), named for the milky white latex it contains, is native to North America but invasive in Europe. Due to its toxic effects, livestock don't eat milkweed, but rabbits are able to eat the milkweed that has been overtaking Kiskunság National Park in Hungary. The challenge is that rabbits give birth in winter and spring, while milkweed dies back in fall and doesn't begin growing again until May. Ostensibly, milkweed molecules have no way to get into the mother rabbits' amniotic fluid, milk, or edible feces, so there is no way for wild baby rabbits to learn that milkweed is a potential food source.

To mimic the milkweed-less interval between the end of the milkweed growing season and the beginning of the breeding season for rabbits,

Altbäcker tested the preferences of kittens born to mothers taken off milkweed three months earlier. While kittens didn't eat as much milkweed as those in the earlier experiments, they liked it better than did control litters. He theorizes compounds in mother's diet remain in her body, perhaps in her fat from the previous season, and when they are metabolized in winter, they are transcribed into amniotic fluid and milk as a message to offspring.

Rabbits illustrate how wild animals learn food selection behaviors. Over time, those behaviors become part of local cultures. Culture also links wild animals with different parts of the environment, as illustrated with moose. Some populations of moose in Norway migrate from high elevations in summer to low elevations along the seashore in winter.[9] Oddly, though, other populations persist in living year-around at high elevations. Why?

Moose who endure the harsh elements during winter at high elevations do so based on a tradition initiated to avoid being hunted by humans along their former migratory route. Moose who weren't hunted continued to migrate and thus spend their winters at more hospitable lower elevations along the seashores. Archeological evidence shows these different migratory patterns have been in place for 5,000 years. Nowadays, moose aren't hunted in Norway, yet they continue to spend their winters at high elevations, their behavior a slave to tradition. Compared with migratory moose, the nonmigratory moose live in a more restricted range of environments, which affects form, function, and behavior. That is reflected in observations that moose that reside at high elevations are smaller and they produce fewer calves compared with migratory moose. Their smaller body size and lower reproductive rate enables them to survive on sparse forage under harsh winter conditions at high elevations.

That wild animals learn to use different environments is also shown by management that changed the culture of elk at Deseret Land and Livestock. In 1984, the managers at the Deseret Land and Livestock (DLL) Ranch in northern Utah began feeding hay to elk during winter to prevent the elk from spending the winter eating forage on their neighbor's ranches. DLL was benefiting from high fees people paid to hunt elk on DLL, but their neighbors were sustaining the costs of feeding the elk during winter. Managers at DLL put the elk on welfare, but by 2004 the costs of feeding 1,000 elk were so high, they decided to wean the "welfare elk."

Rather than just stop feeding, which had proved to be a disaster elsewhere, the managers at DLL decided to use "carrots" (cattle grazing and supplementation) and "sticks" (herding and hunting) to gradually change the behavior of elk.[10] Areas grazed by cattle in spring are attractive to elk in fall and winter due to a nutritious mix of regrowth of grasses and forbs (stimulated by grazing early in spring) with mature forage. Cattle were grazed strategically in areas that would be suitable winter range for elk. Elk were also provided supplements to enable them to better use sagebrush, a shrub abundant on winter ranges but not fully used due to its high concentrations of terpenes. When these measures were in place, managers stopped feeding hay on the traditional winter-feeding area, and they began to use stockmanship to move and place elk in the desired locations. Finally, elk were hunted in areas where they were previously fed, but not in areas the managers now wanted them to overwinter. As the example with moose shows, hunting can be a very effective way to permanently alter the habitat preferences of animals.

During the first winter of the project, the weather was relatively mild and the elk weren't fed any hay. As the managers noted with excitement at the end of that winter, that was the first time in twenty years that no calves had learned to eat hay on the winter feeding area. They were beginning to break the transgenerational bonds to winter feeding in the "welfare elk." Since the project began in 2004, elk have been fed hay only in the severe winters of 2005 and 2010. By 2010, the elk had to be pushed to eat the hay, as no former "welfare recipients" resided in the herd. One generation at a time, the managers of DLL had changed the culture of the elk herd.

As a winter storm blankets our woodlot today, I'm pondering how good intentions alter cultures and adversely affect the ability of animals to survive. Chickadees are dining at our feeder in small groups. Rosy finches flock to the feeders in a frenzy, tossing seeds to and fro. Individuals who can't get to the feeders enjoy a meal of dropped seed on the snow below. Our well-meant welfare may do them more harm than good. Over generations, the knowledge their ancestors had of what and what not to eat, where and where not to go, may be lost as successive generations of birds come to rely on seeds we provide rather than the seeds nature provides.

I recall stories about grizzly bears that learned to feed in garbage dumps in Yellowstone National Park. Tourists loved the spectacle, just as Sue and I delight in watching the chickadees and rosy finches. However, over time the impacts on bear behavior and bear-tourist encounters became a growing concern. At that point, biologists John and Frank Craighead, who spent their lives studying bears, recommended gradually weaning the bears off of the dumps so they'd have time to learn new foraging behaviors. However, the National Park Service ignored their advice and closed the dumps quickly, exacerbating the problem. Unfortunately, the bears had lost their relationships with the foraging environments their ancestors once relied on for food, and with no time to gradually transition back to natural foods, bears (and tourists) suffered greatly.[11]

## How Palates Link Livestock with Landscapes

During the second summer I worked for rancher Henry DeLuca, I was pondering two puzzling observations. I'd just taken a course in genetics and learned about the value of crossbreeding and heterosis, the tendency of a crossbred individual to show qualities superior to those of its parents. I knew, however, that Henry kept his own heifers as replacement stock to maintain his herd, so I asked him why he didn't buy replacements of different breeds from other ranchers.

He began by telling me a story about the time he bought replacement heifers and put them on the Mount Shavano grazing allotment. He had a devil of a time finding them on the mountain during the summer, they were in poor condition when he brought them back to the ranch in fall, and they weren't pregnant. He then told me of the time he had to move his cattle to South Park for "the summer from hell." That summer, too, he had a hard time finding his cattle, they came off the range in poor condition, and they had low reproductive performance. He summed up his stories by saying, "The problem is the cows just don't know the range."

Animals moved to unfamiliar environments often spend inordinate amounts of time walking around the perimeter of fenced pastures. These movements, often referred to as "exploratory behavior," may indeed be learning excursions, but there is another way to explain the excessive amount of walking. Naive animals don't know where they are and they are looking for a place to escape to find a way back home. In Iceland, where livestock are not fenced, the agricultural extension service recommends not putting sheep in unfamiliar areas because the animals will travel for hundreds of miles in search of familiar haunts.

Henry's knowledge came to his brain through his hands and eyes while working with his cattle, but his insights are similar to results of studies that show genes are being expressed in his particular environment to influence form, function, and food and habitat selection behaviors, though he would never have used that verbiage. Like Henry's cattle on Mount Shavano, cattle placed on the same summer range pastures in Oregon and Idaho separate into subgroups that show fidelity to particular areas within pastures.[12] As adults, they prefer to live in locations they learned to use as calves from their mothers.[13] Sheep, too, show fidelity to areas where they were born and raised.[14] Shepherds in the United Kingdom handpick different subgroups of sheep, who show fidelity to different habitats, to disperse animals more broadly across landscapes.

Like rabbits, livestock begin to learn about the flavors of forages in their mother's diet in the womb and through mother's milk.[15] Flavors of onion and garlic, for instance, are transferred in utero and in milk, preparing lambs to eat onion and garlic plants growing in the environment. After birth, offspring quickly learn what to eat from their mother. Six-week-old lambs fed whole-grain wheat with their mothers for as little as one hour a day for five days ate more wheat than lambs exposed to wheat without their mothers. Three years later, with no exposure to wheat beyond the five hours early in life, intake of wheat was nearly ten times higher for lambs that were exposed to wheat with their mothers compared with lambs exposed to wheat without their mothers or not exposed to wheat at all.[16] Lambs also learn to avoid toxic foods from their mother.[17] By eating the same foods as the mother eats and cautiously sampling novel foods the mother avoids, young animals are unlikely to eat poisonous plants to an extent that could be fatal. When a lamb samples a plant that its mother has avoided and then becomes sick, the lamb strongly avoids the food thereafter.[18]

Experiences in utero enable animals to better use poorer quality diets. For example, the offspring of pregnant cows who eat high-fiber diets in winter are better able to eat high-fiber diets. Compared with naive offspring, calves exposed to a high-fiber diet in the womb eat more high-fiber forages and digest them better, and they grow better.[19]

In a related study, young calves were exposed to high-fiber straw for two months with their mothers. Five years later, they were fed straw as a major part of their diet from December to May for three years.[20] Throughout the study, experienced cows ate more straw, gained more weight, maintained better body condition, produced more milk, and conceived sooner than cows not exposed to straw early in life. Genetic analysis shows that heritability of dry matter intake and digestibility of straw is only 20 percent, so experiences in utero and early in life account for 80 percent of the response in these studies.[21]

Using sagebrush as a winter forage enables ranchers to cut feed costs, enhances the abundance of grasses and forbs, maintains sagebrush as part of biodiversity, reduces the use of herbicides, and creates mosaics of habitat for diverse species, including sage grouse, deer, and elk. Two factors help livestock better use sagebrush.[22] As discussed in chapter 4, providing supplemental energy and protein enables livestock to detoxify potentially toxic terpenes in sagebrush. In addition, cattle exposed to sagebrush in utero, early in life, and as adults consistently eat more sagebrush and maintain better body weights than their naive counterparts.

Social groups are a blend of stability (mother) and creativity (offspring) as offspring explore foods and areas not used by mother. When nanny goats from different islands in the French West Indies were moved to an island with a smorgasbord of forages found on each of their home islands, the foraging behaviors of adults didn't change.[23] While the nannies originally affected foraging behaviors of their offspring, as time passed, their offspring were increasingly affected by peers from other islands. Over four generations, peers converged on more similar diets.

When sheep were used to graze forest plantations in Washington State, the first year the ewes and their lambs ate the grasses and forbs and ignored the young tree seedlings, enhancing the growth of the tree seedlings. During the next summer, lambs began eating tree seedlings, and by the end of the summer, their mothers were eating them, too. The third year, mothers taught their new crop of lambs to eat tree seedlings, which ended the contract for the grazing.

We often consider bison and cattle to be grazers, sheep to be mixed feeders who eat large amounts of forbs and shrubs, and goats to be browsers. Nonetheless, any of these herbivores can learn to eat large amounts of grasses, forbs, and shrubs depending on the social and physical landscapes where they are reared.[24] Cattle and bison can browse, and goats can graze. For example, goats in Israel once used large territories, but as the Old World became more populated, cattle (used for plowing) and sheep (used for wool) became more important than goats, which were fenced out of desirable pastureland and placed instead on brush-dominated landscapes.[25] In the process, their diets evolved from less browse (41 percent) and more herbs (51 percent) to more browse (81 percent) and less herbs (19 percent).

The ability of goats in Israel to learn to use browse high in tannins is illustrated by cross-fostering studies with two breeds of goats. Damascus goats prefer high-tannin species of woody plants while Mamber goats do not.[26] Offspring from one breed (Damascus) were reared from birth by females from the other breed (Mamber) and vice-versa. Mamber kids, whose Damascus foster mothers preferred high-tannin browse, ate much more high-tannin browse than did Damascus kids whose Mamber foster mothers did not eat high-tannin browse.

Like goats in Israel, goats in the United States also showed the advantage of experience with the nutritionally challenging shrub blackbrush, which is low in energy and protein and high in tannins. Goats reared from one to four months of age with their mothers on blackbrush-dominated rangeland ate two and a half times more blackbrush than goats that had been reared on an alfalfa-based diet of much higher quality.[27] When allowed to choose between blackbrush and alfalfa pellets, experienced goats ate 30 percent more blackbrush than did naive goats. To better meet needs for energy and protein, experienced goats developed much larger gut (rumen) volume and better ability to cope with tannins compared with goats who were naive to blackbrush. Ranchers often tout the superior genetics of animals that finish well on grass, in part due to their large gut capacity compared with other animals, and while that may be true, they ignore epigenetic influences of experiences in utero and early in life to alter form and function.

Experiences in utero and early life cause changes neurologically, morphologically, and physiologically, and these changes help explain an animal's enhanced ability to cope with different forages. For example, lambs exposed to saltbush in utero grew faster and handled a salt load better than lambs whose mothers grazed on pasture without saltbush.[28] They excreted salt more rapidly and drank less water due to changes in kidney form and function. These changes better enabled lambs to eat saltbush. Likewise, lambs reared on poor-quality grass eat more grass and digest grass better than naive sheep. They do so in part because they can better recycle nitrogen, which helps microbes in their gut digest grass, thus providing the energy and protein they need.[29]

Maternal nutrition also affects the accumulation of intramuscular fat (marbling) and development of skeletal muscles (steaks) in the developing fetus.[30] Nutrient deficiency during mid- to late gestation decreases muscle fiber size and the number of intramuscular fat cells (adipocytes). Poor neonatal nutrition permanently reduces muscle mass and negatively affects animal performance, which is an economic concern for growers because animals won't develop the muscle or fat mass to finish well within a reasonable timeframe.

All of these changes, referred to as *predictive adaptive responses*, act through developmental plasticity in utero and early life to modify form, function, and behavior in ways that confer survival advantages when the environment of rearing matches the environment where an animal will live.[31] That will be so if the postweaning environment doesn't change radically during the lifetime of the offspring. Moving animals to unfamiliar settings severs links with landscapes and explains why animals placed in unfamiliar social and physical settings suffer more from predation, malnutrition, overingesting poisonous plants, and poor reproductive performance compared with animals born and reared in the environments.[32]

Adult livestock typically require three years or longer to adapt to unfamiliar haunts or to radical changes in management in a familiar haunt. That was illustrated when Ray Bannister, an innovative rancher who lives near Wibaux, Montana, changed grazing management. His cows took three years to learn to "mix the best with the rest" as opposed to "eat the best and leave the rest." Likewise, ecologist-rancher-author Bob Budd's cows took three years to learn to use uplands as opposed to riparian areas on Red Canyon Ranch near Lander, Wyoming. In both cases, as the culture of the herd changed, so did ecological, economic, and social benefits, as I described in *Foraging Behavior*.

In *For the Love of Land*, Savory Institute cofounder Jim Howell describes how it would feel for a human to be moved to a strange land.

> *What would you do if you were unwillingly plucked off of your pretty farm in the green hills of Missouri, transported to a new ranch in the badlands of Wyoming, given a brand new set of friends, all new food, different weather, a novel landscape, and salty water? You most likely would protest and perform below your potential, at least initially. . . . What if you had been on that same Wyoming ranch your whole life and had been in charge of the winter country in the Red Desert all that time. You know every square foot of that place, where all the best grass patches are, the good places to take shelter in blizzards, how far you can ride out and still get back before dark, etc. You are intimate with the land.*

## Value of Social Organization and Culture

Mule deer does and their fawns made the rounds through our neck of the woods last night. It had been a week since they had last come by here. We watched until we could see only ethereal shadows nibbling on the many plants on offer at our bistro. Antelope and elk, too, come in groups mostly at night. They roam in small groups in summer and larger groups in winter.

In *The Buffalo Harvest*, frontiersman Frank Mayer describes these "small orbits" for bison. He begins: "Do you remember reading about buffalo herds millions strong, moving in a solid mass, and stopping trains and wagons? Isn't that about the picture as you have it in your mind?" He points out no buffalo herd he saw ever numbered more than 200 animals, and most of them were much smaller. Most herds had from three to sixty animals, with an average of around fifteen. In these small groups, the buffalo traveled and fed, scattered over the plains, but each one distinct and separate from the other herds. He notes,

> *Do keep these small herds in mind . . . they were important to us in our hunting; in fact formed the basis of our attack.*
>
> *At the head of each of these little herds would be its leader. But the leader wasn't a courageous, old bull, ready and willing to whip the universe. It wasn't a bull at all. It was a cow, a sagacious old cow who by the power of her intellect had made herself a leader. Buffalo society, you see, was a matriarchy, and the cow was queen. Wherever she went, the others, including the big bulls who should have known better than to follow a woman, went. When she stampeded, they stampeded. When she got into trouble, they didn't know what to do. And our job as runners was to get her into trouble as soon as we could. Then the rest was easy.*

The issues Mayer raises are significant for three reasons. First, most people believe bison lived in huge herds whose numbers and movements were influenced by predators. Yet, predators such as wolves have little influence on bison numbers or movements, chiefly because bison are dangerous prey for predators, and the same is true for many large herbivores in Africa.[33] Some ranchers now encourage the "natural herd instinct" of cattle, unaware their natural herd instinct is to live in extended families. The trend toward grazing at high stock densities may benefit soil and plants and produce good performance in livestock, but it isn't mimicking natural behavior.

Second, we interpret snapshots of times past – the way things used to be – to fit agendas. A person who favors grazing by large mobs of livestock can emphasize historical accounts of bison herds millions strong moving for days across the prairie, enhancing the health of soil and plants. Or one can contend, as does Frank Mayer, those herds were anomalies – bison formed large herds only during breeding season, mass migrations, and when they were frightened – "Whenever they stampeded they did come together and charged as one vast, solid herd. But when the fright had passed they'd separate into their peculiar small herd formation."

Prior to the arrival of Europeans in North America, Native Americans were devastated by diseases such as smallpox. It is postulated that at least three pandemics decimated native people by 90 percent or more before Pilgrims set foot on Plymouth Rock.[34] With the decline in humans, the number of bison increased, but only from the late 1600s to the mid-1800s when Europeans began to wipe them out. So, depending on which point in time one chooses to emphasize, a person can make a case for large or small populations of bison. This is typical of most natural resource issues. We don't appreciate the changing nature of plants, herbivores, or peoples who formerly inhabited the landscapes we've just recently come to inhabit.

Third, Mayer appreciates that bison lived in philopatric matrilines. A matriline is a line of descent from a female ancestor to offspring in which the individuals in all intervening generations are mothers. In a matriline, an individual belongs to the same descent group as her or his mother. Philopatry is the tendency of animals to stay in or habitually return to a particular area. Retention of daughters and young sons within the matriline, along with dispersal of older males into bachelor groups, is the crux of sociality in many mammals. Many large herbivores live in philopatric matrilines, including bison, white-tailed deer, bighorn sheep, red deer, African elephants, and water buffalo, as well as domestic sheep, feral cattle, and zebu cattle.[35]

A matriarch's knowledge of biophysical haunts is key to the health of her family.[36] In elephants, families led by older, more knowledgeable matriarchs have better reproductive success and survival of their offspring than families led by younger matriarchs.[37] That's why changes in leadership when older females are culled from the herd can dramatically alter how families use habitat, including food resources and migratory routes. The breakdown of family structure is illustrated by comparing elephants in different parts of Kenya – Samburu and Amboseli. Samburu groups are composed mostly of nonrelatives due to the poaching that has led to 85 percent population decline and death of older females targeted by poachers.[38] The loss of social organization and cultural knowledge adversely affects families. Conversely, the largely undisturbed extended families of Amboseli elephants are still kin-based.[39]

This social organization is also illustrated with aquatic mammals. Old female killer whales use knowledge of seascapes to help their families through tough times.[40] Females who are past the age of last reproduction are more likely to lead their families as they travel around feeding grounds, especially in years when Chinook salmon, their main food resource, are in short supply. Menopausal females boost survival of their relatives through the transfer of ecological knowledge, which helps explain why female killer whales live long after they stop reproducing.

The practice of culling livestock that no longer produce offspring has become common in the United States, with little thought given to how it might affect the social organization of livestock herds. The consequences can be severe when herds without matriarchs must fend for themselves in harsh environments such as the deserts of Nevada.[41]

Ironically, we rarely consider how social organization and culture affect the structuring and functioning of ecosystems, including human societies. Traditionally, hunter-gatherers lived beyond sixty years of age, and women with a prolonged postreproductive life span were key to the inclusive fitness of extended families and clans.[42] Their knowledge of where to find and how to use plant resources, along with their roles in food gathering and preparation, improved inclusive fitness. The oldest, most experienced women also knew where to find food during perilous and rare conditions such as a drought. Nowadays, we no longer forage for ourselves, and living in extended families is mostly a thing of the past. No surprise, we manage wild and domestic animals in ways that thwart the development of culture, perhaps to our detriment.[43]

Historically, bison living in extended families played a key central role in the structuring and functioning of ecosystems in North America. There is an opportunity for social species such as bison, cattle, sheep, and goats to play a similar role today. To best realize this goal, we must understand not only how behavior is influenced by grazing management techniques – strategic placement of water, salt, and fences – but, more important, we must also understand how social organization and culture influence use of landscapes. With livestock, for example, we've come to rely on fences and grazing systems rather than culture to influence diet and habitat selection.[44] Could social organization in bison and cattle lead to rotational grazing without fences?

While that may sound crazy, the notion is based on key observations. Social herbivores live in extended families that maintain their cohesiveness by avoiding prolonged contact with other families. Interactions among families likely increases movements across landscapes as different families avoid prolonged contact with one another and with areas their family and other families have temporarily fouled by removing forage and by depositing urine and feces. In addition, individuals within families differ in their preferences for foods and habitats. Not unlike a shepherd in France, a matriarch can maintain the health of her bison family by meeting the needs of different family members through grazing circuits. By knowing when and where to move, she can enable individuals with different needs – young and old, males and females – to eat a broader array of plants and to forage in a greater variety of locations, thus enabling health. Such rotational grazing without fences would help to maintain the biodiversity of landscapes.

If we encouraged extended families, could we improve animal welfare and profitability? Male bison are typically weaned and placed in feedlots at about the time they would be moving from matrilines into bachelor groups. Bison in feedlots experience high stress due to social disorder and aggressive interactions, which adversely affect welfare and increase costs of production.[45] Providing adequate space, offering a choice of foods, and leaving young bison with herd mates on pasture prior to slaughter lowers stress, costs of gain, illness, and death.[46] While bison on rangelands require longer to finish, they are less costly and hence more profitable.

## Our Cultures, Our Families, Ourselves

I once had a PhD student from Iceland, Anna Thorhallsdottir, who treated her graduate committee to an Icelandic buffet just before her comprehensive exams. The foods were unknown to any of us. The centerpiece at the table was a sheep head, hair and all, that appeared to have been cooked with a blowtorch. The various dishes included shark, which had been buried under sand for three months – the stench of ammonia was beyond belief. There were meats and cheeses that resembled nothing we'd seen – let alone eaten – before. We were even treated to lichens for dessert. Needless to say, nobody ate much. We took the smallest amount possible to be polite. My daughter didn't eat at all. She asked: "Anna, don't you have any American foods, you know, like hamburgers?" None of the foods Anna served us were harmful. They were all highly nutritious – considered delicacies in Iceland – but they were unpalatable to us because we were so unfamiliar with what we considered to be strong and unpalatable flavors.

Psychologist Paul Rozin once asked: "Suppose one wishes to know as much as possible about the foods another person likes and eats and can ask only one question. What should that question be? There is no doubt about it, the question should be, 'What is your culture or ethnic group?'" No other question can begin to approach the amount of information as the answer to that question. Consider the differences between the carnivorous diets of Eskimos and the herbivorous diets of many tropical cultures, or between the elaborate cuisines of India or France and the relatively limited amount of food processing carried out by some hunter-gatherers.[47]

Scientists and practitioners don't agree on what constitutes a *suitable* diet for humans. It's possible to find support for any flavor of diet a person may want to consume. Some argue we must eat both plants and animals (omnivorous), others argue we are healthiest when we eat a diet of plants (herbivorous), and still others contend we are most fit when we eat mostly meat and animal products (carnivorous).[48] Authors of diet books view diet as predetermined evolutionarily to be the diet *they* advocate. Nonetheless, flexibility is the key to thriving in foodscapes that change in time and space. Indeed, as Rozin contends, human diets vary from plant-based in deserts and tropical grasslands to animal-based in northern coniferous forests and tundra.[49]

Traditionally, herbivores and hunter-gatherers ate foods in a manner that ensured survival through cycles of abundance and scarcity. Their dietary habits, transferred across generations, became part of the culture.[50] As with rabbits and livestock, human learning begins in utero, as the flavors of substances the mother eats (for example, fruits, vegetables, spices) or inhales (for example, onion, garlic, tobacco, perfumes) are transferred into amniotic fluid to be savored by her fetus. After birth, humans learn about foods through flavors transferred in mother's milk. Amniotic fluid and breast milk share flavor profiles with foods the mother eats. The flavors experienced are unique for each infant and characteristic of their family's dietary culture.

In humans, the flavor of breast milk varies from one mother to another and from one day to the next, depending on the foods eaten in a meal. When mother eats a variety of foods, breastfeeding – unlike monotonous formula feeding – can provide an infant with a rich source of sensory variety.[51] As a result of early and varied flavor experiences, breastfed infants are more willing to eat similarly flavored foods at weaning. Assuming the mother's diet is diverse, varied sensory experiences with food flavors in mother's milk may explain why children who are breastfed tend to be less picky and more willing to try new foods during childhood. Because flavor variety is associated with greater variety in the nutritive content of foods, an infant's preference for varied flavors should ultimately increase the range of nutrients consumed and the likelihood of achieving a well-balanced diet. In other words, the variety effect reflects an important adaptive mechanism in the regulation of food intake among humans.

Traditionally, humans developed social norms and rituals around food gathering, cooking to increase digestibility of fiber and decrease adverse effects of secondary compounds, and preferences for different foods.[52] Nowadays, most people in the United States are descendants of ancestors who came from different parts of the world. They evolved in different habitats, eating different diets. In *Why Some Like It Hot*, ethnobotanist Gary Nabhan describes how interactions among genes, organisms, and environments enable people to use different foods, unique to landscapes they inhabit. Many food sensitivities, once misunderstood as genetic disorders, are adaptations our ancestors evolved in response to the dietary choices and diseases they faced in landscapes where they evolved. They are liabilities only when people are "out of place," eating globalized diets that lack particular mixtures of phytochemicals that activated adaptive responses in their ancestors. Today, we are experiencing what happens when peoples of different ethnicities move to different foodscapes and create new food cultures. And with any diet, there is a broader context that isn't imported with the diet, as Nabhan points out:

> *I realized just how interconnected all the elements of the Cretan cuisine really are. The mix of plants at the islanders' disposal — we cannot replicate it in too many places around the world. The powerful antioxidant combination of olive oil and greens — we cannot ensure that other ethnic populations will take to these foods culturally as the Cretans do. The dedication to religious fasting, a time-tried way to get clean with God, and clean with our bodies — not many Americans have the spiritual will to fast more than a hundred days a year. And finally, the gene frequencies the Cretans carry — Cretans did not gain their tolerance to imbibing large doses of olive oil, or to resisting malaria through fava beans . . . overnight.*

## Lessons from Clara's Kids

Clara Davis's studies of humans illustrate how culture, availability of wholesome alternatives, and flavor-nutrient learning can enable health through nutrition.[53] During the initial course of sampling each of the many foods, the infants in her studies quickly developed preferences. They ate several foods in any meal and often preferred brains, raw beef, bone jelly, and bone marrow – foods disgusting to adults who have not learned to eat them.

No two children ever ate the same combinations of foods and no child ever selected the same mix of foods each day. Davis noted,

> *For every diet differed from every other diet, fifteen different patterns of taste being presented, and not one diet was the predominately cereal and milk diet with smaller supplements of fruit, eggs and meat that is commonly thought proper for this age. To add to the apparent confusion, tastes changed unpredictably from time to time, refusing as we say "to stay put," while meals were often combinations of foods that were strange indeed to us, and would have been a dietitian's nightmare – for example, a breakfast of a pint of orange juice and liver; a supper of several eggs, bananas and milk. They achieved the goal, but by widely various means, as Heaven may presumably be reached by different roads.*[54]

Davis suggests their patterns of selection developed due to "sensory experience, i.e., taste, smell, and doubtless the feeling of comfort and well-being that followed eating, which was evidenced much as in the breast-fed infant." Selected in appropriate combinations, these foods provided biochemical richness children needed to thrive, though Davis's kids could have eaten themselves sick even with this assortment of foods. Had they selected only meat, fish, and eggs, they'd likely have come down with scurvy. Had they been fanatical vegans and eaten only fruits and vegetables, they'd likely have experienced a vitamin $B_{12}$ deficit and megaloblastic anemia. However, no child ever developed a deficiency. Actually, they selected foods that ameliorated diseases such as rickets, which some of them had when they entered the orphanage. Indeed, they all grew normally and were in fine health, as verified by the attending physician.

In a commentary on Davis's studies published in the *Canadian Medical Association Journal*, investigative journalist Stephen Strauss points out that early in the twentieth century a battle was raging between science-infatuated pediatricians and defiant and ostensibly unscientific children.[55] Armed with data from nutrition science, doctors were prescribing what, when, and how much a child should eat, advice most children couldn't stomach. One physician estimated 50 to 90 percent of visits to pediatricians' offices were by mothers worried about their children's refusals of the prescribed diets. Some doctors answered hunger strikes by declaring war on kids. In the 1926 edition of his popular book *The Normal Child*, Alan Brown, at the time the head of pediatrics at the Hospital for Sick Children in Toronto, advised mothers to put children on a starvation diet until they acquiesced to doctor-approved regimes. Over time, doctors began recommending babies be allowed to choose their own diets. The world's most famous baby doctor, Benjamin Spock, argued in *Baby and Child Care* that Nature's way – allowing baby to choose from wholesome foods – was easier on mothers and babies.

Davis planned to do a follow-up study comparing self-selection by children offered a choice of natural and processed foods, but the Depression dashed this hope after a lack of funding ended the original study in 1931. Unfortunately, as we now know, the experiment that Davis envisioned was conducted en masse during the latter half of the twentieth century, with dire effects on public health. As a result, nowadays, many offspring are "prepared" in utero to eat a highly processed diet, even to the extent of being born with a suite of metabolic disorders.[56] Maternal obesity and diabetes during pregnancy increase the risk of obesity and type 2 diabetes in offspring.[57] Insulin-producing cells in the pancreas of a fetus of a diabetic mother are stimulated to grow in size and number due to high levels of blood sugar in mother's diet. That, in turn, causes the fetus to produce more fat, which causes fat babies in a malicious cycle. The greater a woman's weight gain during pregnancy, the higher the risk her child will be overweight by three years of age and will continue to be overweight into adolescence and adulthood.[58] With each passing generation for the last century, these maternal effects have accumulated, and they have led to the twin-epidemics of obesity and type 2 diabetes.

That's happening globally, as illustrated by the Tswana peoples of the eastern Kalahari. They thrived on wild plants during eight years of drought, from 1965 to 1973, when a drought of similar intensity and timing killed more than two million people in the West African Sahel.[59] Tswana elders know more than 200 edible wild plants, and they use an additional 100-plus species in medicine, magic, construction, thatch, dye, and even as props to teach moral lessons. Sadly, their children no longer know the plants in their tribal lands. Children and young adults express little interest in plant lore, so knowledge of wild plants is not being passed to future generations. Most children are eager to adopt Western cultural behaviors, dress, and fast food. A time will come when a drought of severe magnitude will devastate Tswana peoples, for the same reasons it had the peoples in the West African Sahel. Many Tswana will starve in the midst of food plenty, unable to recognize which plants to use for food and medicine.

Historically, people in higher income nations ate greater quantities of refined carbohydrates than people in lower income nations. In those richer nations, people with lower incomes ate more processed foods than people with higher incomes.[60] In the short term, people with lower incomes maximize intake of energy-rich foods at low dollar cost. In the longer term, though, society pays the price. Obesity and diet-related diseases now cost $123 billion annually in the United States, half of which is paid by tax dollars through Medicare and Medicaid. These patterns and associated costs are increasingly common globally. Overweight and obesity, estimated to affect 1.5 billion adults worldwide in 2008, are projected to be 2.16 billion (overweight) and 1.12 billion (obese) by 2030.

Because obesity rates reflect unequal distribution of incomes and wealth, ending obesity is linked with curtailing poverty. Nourishing foods can be inexpensive, but people on a limited budget will struggle to eat healthier foods unless they learn to embrace unfamiliar foods and eating habits. To do so, they will have to eat foods that seem less palatable to them; foods that due to past experiences can score low on taste, variety, enjoyment, and convenience. Because humans learn what to eat in utero and early life, policies must include ways to change people's acquired preferences. I explore this topic of policy more in chapter 12.

Not unlike the Tswana elders, Davis became "the culture" as she enabled "her" kids to thrive by offering them wholesome foods. As she wrote,

> *By this time you have all doubtless perceived that the "trick" in the experiment (if "trick" you wish to call it) was in the food list. Confined to natural, unprocessed and unpurified foods as it was, and without made dishes of any sort it reproduced to a large extent the conditions under which primitive peoples in many parts of the world have been shown to have had scientifically sound diets and excellent nutrition. . . . The results of the experiment, then, leave the choice selection of the foods to be made available to young children in the hands of their elders where everyone has always known it belongs.*

Unfortunately, the foods in supermarkets and the choices people learn to make are nowadays under the control of "elders" in academic, corporate, and political establishments. These institutions have inhibited discourse, shaped and skewed the scientific literature, manufactured and magnified scientific uncertainty, and influenced government policies to advance dangerous "food" products. The emphasis in science and marketing has enabled some in science, industry, and politics to influence the masses and reap the profits. Is it any wonder why something as simple as eating, something bacteria, fish, birds, mammals do each day – without a bit of thought – has become so problematic for humans?

In coming to rely on advice from authorities, we fail to consider a crucial point, one the body of every wild insect, bird, fish, and mammal who ever roamed the planet comprehends from personal experience: The body was the first molecular biologist, geneticist, physiologist, nutritionist, pharmacist, and physician. A healthy body knows what to do with wholesome foods. Appreciating that simple insight could change recommendations – from an endless stream of latest advice on what and what not to eat to creating cultures that know how to grow and combine wholesome foods into meals that nourish and satiate, as Clara Davis did as "mother" to those fifteen orphaned infants in Canada nearly 100 years ago.

# PART IV

# Grappling with Uncertainty

# CHAPTER 12

# How to Poison a Rat, Cow, or Human

It's monsoon season in New Mexico and Arizona. Here in the mountains of Colorado, that translates into afternoon and evening thunderstorms. We went to sleep last night to the patter of rain, flashes of lightning, and thunder rumbling through the mountains. We awoke to the fresh scent of rain-drenched earth and the sight of raindrops perched on the leaves and flowers of red paintbrush, cyan blue penstemon, orange hymenoxys, yellow rabbitbrush, and blue-green sage, as well as the less showy, but no less remarkable, purple flags of blue gamma grass, green florets of Junegrass, and twisted stalks of oat grass and squirrel tail.

The antelope, dining on this feast for our eyes and their bodies, are well adapted to eating the mix of plants now on display, a smorgasbord of more than fifty species of grasses, forbs, and shrubs. The body of the antelope turns phytochemicals into cells, with little concern about toxicity. Their bodies can detoxify phytochemicals in this potentially deadly mix of plants. No doubt, the antelope have also learned how much of each plant to eat and which combinations of plants to eat – for appetizers, main courses, booster phases, and desserts – to best savor a meal.

That's not the case with stocker cattle that summer here. These year-old animals are shipped to the backwoods from all over the United States. Their bodies are not adapted anatomically or physiologically to living on these foods nor have they learned from their mothers how to forage in these landscapes. Thus, they are disadvantaged when it comes to dealing with this attractive but potentially deadly mix of plant life. Take, for instance, the creamy white flowers perched above the pale green leaves of locoweed, a beautiful but potentially deadly herb.

Locoweed is the name for a large number of species that produce an alkaloid called swainsonine, a phytochemical that can harm livestock and wildlife.[1] Poisoning from locoweed is most likely to occur in spring, when locoweed is one of the first, and most nutritious, plants to appear. Animals poisoned from eating locoweed develop a medical condition known as locoism, the most widespread poisonous plant challenge in the western United States. Locoism is reported most often in cattle, sheep, and horses but has also been reported in elk and deer.

Locoweed is named for the neural damage it causes and the ensuing crazy behaviors animals exhibit. Molecules normally processed by the enzyme alpha-mannosidase accumulate in cells when herbivores eat locoweed. This buildup creates holes (vacuolation), especially in neurons in the brain and epithelial cells that line the cavities and surfaces of organs and glands all over the body. The vacuolation resolves once animals stop eating locoweed, if they haven't eaten too much. If they have, vacuolation destroys cells and the damage is irreversible.

Giving livestock a mild dose of lithium chloride to induce nausea quickly teaches them not to eat locoweed, just as it taught sheep and goats in my early research studies to avoid various forages.[2] The puzzling question is why, given the harmful effects of swainsonine, don't cattle, sheep, and horses, as well as elk and deer, learn not to eat locoweed?

## How to Poison in Principle and Practice

Imagine you want to poison a creature through the food it eats. You face two challenges. First, the poison must be tasteless and odorless, not alerting the victim to any possible danger lurking in the food. Animals are cautious of familiar foods that have unfamiliar flavors, and they sample only small amounts of unfamiliar foods. If they get sick after a meal of unfamiliar and familiar foods, they subsequently avoid eating the unfamiliar food, but not the familiar foods.[3]

Second, the poison will work better if the immediate consequences of eating the poisoned food are positive – due to feedback from nutrients in the food – and the lethal effects are delayed for many hours or even for days or weeks. Turns out, that deadly combination is pervasive.

Locoweed meets both these challenges, resulting in the self-destructive foraging behavior manifest in the stocker cattle. Cattle are enticed to sample locoweed early in the growing season when it is one of the more abundant green forages to grace the landscape. When cattle first sample small amounts of locoweed, they aren't deterred by any aversive feedback. In fact, the immediate positive consequences from energy and protein in the lush spring growth condition a strong preference for locoweed. The harmful side effects don't occur for many days or weeks, which makes associating food ingestion (locoweed) with aversive consequences (neural damage) very difficult. By the time the harmful effects are apparent, it's too late for animals to learn that locoweed was the cause of their condition. They are caught in a feedback trap.

Warfarin is a well-known poison that also meets the challenges. Warfarin is used as an anticoagulant (blood thinner) in human medicine, and it is also used for controlling rats and mice in residential, industrial, and agricultural areas. Odorless and flavorless, warfarin is effective when mixed with nutritious food bait because rodents return to the bait and continue to feed over a period of days until a lethal dose accumulates in their bodies. Warfarin can also be mixed with talc and used as a powder, which is applied in areas where the animals are active. The powder accumulates on skin and fur, and the animals consume it during grooming.

As is often the case with plants and animals alike, rats have become resistant to warfarin, so humans have created new poisons of much greater potency. Other 4-hydroxycoumarins used to kill rodents now include coumatetralyl and brodifacoum, referred to as super-warfarin. They are more potent and longer acting, hence more effective – at least for now – even in rat and mouse populations resistant to warfarin. Unlike warfarin, which is readily excreted, newer poisons accumulate slowly in the liver and kidneys, enhancing their delayed effect.

In his diaries, Napoleon Bonaparte's valet, Louis Marchand, wrote detailed descriptions of Napoleon in the months before his death. Those descriptions led Sten Forshufvud to argue that Napoleon died from chronic arsenic poisoning. Forshufvud, a Swedish dentist and physician, was an amateur toxicologist who formulated the controversial theory that Napoleon was poisoned by a member of his entourage while in exile. Arsenic was the poison of choice historically because it is a harmless-looking powder that resembles flour or sugar at a quick glance. As a result, it is virtually undetectable in hot food and drink – and fatal in even small doses.

The success of criminal poisoning depends on imitating the symptoms of a natural disease. The clinical signs of acute arsenic poisoning – vomiting and diarrhea – were easily mistaken for those of common diseases such as food poisoning, dysentery, and cholera. If you are eating a meal and begin to feel ill, how do you know you aren't being poisoned? Your friends and relatives all smile kindly upon you. The meal looks fine, so how can you possibly know there is arsenic in the curry? Of course you can't and the notion was terrifying historically. The analog to arsenic nowadays is refined carbohydrates such as sugar. Eaten in small amounts over time (chronically), both arsenic and sugar result eventually in systemic organ failures.

In an article titled "How Chris McCandless Died," Jon Krakauer discusses further details of McCandless's death.[4] Krakauer had first written about Chris in his well-known book *Into the Wild*. In April of 1992, Chris ventured alone into the Alaskan wilderness, just outside the northern boundary of Denali National Park. He hoped to live simply in solitude. He had little food or gear. By mid-August, he was dead. On September 6, 1992, Butch Killian, a local hunter, found his body in his sleeping bag inside the bus he used for a home. He'd been dead for over two weeks. He weighed 66 pounds (30 kg); he'd lost 74 pounds (34 kg) since arriving in Alaska.

Krakauer originally suggested Chris may have died from "rabbit starvation" due to the leanness of the game he was hunting. He also may have suffered from eating toxic seeds. In mid-June of 1992, Chris began eating roots of *Hedysarum alpinum*, a species of plant in the legume (pea) family known by the common name alpine sweetvetch or potato, as Chris called the plant. On July 14, he began to eat the seeds of sweetvetch as well. Two weeks later, on July 30, he wrote in his diary that he was extremely weak, having considerable difficulty standing up, and that he was in great jeopardy, due to eating the seeds of sweetvetch. Two questions arise. First, how did he know the seeds of sweetvetch were causing his maladies? Second, if he associated the maladies with the seeds, why didn't he stop eating the sweetvetch?

On the one hand, Chris knew the seeds were the culprits because bodies associate internal malaise with novel foods, and the seeds of alpine sweetvetch were the new food in his diet. Like goats eating more of one twig type than another, he was also eating these seeds in large amounts relative to other foods, another clue for his body. On the other hand, he was confused by opposing signals. The immediate positive consequences from the energy and protein provided by the seeds were conditioning a strong preference. But the effects of other toxins – including perhaps the neurotoxin beta-N-oxalyl-L-alpha-beta diaminopropionic acid – in the seeds were delayed in time. Chris was caught in a feedback trap, the dire consequences manifest only after the toxin had its effect. After eating enough sweetvetch, a person becomes crippled. The more seeds eaten, the poorer a person's physical condition. No different from herbivores, the less nutritious the diet, the more adverse the toxicological consequences. Chris eventually figured out the source of the malaise, but it was too late.

Beyond these issues lurked more fundamental cause. Chris lacked the cultural foundations that would have guided his selection of appropriate foods to nurture his body in this landscape. He was conceived and reared in utterly different settings. In form (expressed morphology), function (expressed physiology), and behavior (which foods to select), his body was ill equipped to deal with the challenges he faced in the unfamiliar backcountry of Alaska.

## Mixed Signals from the Skin and the Gut

When I began to study the responses of livestock to foods, I wondered what they learn from different experiences. Imagine a calf in a pen of cattle that has just been fed locoweed. I suddenly walk up to the calf, grab it, and put a balling gun down its throat to deliver a capsule of lithium chloride into its stomach. The calf soon will experience nausea, but will it associate the nausea with the locoweed or with me? If I have a shock collar on the calf, and I shock the calf when it eats the locoweed, will it associate the shock with the locoweed or with me?

Animals have two very basic defense systems – gut defenses and skin defenses. As preeminent psychologist John Garcia wrote, "All animals have evolved coping mechanisms for obtaining nutrients and protective mechanisms to keep from becoming nutrients."[5] How these two defense systems work is illustrated when hawks are fed distinctively colored or flavored mice. When hawks normally fed white mice are given a black mouse and then injected with a toxin, the hawks afterward avoid eating both black and white mice. Why? Because black and white mice *taste* the same, and hawks don't discriminate between mice as a food item based on color. On the other hand, if a distinctive taste is added to black mice, but not to white mice, hawks learn to avoid black mice on sight after a single black mouse–toxicosis event. Hawks' food preferences are based on taste and postingestive feedback (nausea).

Garcia and his colleagues showed that animals made ill following simultaneous exposure to taste and audiovisual stimuli show much stronger aversions to taste than to audiovisual stimuli. If you eat a food, watch a movie, and then get nauseous, you will associate the nausea with the food, not the movie. The taste of food and sensations of nausea and satiety are part of the gut-defense system evolved in response to toxins and nutrients in foods.

On the other hand, if animals receive foot shock following concurrent exposure to taste and audiovisual stimuli, they show stronger aversions to audiovisual cues than to taste cues. If you eat a food and watch a movie that makes you sad, you will associate sadness with the movie, not the food. Auditory and visual stimuli and sensations of pain (or satisfaction) are associated with the skin-defense system, evolved in response to predation and other aversive (or positive) external stimuli.

Odors are readily associated with both skin- and gut-defense systems. The odors of different foods serve as cues for the gut-defense system, while the odors of different predators forewarn the skin-defense system of possible danger.

The same kinds of responses have been demonstrated for food aversions and place aversions. Nausea caused by toxins in foods decreases palatability, but that doesn't necessarily cause animals to avoid the location where they ate the food. Conversely an attack by a predator will cause an animal to avoid the place where it had been eating when attacked, but that doesn't decrease the palatability of the food it was consuming at the time.

Like hawks and mice, cattle, sheep, goats, and horses given a capsule of lithium chloride with a balling gun will associate the attack to its skin-defense system with the person who gives them the capsule, but they will associate the food it just ate with the lithium chloride-induced toxicosis and its gut-defense system. The automatic pairing of foods with postingestive feedback means that even if I could have explained to that calf that the lithium chloride – and not the food – was the cause of its toxicosis, the animal would still be averse to the food. As we discussed in chapter 4, the gut-defense system links food with feedback regardless of what an animal "thinks" caused the illness.

Mixed signals between skin defenses and gut defenses are why humans don't acquire aversions to foods that cause allergies. The most common food allergens – milk, eggs, tree nuts, fish, shellfish, soy, and wheat – account for nearly 90 percent of all allergic reactions. On the one hand, people derive immediate positive postingestive consequences from the rich sources of nutrients in these foods, which cause satiety, not nausea. People continue to eat foods that cause allergies because allergic responses consist of dermatitis and respiratory distress, including life-threatening anaphylactic responses, maladies a body doesn't associate with food. These responses occur when the body's immune system mistakenly identifies a protein in food as harmful. When some proteins or fragments of proteins aren't digested, they are then tagged by immunoglobulin E (IgE). These tags trick the immune system into thinking the protein is harmful. The immune system, thinking the body is under attack, triggers an allergic reaction.

## Hazards of Foraging in Supermarkets

The challenges herbivores face when toxicity is delayed days, weeks, or even years aren't that different from those encountered by people foraging in the aisles of supermarkets with 600,000 items all attractively displayed. While many of these foods appear to be nutritious, they are slow-acting toxins – as deadly as arsenic, locoweed, warfarin, or sweetvetch – that deliver mixed signals: immediate positive flavor-feedback consequences but delayed illness.

The instant gratification and delayed consequences are catching up with us. The Centers for Disease Control and Prevention estimates 70 percent of people in the United States will die of a diet-related disease. We are slowly killing ourselves with the foods we've learned to eat in utero and early in life – no different from stocker cattle or rats slowly dying from eating foods laced with swainsonine, warfarin, or arsenic, or Chris McCandless slowly dying from chronic toxicity.

Consider the immediate benefits versus the delayed costs of eating meats preserved with sodium nitrite. Sodium nitrate is used to cure meat because it prevents bacterial growth. It also gives meat a desirable dark red color, a result of a reaction with myoglobin. Due to its toxicity – the lethal dose is about 22 mg/kg/body weight – the maximum allowed nitrite concentration in meat products is 200 ppm. Under some conditions, especially while cooking, nitrites in meat can react with degradation products of amino acids, forming nitrosamines, which are carcinogens. The immediate positive consequences (nutrients) of eating meats preserved with nitrite are offset by delayed aversive effects (cancer). To further complicate the issue, beets, cabbage, carrots, celery, radishes, and spinach are all high in sodium nitrate. People who eat large amounts of these vegetables as well as cured meat have a greater risk of developing cancer.[6]

Processed foods also provide immediate positive consequences because the fats and refined carbohydrates they contain give us an energy boost. However, overconsumption of fats and refined carbohydrates can lead to obesity, diabetes, and diet-related diseases over time. When we extract and purify compounds to create foods that are highly reinforcing, we thwart the ability of a body to eat in moderation: We put the accelerator to the floor and remove the brakes. The overwhelming reinforcement from amplifying feedback signals leads to addiction with any kind of behavior, including ingestion of energy-dense foods that are rapidly digested and absorbed into the body. In that sense, energy-dense processed foods share pharmacokinetic properties – concentrated dose and rapid rate of absorption – with drugs of abuse, a result of the addition of fat or refined carbohydrates and the rapid rate that refined carbohydrates are digested and absorbed into the body, as indicated by glycemic load.[7]

Processed food products such as sugar-laden chocolate bars and ice cream, as well as fat-laden French fries, pizza, cookies, chips, cake, buttered popcorn, muffins, and sodas, have high concentrations of refined carbohydrates and fats. Homemade versions of these are often equally loaded with carbohydrates and fats. Many of these products also contain ingredients, such as white flour, from which the health-promoting and satiating phytochemicals and fiber have been removed, which further accelerates the rate at which they are absorbed by the body. For instance, the sugar in a highly processed milk chocolate bar (high glycemic load) will be more quickly absorbed into the body than the natural sugars in a banana (low glycemic load). This is because the banana is unprocessed, and while it contains sugar, it also has a host of phytochemicals along with fiber, protein, and water, which slow the rate sugar enters the bloodstream.

In 2008, people in the United States consumed over 28,000 grams (60 pounds) of added sugar per person per year, and this didn't include fruit juices.[8] The average intake per person was 76.7 grams per day, which is equal to 19 teaspoons of sugar (assuming 4 grams of sugar per teaspoon) or 304 calories (4 calories per gram of sugar). Sugar intake declined by 23 percent from 2000 and 2008, mainly because people drank less sugar-sweetened beverages. Those levels of intake (59 grams per day) are still more than the 24 to 48 grams per day some recommend.

This brings us back to one of the phenomena that underlies nutritional wisdom: The presence of phytochemicals in whole foods limits how much of any one food we can eat. That enhances health through nutrition by exposing a body to a wide array of phytochemicals, thus enabling cells to pick and choose from a variety of compounds. With energy-dense processed foods, the appetite-stimulating effects of variety aren't offset by the appetite-moderating effects of phytochemicals. Thus, a person can easily devour sixty grams or more of sugar in a processed food. With *unlimited access to vast arrays* of foods and drinks high in refined carbohydrates, *people never satiate.* A twelve-ounce can of Mountain Dew alone has 46 grams of sugar. Few of us will consume the equivalent of six small apples in a day, let alone in a meal.

## An Example of Delayed Onset of Poisoning

Diabetes was one of the first diseases to be described. An Egyptian manuscript from 1500 BCE mentioned "too great emptying of the urine," and Indian physicians at that time classified the disease as *madhumeha* or "honey urine," noting the urine attracted ants. The term *diabetes* (to pass through) was first used in 230 BCE by the Greek Appollonius of Memphis. The disease was rare in the Roman Empire. The physician Galen saw only two cases in his career.

The classical symptoms of high blood sugar are increased hunger (polyphagia) and thirst (polydipsia) and frequent urination (polyuria). The first described cases are believed to be type 1 diabetes, considered an autoimmune disease, resulting from an immune system attack on the pancreatic beta cells that produce insulin – the hormone that helps cells absorb glucose. Type 2 diabetes, formerly called adult-onset or non-insulin-dependent diabetes, can develop at any age. It often becomes apparent during adulthood, but type 2 diabetes is rising in children.

Diabetes is a group of metabolic diseases that occur when a person has high blood sugar (glucose), either because the body doesn't produce enough insulin or because cells have become resistant to insulin. Diabetes is chronic and can rarely be cured. Managing diabetes means maintaining blood glucose levels as close to normal as possible, without causing hypoglycemia. This can best be accomplished with diet and exercise. Ironically, and in contrast to most humans, rats rendered diabetic voluntarily choose diets devoid of carbohydrates, consuming only protein and fat, as discussed in chapter 1.[9] Why can't a human who has diabetes figure that out?

Type 2 diabetes develops gradually and the symptoms are typically subtle or absent. According to a Center for Disease Control and Prevention report issued in 2015, 30.3 million Americans have diabetes and another 84.1 million have prediabetes. More than 50 percent of American adults are prediabetic or diabetic, but only about one in eight people with prediabetes knows they have a problem. Nearly one in four adults who are living with diabetes – 7.2 million Americans – don't know they have the condition.

The incidence of type 2 diabetes in developing countries is following the trend referred to as the Western diet and lifestyle.

Like other feedback traps, people are unlikely to relate the slowly evolving symptoms of diabetes with their eating choices. They've learned to like the flavors and feedback from the foods high in fats and refined carbohydrates that they've been consuming. Their body no longer suspects the familiar food as the culprit, unaware of the consequences, no different from the poor old cow eating locoweed or Chris McCandless ingesting sweetvetch.

All forms of diabetes increase the risk of long-term complications that typically develop after ten to twenty years. They usually affect the skin defense system more than the gut defense system. Diabetes doubles the risk of cardiovascular disease. Diabetic retinopathy, which affects blood vessel formation in the retina of the eye, impairs vision and leads to blindness. Diabetic nephropathy, the impact of diabetes on the kidneys, scars kidney tissue. That leads to loss of progressively larger amounts of protein in urine and chronic kidney disease, requiring dialysis. Diabetic neuropathy, the impact of diabetes on the nervous system, causes numbness, tingling, and pain in the feet and also increases the risk of skin damage due to altered sensation.

## Pathways to Addiction

Let's return to the question of why humans who have diabetes don't instinctively stop eating sugar. People once chewed raw sugarcane for its sweetness. But chewing sugarcane, or eating other fruits with a rich array of phytochemicals, is markedly different from eating foods high in sucrose, high-fructose corn syrup, or fructose-containing sugars. Fructose in fruit is bound up in a complex matrix of primary and secondary compounds that are digested and absorbed much more slowly than the fructose in high-fructose corn syrup in a soda.

Like munching on sugarcane or eating fruit, chewing coca leaves is mildly reinforcing, but people don't become addicted to coca leaves or to fruits.[10] However, when the alkaloid cocaine is extracted from coca leaves and then purified, the dose and rate of release of cocaine are both greatly magnified.[11] In this way, drugs like cocaine, heroin, morphine, and methamphetamine hijack reward systems in the brain by acting directly on opioid and dopamine receptors. Dopamine is a neurotransmitter that is active in reward and pleasure centers in the brain. It appears to mediate desire and motivation, however, more than pleasure.

Excessive intake of highly palatable foods shares similarities with the effects on brain and behavior seen with some drugs of abuse.[12] In recognizing these parallels, some scientists are cautious about the notion of "food addiction" because appetite and eating food are necessary to human survival, whereas drugs of abuse are not.[13] However, while humans need food to survive, they don't need excessive amounts of the highly palatable foods now common in our diets.

Four components of addiction – bingeing, withdrawal, craving, and cross-sensitization – have been demonstrated behaviorally with sugar.[14] These behaviors are related to neurochemical changes in the brain that occur with addictive drugs and include changes in dopamine and opioid receptor binding, certain kinds of RNA (enkephalin mRNA) expression, and dopamine and acetylcholine release in the nucleus accumbens, a site involved in addictive behavior. When rats binge on foods high in sugar and fat, they experience changes in the dopamine system akin to drugs of abuse, but they don't exhibit signs of opiate-like withdrawal, which suggests sugar and fat may play distinct roles in addictive potential of processed foods.[15]

For better or worse, addiction to food may have an epigenetic basis.[16] When pregnant rats are maintained on a highly palatable processed diet, the offspring show an increased preference for sugar and fat and changes in dopamine and opioid gene expression in reward-related brain regions. Diets high in sugar and fat alter mesolimbic reward pathways in adults, and in utero exposure to a highly palatable diet results in similar alterations in brain-reward functioning in offspring.

Researchers once believed people who become addicted to drugs, alcohol, and foods and to activities like sex and gambling are *more sensitive* to the highly rewarding consequences of dopamine release in the brain. The view was that the more dopamine released, the more desirable (pleasurable) the activity. However, the opposite is true. People who become addicted have *fewer* dopamine receptors so they are *less* sensitive. Their brains have decreased (down regulated) the number of dopamine receptors so that *more* dopamine is required to produce an effect. People who are *initially overly sensitive*, and hence more likely to become addicted, require ever-increasing amounts of a drug or activity to achieve the same level of reward because they have *fewer* dopamine receptors. So, some people may initially overeat because they are born with a highly sensitive dopamine system that gradually becomes less sensitive as they overeat.

That's one reason why addicted people gradually lose the ability to moderate their behaviors. Repeated exposure to addicting drugs and foods disrupts the balance between the limbic system (the part of the brain involved in reward-seeking behaviors) and the frontal cortex (the part of the brain responsible for moderating behaviors). Neurons whose primary neurotransmitter is dopamine are present in various parts of the limbic system. Those neurons interact with neurons in the frontal cortex, which exerts control over behaviors. The frontal cortex, which contains most of the dopamine-sensitive neurons in the cerebral cortex, helps us anticipate the consequences of our actions, choose between different actions, and override and suppress unacceptable social behaviors. The frontal cortex also plays a key role in retaining longer term memories associated with emotions derived by input from the limbic system.

What we refer to as self-control is mediated by a complex network in the brain. Self-control can be impaired by disturbances in any number of regions in this neural network. The ventromedial prefrontal cortex (vmPFC) helps the brain calculate values for different options. As we choose among foods, our brain integrates characteristics of each option. We then choose the one with the highest immediate value. To know what matters most at the moment when we make the choice, the vmPFC exchanges information with other areas in the brain that are also responding to the different aspects of the options we are contemplating.

When we're stressed, we choose short-term gains over long-term costs. In a study of fifty-one people who were trying to maintain a healthy lifestyle, twenty-nine were subjected to moderate stress.[17] People were then asked to choose between eating a tasty but unhealthy food or a healthy but relatively less tasty item. Those who were stressed were more likely to select an unhealthy food, despite being reminded which foods were better for them during the experiment. Consuming more palatable food may diminish stress in humans.[18] The decision to risk long-term gain for a short-term benefit is best understood from an evolutionary perspective: Stress prepares the body for fight or flight. Long-term goals take a backseat in stressful situations.

During a stressful situation, communication increases between reward regions of the brain – the striatum and amygdala – and the vmPFC. As stress increases, regions of the dorsolateral prefrontal cortex involved in goal maintenance and long-term planning communicate less effectively with vmPFC. While people still care about eating healthily, when stressed, they can't resist if presented with a strong temptation such as their favorite chocolate bar.

We've enabled pathways to addiction by the food systems we've created. For example, the ecological role fruit plays in farming and food systems is different now from in our hunter-gatherer days. By providing fruit, plants help animals store energy, as the liver transforms fructose into the fat they will use during the dormant season when energy is not readily available. By eating fruit, animals help plants disperse seeds across landscapes and their feces provide safe sites for seeds to germinate. The fruits hunter-gatherers and wild animals procure in nature differ from those humans purchase in grocery stores because plants package and dispense fruit differently from humans. Nature provides fruit seasonally; grocers sell fruit year-round. Animals must work to obtain fruit in nature; fruit is easily obtained at the grocer. Nature grows fruit rich in phytochemicals; people grow fruit in ways that promote growth and ease of shipping to the detriment of phytochemical richness. Finally, the sugar fructose in plants is much diluted compared with that people add to processed foods and beverages we have come to crave.

Compared with eating foods or drinking beverages high in glucose, drinking beverages high in fructose *lowers* circulating levels of satiety hormones, such as insulin and leptin, and *raises* levels of hunger hormones like ghrelin.[19] *Decreases* in circulating levels of insulin and leptin and *increases* in ghrelin can increase calorie intake and contribute to weight gain and obesity during chronic consumption of foods or drinks high in sugar or high-fructose corn syrup. Moreover, fructose, unlike glucose, doesn't cross the blood-brain barrier and may contribute to increased energy intake because it doesn't activate glucose sensors in the brain. The 50-50 mix of glucose and fructose in table sugar and high-fructose corn syrup encourages fat synthesis. Fructose stimulates the liver to produce triglycerides, which are stored in fat cells, while glucose stimulates insulin secretion and fat storage. Unlike glucose, fructose doesn't stimulate the pancreas to produce insulin. Fructose appears to behave more like fat than carbohydrates with respect to insulin secretion, leptin production, and postprandial triglyceride levels.[20]

## Prisoners of an Unhealthy Food System

As illustrated throughout this chapter, many of our unhealthy behaviors are shaped by immediate positive consequences and delayed negative consequences. Munching a few more brownies, sipping a sixty-four–ounce soda, and buying them on credit all lead to immediate positive consequences – it's just a few extra calories and a little more money. The immediate gratification that comes from having it all now enables food and credit card companies to make fortunes to the detriment of people who pay massive "interest rates" on "short-term loans."[21]

Sadly, our impulses come to own us. As we ingest processed foods (and pesticide-laden produce), immediate benefits are followed by costs that occur weeks, months, or years later for individuals and societies. Most of the social, economic, and environmental challenges we face arise because we aren't able to learn effectively when the consequences of our actions are *delayed in time* (eating) or *distant in space* (climate change).

Most people don't have a clue about the many ways the food industry controls the contingencies that shape their behavior or how industry hijacks their preferences for wholesome foods as it designs energy-dense, nutrient-poor processed foods. Nothing is left to chance – from the artificial flavors that mimic fruits to the refined carbohydrates that provide reinforcing feedback; from the packaging that entices us to buy the product to the shelf positions and locations in the grocery store aisle. Scientists in the food industry well know how to influence the choices we make. They understand that contingencies shape behaviors and our so-called free will has nothing to do with any of our behaviors, including food selection and ingestion.

As people fall under the spell of processed foods, in many cases now beginning in utero and early in life, they are likely to eat diets high in refined carbohydrates and come to crave carbohydrates as a source of energy. Carbohydrate cravings are established and maintained through chronically high levels of insulin, which signals a body to use glucose, rather than fatty acids, for energy.[22] The homeostatic system that once maintained blood sugar in a healthful range now establishes an environment in which cells are primed to use glucose for energy, and only glucose can satisfy demands. High-glycemic index meals decrease plasma glucose, increase hunger, and stimulate brain regions associated with reward and craving following a meal.[23]

Without knowledge of what's happening, we are helpless in the face of foods that behave as slow-acting poisons, which is bad enough in itself, but then add a compounding effect that some of those same foods also have an addictive quality. So we really do end up as prisoners within a system feeding us a diet that will probably shorten our life spans and definitely will cause us poorer health and lead to even greater financial burden in health care.

Unlike other causes of preventable death and disability – tobacco, infectious diseases, and injuries – the obesity epidemic has yet to be addressed or reversed by public health measures in any populations.[24] This helplessness increases the urgency for policies that create contingencies (incentives) to change supply-side behaviors. When shaping policies to prevent "poisoning," government officials must admit that the occurrence of obesity and diabetes isn't a failure of individual willpower to resist energy-dense foods high in fats and refined carbohydrates. Rather, it is a lack of political will to confront powerful companies that increasingly control human food webs. Agribusiness is a global industrial complex run by a handful of huge multinational corporations that control the food chain from production of seeds, feeds, and chemical fertilizers and pesticides to processing, marketing, and distribution of food products.

To counter obesity and diet-related diseases, legislators would have to prioritize the well-being of the public over that of corporations by changing the laws that incentivize corporations to create harmful foods. An example from history shows how that could be accomplished. In the eighteenth century, ship captains were paid for each prisoner who walked *onto* their ships for the trip from England to Australia. Death losses were high, averaging 12 percent from 1790 and 1792, with a loss of 37 percent on one voyage. Casualties were due to overcrowding and poor nutrition. Ship captains received many pleas appealing to their moral duty to care for prisoners. But these supplications had no effect on death rates. Finally, someone suggested changing incentives: Pay the ship captains for the number of prisoners who walked *off* their boats in Australia. The results were remarkable. Three ships carrying 422 prisoners made the trip from England to Australia in 1793 and only one prisoner died. None of the ship captains had become better people, but in response to better contingencies (incentives), they behaved as if they had.

Legislative solutions to change the behavior of "ship captains" – so prisoners walk *off* their boats – would require the kind of collective courage that's difficult to implement at the national level because America is a pay-to-play political system. Lobbyists from industry pay, so they are at the table and they are serving the meals. The result is a political system that's among the most corrupt in the world, but with a twist of genius: Through their infamous decision in Citizens United v. Federal Election Commission, the US Supreme Court has for all practical purposes legalized the corruption and put American democracy on a descending death spiral.[25] And while legislators talk endlessly about reducing the size of budgets, rarely do they discuss how we could save hundreds of billions of dollars annually: Change what and how we eat. The National Diabetes Information Clearinghouse estimates diabetes alone costs $132 billion annually in the United States. We're paying a fortune to treat obesity, diabetes, cardiovascular disease, and some cancers – all preventable through changes in diet and lifestyle.

Rather than leave our health to "ship captains" at national levels, we must take responsibility for our own behavior. Forget the duplicitous top-down influences; at the individual level there are clear-cut choices we can make that will immediately begin to improve our health. We can begin by first understanding the contingencies controlling our behavior.[26]

Recall from chapter 11 how we changed the behavior of welfare elk at Deseret Land and Livestock? We changed the *contingencies* that were controlling their behavior. Over time, we transformed the culture from a herd dependent on processed foods (hay) to one that ate natural forages. There is no way to stress enough that *contingencies shape behavior*. Like elk, people do what they do because of the consequences they experience for their behaviors. Every behavior has a consequence, and the likelihood of a behavior continuing depends on the consequences of that behavior. Understanding how consequences influence behavior is the key to changing behavior. And understanding that contingencies (contexts) shape behavior is key.

To counter foods and food environments designed by industry to entice us to eat the harmful products they sell, we must learn to forage selectively, choosing only wholesome foods at the grocery store. Forget willpower; it isn't effective for changing your behavior. Forget reading labels. If the food product has a label, the creation probably isn't any good for you. Don't bring toxic junk food home. You don't have the willpower to avoid it. Forget diets, they don't work. Avoid locations such as fast-food outlets because they encourage unhealthy eating habits. Instead, surround yourself with wholesome foods that nourish and satiate and eat to your heart's content – within a restricted time-period each day. These changes all counter the addicting influences of processed foods. Over time, one child and one family at a time, from the bottom up, we can wean ourselves off processed foods. In the process, we can create a culture that knows how to combine wholesome foods into meals that nourish and satiate.

# CHAPTER 13

# When Authority Trumps Wisdom

Though I'd eaten fish every Friday from my time in the womb throughout the first eighteen years of my life on Earth, I never liked fish, regardless of how it was prepared. Nor have I even once craved fish, despite its alleged benefits for my health. I know fish is high in many nutrients including protein, iodine, and various vitamins and minerals. The fatty fish I most despise – such as trout, salmon, sardines, tuna, and mackerel – is high in vitamin D, the fat-soluble nutrient deficient in the bodies of some people. I know fatty fish is high in omega-3 fatty acids, crucial for my body and brain to function optimally. Some folks recommend that to meet my requirements for omega-3s, I should eat fatty fish at least once or twice a week. Despite all of this instruction, my body still can't stomach fatty fish, so I don't eat it. What is my body to make of the conflict between *authority* and *personal experience*?

With the release of the eighth edition of the US Government's Dietary Guidelines in 2015, it seems reasonable to consider – in the throes of an epidemic of obesity that has rendered Americans less healthy than ever – if the guidelines can be trusted. Arguably, they've done more harm than good. The goals of the guidelines, first released by the Departments of Agriculture and Health and Human Services in 1980, were to increase Americans' intake of carbohydrates to 55 to 60 percent of caloric intake; reduce fat consumption from 40 percent to less than 30 percent of calorie intake; reduce saturated fat to 10 percent of calories and increase monounsaturated and polyunsaturated fats each to 10 percent of calories; reduce cholesterol intake to less than 300 milligrams a day; reduce salt consumption by 50 to 80 percent; and reduce sugar intake by 40 percent. Except for sugar, from today's perspective, the goals could hardly be more misleading.

Was the impetus for these guidelines an honest desire to improve health? Or were they tainted by corporate influence from the beginning? Why weren't there any guidelines before 1980? Did the government not care about the American diet before then?

The following examples – vitamin D, fat, and salt – reveal a basic insight into our broken relationship with food. Most of us have little meaningful experience of wholesome foods in health. Authority figures tell us what and what not to eat. We don't listen to what our bodies tell us – at a noncognitive, synthetic, intuitive level – because our cognitive, rational, analytical sides are too busy listening to experts interpret the latest findings about diet.

## How Vitamin D Became the Rage

Vitamin D promotes absorption of calcium and phosphorus from the small intestines. Severely low levels of vitamin D can result in the soft, brittle bones referred to as rickets in children. While millions of people assume taking supplemental vitamin D will also help to alleviate depression, fatigue, muscle weakness, heart disease, and cancer, no broadly accepted evidence supports the premise that vitamin D can be used to prevent or treat these disorders.

Like other fat-soluble vitamins – A, E, and K – vitamin D is stored in the liver and fatty tissues for up to six months and generally poses a greater risk for toxicity than deficiency, especially when consumed in excess. A body needs only small amounts of any vitamin. Eating a balanced diet will not lead to toxicity in otherwise healthy individuals, but taking supplements that contain megadoses of vitamins A, D, E, and K may lead to toxicity.

Michael Holick, professor of medicine, physiology, and biophysics at Boston University School of Medicine, is a chief advocate that nearly everybody needs supplemental vitamin D. He cites as evidence studies suggesting an association between low vitamin D levels and higher rates of various diseases. These correlational studies don't prove cause and effect, but he is persuaded by the fact that many studies suggest low blood levels of vitamin D are hazardous.

In a review published in 2007 in *The New England Journal of Medicine*, Holick notes that after foods in the United States were fortified with vitamin D, rickets appeared to be an issue of the past.[1] Nevertheless, he argues, vitamin D deficiency is common, and levels once considered normal – 21 to 29 nanograms per milliliter of blood (ng/ml) – are associated with increased risks of cancer, autoimmune disease, diabetes, schizophrenia, depression, poor lung capacity, and wheezing. In 2011, Holick led a committee of the Endocrine Society that recommended vitamin D levels be increased to at least 30 nanograms per milliliter, which implies most people are deficient in vitamin D.[2] His committee endorsed a daily allowance of 600 IU up to age seventy, and 800 IU for people who are older than seventy. Diet can't provide that much vitamin D, and a body would require nearly constant exposure to sunlight to reach these levels.

So, the rage to take vitamin D began not in natural food stores but in medical journals with research papers that linked vitamin D levels formerly considered normal to multiple sclerosis and mental illness, then to risk of cancer and poor bone health. People who read these findings in the popular press began to believe they were deficient in vitamin D. Requests for blood testing skyrocketed. Vitamin D tests are now part of routine evaluation of patients.

According to the Centers for Disease Control and Prevention, the number of blood tests for vitamin D levels among Medicare beneficiaries, mostly people sixty-five and older, increased eighty-three–fold from 2000 to 2010. Among patients with commercial insurance, testing rates rose 2.5-fold from 2009 to 2014. Labs performing these tests now report normal levels of vitamin D – 20 to 30 nanograms per milliliter of blood – as insufficient. As a result, millions of healthy people think they have a deficiency. Some are taking supplemental doses so high they can be dangerous.

In the ensuing years, however, studies where people were randomly assigned to take vitamin D or a placebo failed to support Holick's claims for the benefits of vitamin D. Holick argued those studies were too small to be definitive, but larger and more rigorous trials have not supported his hypothesis either. One study with over 5,000 participants found that vitamin D didn't prevent heart attacks.[3] Another study of 2,303 healthy postmenopausal women, randomly assigned to take vitamin D and calcium supplements or a placebo, found no evidence vitamin D protects women against cancer.[4] Despite these findings, many people didn't want to hear that advice. They are happy with routine testing for vitamin D. They want to know their status, so many doctors test for vitamin D. Vitamin D has become a religion.

The authors of one study contend the benefits of vitamin D occur for all women.[5] Yet, the margins of error (95 percent confidence intervals) in their study overlapped for nearly every supplement level. Those wide margins reflect the fact that women in the study differed greatly in their responses to supplemental vitamin D, to the point that the differences among treatments weren't statistically significant. Still, the study authors recommend women over fifty-five years of age take supplemental vitamin D, stating that the average person seventy years of age and younger should take 600 IU, while those over seventy years of age should take 800 IU.

Given the great variation among individuals, how much vitamin D should you take each day? What if you are a woman who won't benefit from supplemental vitamin D? Excess vitamin D can be toxic, a result of the buildup of calcium in blood (hypercalcemia). That can cause poor appetite, nausea, and vomiting, as well as weakness, frequent urination, and kidney problems, all symptoms of toxicity from too much vitamin D. That doesn't happen with exposure to sunlight because the body regulates the amount of vitamin D produced by exposure to sunlight. The only way to know your vitamin D status is to have a blood sample analyzed in the lab. But your vitamin D status will vary over time due to vitamin D in your diet and sunlight. So, a one-off measure of your vitamin D status still doesn't tell you how much vitamin D you need.

As usual, lost in the frenzy is any notion that *a body can select foods* to rectify a deficit of vitamin D. The vitamin-D deficiency rickets was common in big cities in the United States in the 1920s and 1930s. Cod liver oil was given as the cure for rickets. Children deficient in vitamin D like cod liver oil, as Clara Davis showed in her studies with children who had rickets, while well-nourished children find the odor and taste of cod liver oil repulsive. As blood calcium and phosphorus levels normalize and rickets subsides, children no longer desire cod liver oil.

But the ability of a body to compensate for the effects of a deficit of vitamin D on calcium and phosphorus absorption is more subtle than that. People deficient in vitamin D compensate by eating more of foods high in calcium. People prefer cream cheese high in calcium when they are deficient in calcium, but they prefer cream cheese lower in calcium when they aren't deficient.[6] When people were offered cream cheese with low, intermediate, or high levels of added calcium, those who weren't deficient in calcium reported the best tasting cheese was the one low in calcium and the worst tasting cheeses were those higher in calcium. Conversely, hemodialysis patients, who have higher needs for calcium, reported that cheese with higher levels of calcium tasted best to them, and cheese with the lowest level of calcium tasted worst. Intake of calcium is higher in lactating than nonlactating women and when women are nursing twins as opposed to a single child.[7] My aversion to fatty fish high in vitamin D explains, in part, my strong liking for cheeses high in calcium.

Like humans, many species of birds and mammals self-select for foods high in calcium when fed a diet deficient in vitamin D.[8] By selecting a high-calcium diet, lactating rats deficient in vitamin D increase plasma levels of calcium and blood levels of phosphate.[9] Their appetite increases, with a doubling of food consumption, they lose less weight and produce more milk, as indicated by improved growth of their rat pups. The lower milk production by vitamin D–deficient rats is due to their low food intake, which is partially dependent on the hypocalcemia that normally occurs with a deficiency in vitamin D.

Interestingly, a body can temporarily rectify a calcium deficiency by increasing the intake of salt (sodium chloride) because sodium liberates calcium from plasma proteins. Rats and lambs deficient in calcium eat more salt, which reduces their need for calcium.[10] Pregnant and lactating women and children with high demands for calcium also eat more salty foods.

## How Fat Became Toxic

For the past fifty years, people have avoided high-fat foods such as butter because of their alleged role in cardiovascular disease. They were taught to choose margarine and other low-fat foods for their health benefits. Yet, if you ask which they prefer – butter or margarine – most people say butter. The rich flavor of butter made from milk of cows foraging on phytochemically rich pastures and the utter lack of palatability of margarine is the wisdom body's way of telling you butter is better. The same is true for the more wholesome flavors of whole milk, yogurt, cheese, and meat compared with their low-fat counterparts. People's experiences stand in stark contrast to recommendations regarding butter, red meats, and cheeses high in fat.

Historically, margarine was endorsed as a substitute for butter because butter is higher in saturated fat (63 percent) than monounsaturated fat (26 percent) or polyunsaturated fat (4 percent). Margarine is higher in unsaturated fat and contains no cholesterol. But we've learned trans-fats in margarines pack a potent insult to a body.[11] They promote inflammation and may increase insulin resistance and triglycerides, and they lower desirable lipoproteins (HDL) and raise undesirable lipoproteins (LDL). Margarine is twice as bad as saturated fat in its impact on ratios of LDL to HDL. Altered partially hydrogenated fats made from vegetable oils also block the ability of the body to use essential omega-3 and omega-6 polyunsaturated fatty acids, which can cause impaired immune system function and sexual dysfunction. Consuming hydrogenated fats is associated with obesity and diabetes, as well as cancer, atherosclerosis, immune system dysfunction, low-birth-weight babies, birth defects, decreased visual acuity, sterility, difficulty in lactation, and problems with bones and tendons. Compared with butter, margarine is truly bad for us. But isn't that what the wisdom of the body was trying to tell us all along?

Nonetheless, reducing intake of fat has been the focus of dietary recommendations to decrease risk of coronary heart disease (CHD). Two lines of evidence, though, suggest that *types of fat*, rather than *total amount of fat*, in the diet have a more important influence on the risk of CHD.[12] First, metabolic studies suggest the type of fat, but not the total amount of fat, predicts serum cholesterol levels. Second, findings from *prospective cohort studies* and *controlled clinical trials* suggest replacing saturated fat with unsaturated fat is more effective in lowering risk of CHD than simply reducing total fat consumption. Based on these associations, some people now advise replacing saturated fatty acids (SFAs) with polyunsaturated fatty acids (PUFAs) – rather than monounsaturated fatty acids (MUFAs) or carbohydrates.[13]

In *prospective cohort studies*, researchers periodically monitor the diets of a large population (cohort) of people. Participants select their own diet and then report on what they eat. People who eat a large amount of saturated fat typically eat less carbohydrates or unsaturated fat, or both, to maintain energy intake. Conversely, people who eat less saturated fat eat more carbohydrates or unsaturated fats. While *prospective cohort studies* are useful for exploring associations, those associations (correlations) cannot establish causation.

The gold standards for establishing causation are *randomized controlled trials* where people are randomly assigned to different treatments. Researchers conduct two types of trials. In one kind of trial, people in one group are assigned, for example, to a diet lower in saturated fat and taught to replace saturated fat with carbohydrates or unsaturated fats, while people in the other group eat a diet high in saturated fats. In the other kind of trial, researchers give people the assigned diet, either high or low in saturated fat, balanced with a similar amount of energy from other sources. The latter kind of trial is much harder to conduct for any length of time.

Meta-analyses of *prospective cohort studies* that didn't consider energy source found no effect of SFAs on risk of CHD.[14] In contrast, meta-analyses that specifically evaluated the effect of replacing SFAs with PUFAs found a benefit, while replacing SFAs with carbohydrates, especially refined carbohydrates, yielded no benefit.[15] The different responses are due to the source of energy, specifically unsaturated fats or carbohydrates. That's consistent with *randomized controlled trials* that show less risk of CHD when SFAs are replaced with PUFAs.[16] The evidence for reducing SFAs and replacing them with PUFAs or MUFAs has grown stronger as better methods are more widely adopted for analysis of dietary intake in observational studies.

The findings of a Cochrane Systematic Review, one of the most trustworthy kinds of reviews in science, suggest a small decrease in CHD by reducing intake of SFAs.[17] The analyses included fifteen *randomized controlled trials* that used various interventions ranging from providing all food to simply giving advice on how to reduce SFAs. These trials suggest eating less SFAs reduced risk of cardiovascular events by 17 percent, but the effects were less clear for all-cause mortality (3 percent decrease) and cardiovascular mortality (5 percent decrease). The effects of replacing SFAs with MUFAs were uncertain because only one small trial was included in the analysis. The degree of reduction in cardiovascular events was inversely related with serum total cholesterol, and suggestions of greater protection occurred with greater reduction in saturated fat or greater increase in polyunsaturated and monounsaturated fats. Based on this review, it's unclear which type of unsaturated fat is best.

Eating foods high in SFAs raises low-density lipoprotein cholesterol (LDL), a risk factor for CHD, but also raises cardioprotective high-density lipoprotein cholesterol (HDL). Conversely, eating foods high in refined carbohydrates raises LDL and leaves HDL relatively unchanged.[18] Eating high glycemic index carbohydrates increases triglycerides, and the overall effects may be worse than those of SFAs. Replacing SFAs with high glycemic index carbohydrates was associated with a significant increase in risk of myocardial infarction for more than 53,000 Danish adults followed for twelve years.[19] While replacing saturated fatty acids with unsaturated or low glycemic index carbohydrates may reduce the risk of heart disease, so far these changes have not been common for Americans.

To add to the confusion, foods are a mix of SFAs, MUFAs, and PUFAs. Red meat, whole-milk products, nuts, olives, and avocados are high in MUFAs, as are beef fat and pork fat. Tallow (beef fat) is roughly 50 percent SFA, 42 percent MUFA, and 4 percent PUFA. Lard (pork fat) has about 39 percent SFA, 45 percent MUFA, and 11 percent PUFA. In contrast, olive oil is 14 percent SFA, 73 percent MUFA, and 10 percent PUFA.

The most recent *prospective cohort study* – the Prospective Urban Rural Epidemiology (PURE) study – followed individuals aged thirty-five to seventy years in eighteen countries.[20] PURE researchers collected self-reported dietary data from more than 135,000 people in the eighteen countries and grouped them according to the amount of carbohydrate, fat, and protein they consumed. After tracking the participants' health over a seven-year period, researchers found that those with the highest intake of fat (35 percent of daily calories) were 23 percent less likely to have died than those with the lowest intake of fat (10 percent of daily calories). Conversely, for carbohydrates, those with the highest intake (77 percent of daily calories) were 28 percent more likely to have died than those with the lowest intake (46 percent of daily calories). From these findings the authors' main conclusion is that "high carbohydrate intake was associated with higher risk of total mortality, whereas total fat and individual types of fat were related to lower total mortality."

The current American Heart Association / American College of Cardiology guideline is to reduce intake of SFAs to 5 to 6 percent of energy intake for people with elevated LDL cholesterol.[21] The 2015 to 2020 Dietary Guidelines Advisory Committee suggests the general public consume less than 10 percent of calories from SFAs and replace SFAs with unsaturated fatty acids. Average intake of SFAs in adults in the United States is 11 percent of total energy intake, while 30 to 40 percent consume less than 10 percent and roughly 5 percent of adults consume less than 7 percent.[22] Thus, based on these recommendations, most adults should eat less SFAs to reduce risk of cardiovascular disease.

To put all of these studies and recommendations in perspective, the first studies to link fat with heart disease were done mostly in North America and Europe, which *traditionally had the highest intake of fat worldwide.*[23] Under those conditions, eating less saturated fat may reduce heart disease. In other parts of the world, *where carbohydrate intake is excessive*, individuals may benefit from eating fewer carbohydrates, especially refined carbohydrates, and more fats. The key point is that excessive intake of any compound or type of food can be harmful.

The prevalence of low-fat nutrition advice has enormously influenced people in the United States, and yet that advice is based largely on uncontrolled studies that confound many variables, with tragic consequences for our health. Journalist Nina Teicholz, in her book *The Big Fat Surprise*, documents how a handful of researchers – through arrogance, egos, bias, and institutional consensus – allowed hypotheses to become dogma enshrined by expert committees and politicians. Many Americans now consider even trace amounts of fats to be harmful. One survey found that 88 percent of 1,000 mothers believed a low-fat diet was "important" or "very important" for the health of their infants and 83 percent responded they sometimes or always avoided giving fatty foods to their children.[24] That's reflected in the plethora of fat-free foods in supermarkets. Though fat is getting a bit of a reprieve now, finding full-fat foods is challenging.

Surveys by Paul Rozin and colleagues illustrate just how toxic fat has become.[25] In one survey, they asked Americans to assume they were going to be stranded on a desert island for a year with water and one food. They asked them to pick the food they thought would be best for their health: hot dogs, corn, alfalfa sprouts, spinach, peaches, bananas, or chocolate milk. The most popular choice was bananas (42 percent), and the least popular were hot dogs (4 percent) or chocolate milk (3 percent), the two foods that would have best supported survival in that scenario. Rozin and colleagues conclude: "Fat seems to have assumed, even at low levels, the role of a toxin." That conclusion is underscored by another of their surveys that shows one in three Americans believes a fat-free diet is better than a diet containing just a pinch of fat.

In *People of the Deer*, Farley Mowat points out the following:

> *If the Ihalmiut hunter shoots a deer for food when he is on a trip far from the camps, he seldom bothers to go to the trouble of building a fire. Usually his first act is to cut off the lower legs of the deer, strip away the meat, and crack the bones for marrow. Marrow is fat, and an eternal craving for fat is part of the price of living on an all-meat diet. In the case of people like the Ihalmiut, who subsisted on meat, fat was prized due in part to the excess of protein relative to energy in their diets.*

The Ihalmiut knew a lack of fat led to illness. As related in chapter 10, rabbit starvation is a similar condition where lack of fat leads to death. As Mowat points out, the Ihalmiut custom, winter or summer, was to eat not less than one mouthful of fat for every three of lean meat. This was the ideal proportion, but it wasn't always possible to maintain that proportion. When fat became scarce, the resistance of the Ihalmiut diminished and they were most susceptible to disease. Nowadays, this excess-protein condition is dangerous for Atkins' dieters who drastically reduce dietary carbohydrates and, given the prevailing fat phobia, are reluctant to eat fat.

These nuances of a culture in tune with a foodscape became obvious when the authorities attempted to change foraging behaviors of the Ihalmiut. In addition to caribou, other potential sources of food lived in the Barrens – hares, whitefish, trout, graylings, and suckers – but the Ihalmiut didn't rely on them for food. As Mowat notes, "With their usual acumen the authorities have seized upon this evidence of the remissness of the primitive mind. Obviously, they think, the ignorant natives must be unaware of the untouched reservoirs of food in the lakes and rivers, if they are so backward that they have not even learned to make and to use nets. So the authorities would supply nets, and thereby solve the starvation problem in the Barrens."

While not obvious to authorities who lived in cities, two problems arose. Logistical challenges involved setting nets under ten or twelve feet of freshwater ice in winter. Nutritional problems arose because fish couldn't supply even a fraction of the fat the Ihalmiut required each day. As Mowat argues, "In winter, a prolonged diet of fish would be as disastrous as poison to the People, and starvation in the form of fatal deficiencies would smite those whose bellies are distended with fish as violently as it smites those whose bellies are empty."

Other issues are involved in the nuance of a culture that relied on meat and fat for sustenance. Glycogen in meat and fat are important sources of energy for non-glucose-dependent tissues, and fat spares glucose for glucose-dependent tissues.[26] Humans need a consistent source of glycemic carbohydrate to support brain, kidney medulla, red blood cells, and reproductive tissues. Under normal conditions, a glucose requirement of 150 to 170 grams per day is met by dietary carbohydrate and gluconeogenesis from noncarbohydrate sources. When fat is gone, during lengthy fasts or starvation, tissue protein is catabolized to provide amino acids for gluconeogenesis. But gluconeogenesis alone can't meet needs for glucose without carbohydrates.

Eating meat increases renal acid loads and can cause metabolic acidosis, whereas eating fruits and vegetables and adequate calcium maintains bone health.[27] Potassium and bicarbonate in fruits and vegetables increases buffering capacity to maintain acid-base balance and prevent bone resorption. Phytochemically rich diets positively influence deposition, mineralization, and resorption, chiefly by inhibiting bone resorption. So, where did the Ihalmiut get vegetables?

In addition to herbs eaten seasonally, they got them from the fat of herbivores who foraged on phytochemically rich diets, as we discussed in chapter 8. And they got them from the intestines of caribou. Indeed, mountain men got "vegetables" from the intestines of bison, as William Holston describes in the essay "The Diet of the Mountain Man":

> *Perhaps the most unique food eaten by the mountain men were boudins. These were the small intestines of the bison. Cut into convenient lengths, the intestines were roasted on sticks until the heat caused them to become puffy. Boudins were tied sometimes to prevent the fat from leaking into the fire. The intestines quickly became distended, and little clouds of steam escaped from numerous tiny punctures. When crisp, the intestines and their contents were eaten with much gusto.*

Like the mountain men, the Hadza of Africa carve pieces of the stomach into bite-size chunks and eat it sushi-style after they kill a wild herbivore.[28] They also toss the tubular colon on the fire for a minute at best and clearly not long enough to terminate the menagerie of invisible microbes clinging to the inside wall of the colon. They proceeded to cut the colon into chunks and to eat them more or less raw. Little doubt Ihalmiut, mountain men, and Hadza benefited from eating herbivores that benefit from eating an assortment of phytochemically rich plants that positively influence the flavor, color, and quality of meat and fat for human consumption.

## Physiology Trumps Policy: The Case for Salt

Salt is prominent in human history and science. Salt has been the foundation of societies, a currency of trade, and a cause for war. During the past century, sodium chloride has been studied extensively to understand its role in human physiology and its impact on health. The latter has focused on the role of salt in blood pressure, an issue fraught with controversy.[29]

Since the Surgeon General began issuing guidelines on health more than fifty years ago, the recommendation has been for adults to consume less salt.[30] In 2003, the Electrolyte DRI Committee of the Institute of Medicine (a part of the National Academy of Sciences) endorsed a maximum of 2,300 milligrams per day.[31] The 2005, 2010, and 2015 US Dietary Guidelines continued to endorse 2,300 milligrams per day as safe for healthy adults and 1,500 milligrams per day for individuals at risk of hypertension. These panels assumed if adults understood the dangers of excess sodium for health and how to reduce sodium in their diets, and if more low-sodium foods were available, more individuals would reduce intake of salt. But that hasn't happened. Americans consume about 3,500 milligrams per day, while other people around the globe consume from 2,500 to 4,500 milligrams per day.

At the same time that more restrictive guidelines have been introduced over the past thirty years, research has continued to provide new insights regarding both the lack of safety and effectiveness of lowering sodium intake. Two lines of evidence suggest a normal range of sodium intake exists, which enables optimal function of the central and peripheral nervous systems, and they *are not consistent* with the dietary guidelines.

One source of evidence comes from research on central and peripheral mechanisms that control sodium appetite. Studies by Curt Richter, discussed in chapter 4, provided the first evidence that animals expressed ingestive behavior that assured survival when threatened by sodium depletion.[32] Decades of research reveal a complex neural network linking multiple centers in the brain, which are integrated with peripheral input from organ systems via neural and hormonal signals.[33] The sophistication of the central control of sodium appetite is compelling support for the proposition that vertebrates evolved ways to assure their physiological needs for sodium are defended when access to sodium is limited or when excessive amounts of sodium are lost under conditions of stress such as sweating, diarrheal illness, or hemorrhage.

The second line of evidence that bodies regulate intake of sodium in accord with needs comes from measures of salt intake in humans over the past three decades. Self-selection and regulation was shown in six carefully executed surveys, conducted by the British Foods Standards Agency (FSA) on populations in the UK. Several of the UK surveys were included in the landmark Intersalt study that also showed people regulate intake of sodium based on their needs.[34] That assessment was done on more than 10,000 adults at 52 sites in 32 countries. Within the timeframe of the FSA surveys, several other government-sponsored studies in the UK and closely related demographic groups included measures of urinary excretion of sodium (UNaV), which corroborated the findings on salt intake from the other studies.[35]

Studies at six academic medical centers in the United States provide additional evidence that adults naturally seek a range of sodium intake to meet bodily needs.[36] These studies of sodium restriction in mildly hypertensive people were rigorously established randomized, double-blind, crossover trials.[37] Participants decreased intake of sodium when their diet was adequate in sodium and increased intake of sodium when their diet was lacking in sodium. While needs for sodium vary among individuals, all these studies show a "normal range" of sodium intake can be defined in humans. That range is consistent with the neuroscience research that demonstrates salt consumption is homeostatically regulated within a relatively narrow range.

These diverse sources of data provide compelling evidence that people regulate intake of salt in accord with their needs. Yet, US policy promotes population-wide reductions in sodium intake. David McCarron, from the Department of Nutrition at the University of California, Davis, argues the policy is based on flawed assumptions that current intakes are excessive, that the "healthy range" must be below current intakes, and unlike all other essential nutrients, lower sodium intake is always better. Public policy is then based on the mistaken assumption that salt intake can be dictated by reducing sodium content in the food supply.[38]

Those assumptions are not consistent with findings across forty-five societies and five decades that show humans consume a reproducible, narrow range of sodium: approximately 2,600 to 4,800 milligrams per day. This range is independent of the food supply, verifiable in randomized controlled trials, consistent with the physiologic regulators of sodium intake, and not modifiable by public policy interventions. These findings indicate that human sodium intake is controlled by *physiology*, which cannot be modified by public health *policies*. Unlike with vitamin D and fat, with salt the wisdom body trumps policy.

In the case of vitamins and energy (fats and refined carbohydrates), we've come to trust authorities in corporate, political, and academic institutions, while ignoring the one expert who knows the most about us: our own body. We now mistakenly believe taste is not a guide to what should be eaten, nor should we simply eat what we enjoy. We are taught important components of foods are discernible only in scientific labs. And we are taught that science has produced rules of nutrition that will prevent illness and encourage longevity. We've abandoned cultural mores about how to grow and combine wholesome foods into meals that nourish and satiate. We like to think that we think for ourselves, but if that's the case, why are two-thirds of the people in the United States overweight or obese and on a fast track to diabetes, heart disease, and cancer? Is that the kind of life we want to choose for ourselves and for our families?

## CHAPTER 14

# When Beliefs Trump Authority

Sea Gypsies live in the Burmese archipelago of the west coast of Thailand. They spend over half their lives in boats at sea. Sea Gypsy children learn to swim before they learn to walk. As children, they dive thirty to seventy-five feet under water to gather clams and sea cucumbers. They learn to lower their heart rate while swimming at this depth and can stay under water twice as long as most swimmers. They can see without goggles at great depths underwater. Unlike land-dwelling humans, whose pupils get larger under water, the pupils of Sea Gypsies constrict 22 percent. This behavior, previously considered a fixed reflex under genetic control, is learned.

They are so in tune with the sea that when the tsunami of 2004 hit the Indian Ocean, killing thousands of people, they survived. Canadian psychiatrist Norman Doidge describes what occurred in *The Brain That Changes Itself*. Sea Gypsies saw the sea recede in a strange way, followed by an oddly small wave. They saw dolphins begin to swim for deeper water. Elephants began moving to higher ground. Cicadas fell silent. Doidge writes: "The Sea Gypsies began telling each other their ancient story about 'The Wave That Eats People,' saying it had come again. Long before modern science put this all together, they had either fled the sea to the shore, seeking the highest ground, or gone into very deep waters, where they also survived."

What the Sea Gypsies did, which people under the influence of analytical science didn't do, was synthesize all these unusual events to see the whole, using a remarkably wide-angled lens, even by Eastern standards. Burmese boatmen were also at sea when these otherworldly events were occurring, but they didn't survive. As Doidge points out, a Sea Gypsy, who was asked why the Burmese boatmen perished, replied, "They were looking at squid. They were not looking at anything. They saw nothing, they looked at nothing. They don't know how to look."

Historically, scientists assumed that how humans "see" the world was based merely on different *interpretations of our experiences*, not on differences in brain development that influence how people *experience* and *perceive* the world. Today, we know differences in brain development are responsible for how creatures *perceive*, *believe*, and *behave*. People experience the world through a cortex with some thirty billion neurons capable of making one million billion synaptic connections. The number of possible neural circuits in a human being is ten followed by at least a million zeroes. To put this number in context, the total number of elementary particles in the universe is ten followed by seventy-nine zeroes.

As American neuroscientist Joseph LeDoux notes in *Synaptic Self*, we develop habits as "neurons that fire together wire together." Those networks form as organ systems throughout the body, including the brain, become linked with the social and physical environments where we live. We begin to create those networks in the womb, and our experiences early in life influence how we perceive, what we believe, and how we behave. We believe we experience the world objectively, but our *experiences* influence our *perceptions*, which influence our *beliefs*, which feed back to influence our *experiences*: I wouldn't have seen it if I hadn't believed it.

## Diet as Religion

Beliefs about how food is produced alter our perceptions of eating.[1] For example, the presence of an organic label influences our perceived pleasantness and flavor of food. We enjoy food more when we believe it was produced ethically. Savory broth is perceived as more pleasant when the label states "rich and delicious taste" rather than when it's labeled as "contains monosodium glutamate." Coffee labeled "ecofriendly" tastes better to people than identical unlabeled coffee. People perceive that the same wine tastes better when they believe it is more expensive.

Beliefs about meat have changed during my lifetime. When I was a child, grain-fed beef conjured up images of a delicious steak, a dietary item my family couldn't afford and rarely ate. Now, for many people, images of grain-fed beef and factory-farmed chickens are less palatable than pasture-reared beef, pork, and poultry.[2] Given meat of identical origin, people who were told the meat came from a factory farm reported the meat looked, smelled, and tasted less pleasant compared with meat that came from animals raised on a humane farm. People reported that "factory-farmed" meat tasted saltier and greasier, and they ate less of it.

These behavioral responses are mediated by neurotransmitters, peptides, and hormones. For example, areas of the brain associated with reward valuation are more active when people drink a milk shake labeled "regular" compared with an identical milk shake labeled "low-fat."[3] Levels of ghrelin (the hunger hormone) decrease after people drink a 380-calorie milk shake labeled "indulgent, 620 calories." Ghrelin doesn't decrease after people drink an identical milk shake labeled "sensible, 140 calories." Ghrelin, which is produced by cells in the gastrointestinal tract, acts on cells in the hypothalamus to increase hunger, gastric acid secretion, and gastrointestinal motility. Ghrelin is secreted when the stomach is empty. When the stomach is stretched, secretion stops. Ghrelin also functions as a neuropeptide. Besides regulating appetite, ghrelin plays a key role in the distribution and rate of use of energy. Ghrelin also plays a pivotal role regulating reward perception through dopamine neurons that link the ventral tegmental area in the brain to the nucleus accumbens, the site involved in addictive behavior.

Gluten provides another example of how beliefs affect perceptions. Gluten – a protein found in wheat, barley, and rye – gives elasticity to dough, makes bread rise, and gives the final creation a chewy texture most people relish. Despite these benefits, gluten adversely affects 0.5 to 13 percent of people in the United States with gas, bloating, diarrhea, vomiting, migraine headaches, and joint pain.[4] These ailments are due mainly to celiac disease, an autoimmune disorder that adversely affects the digestive process in the small intestine. In addition to gluten, fermentable carbohydrates in gluten-containing cereal grains or nongluten proteins in foods may trigger celiac symptoms.[5] The incidence of celiac disease increased during the past few decades due to increased wheat consumption, greater awareness of the disease, improved diagnostic techniques, and modern varieties of wheat.[6] People with celiac disease experience major improvements in their symptoms when they eat foods made from ancient varieties of wheat.[7]

These issues aside, the perception that gluten causes disease illustrates the power of beliefs to alter people's responses to foods. Using blood samples from a nationally representative sample of 7,800 people, Joseph Murray, a gastroenterologist at the Mayo Clinic in Rochester, Minnesota, estimated the number of Americans who have celiac disease is 1.8 million, but only 17 percent of them know they have it. Meanwhile, 1.6 million Americans are on a gluten-free diet, but most of them don't have celiac disease.[8] They get sick when they eat foods they believe contain gluten because they believe gluten is harmful, as shown in a study in which self-identified "gluten-sensitive" people cycled through high-gluten, low-gluten, and no-gluten diets, without knowing which diet they were eating.[9] During the study, people experienced pain, bloating, nausea, and gas to a similar degree on all three diets, including the no-gluten diet.

Sin, damnation, and salvation are recurrent themes with diet, as illustrated by the grain-free monks of ancient China. In *The Gluten Lie*, Alan Levinovitz, faculty member at James Madison University, contends monks, like all diet gurus, used a time-honored prescription. They ridiculed the food culture of the day, which was based on *wugu*, or "five grains." According to their teachings, grain-laden diets "rotted and befouled" organs, leading to disease and death. A person could achieve health by avoiding the five grains. Naturally, to realize the benefits of the monk's diet, a person had to take proprietary supplements, alchemical mixtures only a select few knew how to prepare. All of this is familiar to anyone who reads popular literature about diet.

Peruse any bestselling diet book and you'll find authors portray their advice as sound science while they make a case for their "special diet" to promote your health and longevity. While their intentions may be good, dogmatic belief in a science-based diet doesn't reflect the fact that scientific understanding is ever incomplete and changing. Science is often a veneer for age-old tactics – used to reinforce narratives of good and evil – meant to control people. While that's certainly the case today, those tactics were familiar to authorities ages ago.

Then, as now, the appeal of dietary fads had much to do with myths, not facts, and people's beliefs. Chief among these is belief in a paradise lost, an appealing narrative about a time when everyone was happy and healthy, until they ate the wrong food and fell from grace. Many factors influence dietary habits, including the interplay of religious, moral, hygienic, ecological, and economic factors, as cultural geographer Frederick Simoons reveals in *Eat Not This Flesh*. Simoons reconstructs the origins and spread from antiquity to present of taboos against eating beef, pork, chicken and eggs, camel, dog, horse, and fish. While no single explanation exists for food taboos, he finds powerful, recurrent themes – maintaining ritual purity, food health, and well-being – underlying dietary habits.

## Nocebo Effects: I Shall Harm

In medicine, a *nocebo* – Latin for "I shall harm" – is a harmless substance that creates harmful effects in a patient who takes it. Conversely, a *placebo* – Latin for "I shall please" – is an inert substance that creates a positive response in a patient who takes it. Nocebo and placebo effects are not caused by responses to biologically active compounds. Rather, they result from a patient's beliefs and expectations about how a substance will affect him or her.

Appreciation of placebo and nocebo effects increased with the advent of brain-imaging tools that show nocebo effects influence areas of the brain that modulate pain reception.[10] The pain, bloating, nausea, insomnia, and fatigue experienced as a result of sham treatments can be painfully real, afflicting about a quarter of those assigned to the placebo treatment group in drug trials. As a result, patients in randomized clinical trials often stop taking their drug, complaining of side effects, even though some of these patients never received any medication.

Expectations influence the number and seriousness of side effects experienced by women who receive hormone therapies such as tamoxifen as part of their treatment for breast cancer.[11] When women have greater expectations of side effects, they experience nearly twice as much joint pain and weight gain and twice as many hot flashes after two years of hormone therapy than women with positive expectations. The women also experienced symptoms that were not directly attributable to their medication including back pain, breathing problems, and dizziness. Women may stop taking their adjuvant hormone treatment if they experience too many adverse side effects, which can affect the success of treatment. Early interventions such as counseling may help women understand the effects of their expectations on their experiences and thus better manage their symptoms and commit to continuing to take their medication.

"Value information," such as the price of a medication, can modulate nocebo effects. While more expensive medications can make us feel better, even when they are no different from less expensive generic drugs, they can also make us feel worse.[12] We are likely to experience more negative side effects when we take a drug we think is pricier. For example, brain imaging shows regions associated with pain were more active in an "expensive-cream" study group. The painful effects occurred as nerves were activated from the prefrontal cortex to the spinal cord, findings that refuted the view that beliefs don't affect nerve sensation in the spinal cord.

In her book *Sleep Paralysis*, Shelley Adler, professor in the Department of Family and Community Medicine at the University of California, San Francisco, set out to explore the phenomenon of the classic nightmare across cultures and history, how it straddles the mind-body divide, and its possible connection with sudden unexplained nocturnal death syndrome. As she writes in the introduction: "The night-mare, poised as it is between the supernatural and the natural worlds, and between the meaningful and the biological, is perfectly positioned to teach us about the seamless connection between our minds and our bodies."

Nightmare nocebo effects can even cause death. During the 1980s, more than 100 healthy Hmong men with a median age of thirty-three years died in their sleep. At the time, their death rate was equivalent to the top five natural causes of death for other American men in their age group. Something was killing them, but there was no obvious cause of death. None of them had been sick, physically. They were united only by a shared culture and their dislocation from Laos to the United States. Adler comes to a striking conclusion about the cause of their deaths: They died during nightmares, killed by their beliefs in the spirit world, though the medical cause of their deaths was ascribed to an obscure genetic cardiac arrhythmia common in Southeast Asia.

In REM sleep, we dream and our minds typically shut off the physical control of our body. During a phenomenon called *sleep paralysis*, however, a person experiences a REM state that is out of sequence. We're supposed to be temporarily paralyzed while we experience REM sleep, but we aren't supposed to be conscious. Yet, that's what happens during sleep paralysis, which is a strange mix of brain states that are normally detached in time.

People who have experienced sleep paralysis often feel a malevolent being is near them. This feeling is consistent across cultures, though it goes by different names, manifest in one's culture. In Indonesia, it is called *digeunton*, "pressed on." In China, it is *bei gui ya*, "held by a ghost." The Hungarians know it as *boszorkany-nyomas*, "witches' pressure." In Newfoundland, the spirit that comes is called the Old Hag, and the experience of sleep paralysis, *ag rog*, is "hag ridden." The Dutch name, which is closest to English, is *nachtmerrie*, the nightmare. The "mare" in question comes from the German *mahr* or Old Norse *mara*, which signified usually female supernatural beings who suffocate them by lying on their chests.

The bewitching nature of sleep paralysis, within the context of its common cultural interpretations, is akin to what Andrew Weil describes as "medical hexing."[13] That occurs when your doctor tells you that you'll be on medication for the rest of your life or that you have a 5 percent chance of five-year survival or that you have an "incurable" illness. Though they don't necessarily mean to harm you, they are cursing you with a form of medical hexing. They think they're telling it to you straight, that you deserve to know, that you should be realistic and make arrangements, if necessary. But when they say such things, they instill in your conscious and subconscious mind a belief that you won't get well, and as long as the mind holds this negative belief, it becomes a self-fulfilling prophecy. If you believe that you'll never recover, you won't, just as the Hmong men who died because they believed that their paralysis was caused by a malevolent being intent on harming them during their nightmares.

Is the bewitching nature of medical hexing, especially within the context of its common cultural interpretations, akin to food hexing? To what degree does all the negative information we receive about foods adversely affect our responses to different dietary items? We are told repeatedly about the harmful effects on our health of gluten, fat, red meat, dairy, and salt, which undoubtedly contributes to our chronic stress about foods and eating. We are bombarded by both supraliminal (aware) and subliminal (below the level of awareness) stimuli that condition our responses.[14] This occurs to the point that many people in the United States now believe food is as much a toxin as a nutrient and eating is nearly as dangerous as not eating.[15]

Our anxieties are fueled by uncertainty regarding unending changes in dietary advice from authorities. Compounds once promoted as unhealthy are found to be healthful, including salt, red meat, and saturated fat, while those once thought to promote health are shown to be harmful, including trans-fats, supplemental vitamins, and antioxidants. More than any other group, people in the United States worry about and modify their diet in ways they believe to be healthful, based on the latest findings of what and what not to eat and which supplements they should be taking. Yet, we are least apt to consider ourselves healthy eaters.

Historically, cardiovascular disease was higher in the United States than in France, and the angst over eating in the United States was based in part on the supposed longer life of people in southern Europe who eat a traditional Mediterranean diet.[16] Ironically, 86 percent of a sample of French adults derived more than 30 percent of their energy from dietary fat and 96 percent ate a diet that exceeded recommendations in the United States for less than 10 percent of calories from saturated fats.[17] Aligned with these findings, people in central and northern Europe who traditionally ate high-fat diets lived *longer* than people in southern Europe.

Consistent with a belief that most health differences can be traced to particular compounds or foods, searches by scientists for the answer to the "French paradox" focus on finding protective components such as red wine in the French diet.[18] The scientific literature has emphasized features of food suspected to influence longevity, rather than social norms and attitudes toward food. Little regard is given by scientists for alternative explanations, including different patterns of food intake and stress. French culture values cuisine – food freshness and variety, a balance of foods, and quality as opposed to quantity – and a pleasure-oriented attitude toward food.[19] Pleasure is related with good health, as stress is related with poor health.[20]

## Placebo Effects: I Shall Please

Many patients suffering from gastrointestinal disorders seek complementary and integrative medicine (CIM) therapies. While the therapies can certainly be effective, research indicates that the quality of patient–provider interactions can also improve health outcomes for irritable bowel syndrome.[21] Researchers also find that an expanded patient–provider visit improves symptoms of gastroesophageal reflux disease (GERD).[22] Symptom severity doesn't differ between an over-the-counter homeopathic product, Acidil, and placebo groups. However, people who received the expanded visit were significantly more likely to report a 50 percent or greater improvement in symptom severity compared to patients who received the standard visit.

A paper published in *The New England Journal of Medicine* in 2011, praised by scholars as one of the most carefully controlled and definitive placebo studies ever done, describes a study of forty asthma patients given four treatments: inhalers that delivered albuterol, placebo inhalers that delivered no albuterol, sham acupuncture, and no treatment.[23] The patients made twelve visits to the study site; they received each of the four treatments three times during the course of those twelve visits. While the researchers hoped to find improved lung function with both the real and placebo treatments, they found only patients given albuterol showed significant physical improvement. Yet patients' self-assessments of their perceived improvement did not differ for the albuterol, placebo, or sham acupuncture treatments: Their subjective responses contradicted their objective physical measures.

Remarkably, these effects can occur even when people are told they are receiving a placebo, as illustrated in studies of irritable bowel syndrome (IBS). One study compared two groups of IBS sufferers: One group received no treatment, while the other group was told they'd be taking an inert drug delivered in bottles labeled "placebo pills."[24] They were told placebos often have healing effects. The results surprised the researchers. Patients who knew they were taking a placebo reported twice as much symptom relief as the no-treatment group. That difference was comparable to the improvement seen in trials for the best IBS drugs.

A study of migraine headache confirms that a patient's expectations influence the effects of both the medication and a placebo.[25] While Maxalt – a drug that relieves migraine symptoms, including nausea, vomiting, and sensitivity to light/sound – was superior to a placebo for pain relief, information about Maxalt and the placebo, which ranged from negative to neutral to positive, altered the effects of both Maxalt and the placebo. The effectiveness of the placebo was superior to that of no treatment, accounting for more than 50 percent of the drug effect.

Evidence of a physiological basis for placebo effects comes from various sources, including studies of dental patients that show chemically blocking the release of endorphins – the brain's natural pain relievers – also blocks placebo effects.[26] Several neurotransmitters are involved, some of which use the same pathways as opium and marijuana. People suffering from depression who are given placebos show changes in electrical and metabolic activity in several regions of the brain. These and other studies suggest placebos change chemical responses in the brain, similar to those of active drugs, a hypothesis confirmed by brain-scan technologies such as positron emission tomography (PET) and functional magnetic resonance imaging (fMRI).

The power of the mind to influence health is illustrated in the story of John Whitley, a 60-year-old driver from Nyack, New York.[27] Whitley was diagnosed with stage 4 pancreatic cancer in July 2011. The cancer had spread to his liver, where doctors found an inoperable tumor. He was told he had less than a year to live. Determined to beat the odds, he entered a trial where some participants got an experimental drug and others got a placebo. Every afternoon, as he sat in his apartment and took the drug, he told himself, "This is a miracle drug that is going to save my life." That September, his oncologist called him in shock: Scans showed that his tumor had vanished, but his miracle remission wasn't due to the experimental drug. He received the placebo, not the drug. Nearly seven years after his diagnosis, Whitley is disease free.

He now counsels others through PanCan.org, the website where he first learned of others who had survived cancer. Some experts believe "patient activism" helps people who are facing what seem like overwhelming odds. Kelly Turner, an independent cancer researcher and author of *Radical Remission*, has found that reducing stress and inflammation helps bolster the immune system to better fight the disease. Turner's work and Whitley's story are supported by research that shows how stress and doubts, or a lack of them, can affect the spread of cancer.[28]

More than 3,500 such case studies have been documented in the medical literature in the Spontaneous Remission Project, compiled by the Institute of Noetic Sciences. That review suggest there's no such thing as an incurable illness, which is the point author Lissa Rankin makes in *Mind over Medicine*. Trained as a medical doctor in the classical sense and working in the profession, Rankin became interested in the outlier patients – the ones who defied all odds. Rather than anomalies to be disregarded, she came to see them as folks who could provide deep insights into how what we believe influences our health. To better understand what was happening, she studied the peer-reviewed scientific literature and amassed evidence that the medical establishment had been discovering for the past fifty years: The body can heal itself.

Our beliefs hold more power than medicine. If you or someone you love is suffering from an allegedly chronic, incurable, or terminal illness and you want to optimize the chance for remission, you have to start by cleansing your mind of negative beliefs that will sabotage your self-healing efforts. Healing comes as our thoughts, feelings, and beliefs alter our bodies in form and function. The chronic stress of loneliness, pessimism, depression, fear, and anxiety harm a body. Intimate relationships, gratitude, and authentic self-expression activate self-healing processes. Health comes from loving, nurturing, and creative relationships with ourselves and the natural world, including what we eat, manifest through professional and personal associations.

That dynamic is likely at play with faith healers. As Rankin points out, "People come from far and wide to be in the presence of someone they *believe* will heal them. Add to that other people in need of healing who have made a pilgrimage and who share the same positive belief. Throw in rituals and practices that reinforce the belief – the loving hug, the laying of hands, the meditation, the herbs, the holy water – and you've got a recipe for elicitation of relaxation responses and self-healing that scientists would dub the mega-placebo effect."

## Dying to Be Me

In *Dying to Be Me*, author and consultant Anita Moorjani tells how from the time she was a little girl growing up in Hong Kong, she'd been pushed and pulled by different cultural beliefs. Raised in a strongly traditional Hindu family, she lived in a largely Chinese and British society. From her nanny, she learned Chinese customs and beliefs, including Buddhism. While attending British schools, she was told she must read the Bible and go to a Christian church every Sunday or she was going to hell. Still conflicted, she had been working in the corporate world for years when in April 2002, at age forty-two, she was diagnosed with lymphoma. For four years, her body was ravaged by cancer and her dread of the disease. During that time, as her weight dropped to eighty pounds, she became too weak to walk, and she began to give up hope that she would heal.

After losing two friends to cancer and her own diagnosis of lymphoma, Moorjani began to study everything she could about holistic health, including Eastern healing systems. She tried hypnotherapy, meditating, praying, chanting mantras, and taking Chinese herbal remedies. When none of these worked for her, she traveled to India to follow the healing system of Ayurveda. She told her yoga master that medical tests confirmed she had lymphoma. As she recounts in her book, he replied: "Cancer is just a word that creates fear. Forget about the word and let's just focus on balancing your body. All illnesses are just symptoms of imbalance. No illness can remain when your entire system is in balance." He put her on a regimen that included a specific diet of vegetarian food and herbal remedies, along with a routine of yoga poses that she did at sunrise and sunset. She spent six months in India, and during that time she regained her health.

At the end of six months, she and her yoga master were convinced she was healed, so she returned to Hong Kong. Her friends there told her how good she looked. When she told them about her Ayurvedic regimen, though, she received mainly fear-based and negative responses. While her friends were well meaning and concerned, they'd been taught to believe cancer could not be cured with Ayurveda. Anita felt her old doubts and fears creep back into her psyche.

At that point, she began experimenting with other ways of healing. She attempted Traditional Chinese Medicine (TCM), but because of its conflict with Ayurveda, she was confused. In TCM, you are encouraged to eat meat, especially pork. In Ayurveda, you are encouraged to be vegetarian;

meat is the worst thing you can eat. In desperation, she turned to Western naturopathy, but that just added to her confusion and fears. In that system, sugar and dairy are absolutely to be avoided, as they are thought to fuel the growth of cancer cells. In Ayurveda, dairy is a must and sugar and sweet foods are required as part of a balanced diet.

As she became overwhelmed with doubt, fear, and despair, she watched helplessly as her health deteriorated. Finally, her organ systems failed and she slipped into a coma, death imminent. While her body was being cared for by doctors, she had a near-death experience that saved her life. On that journey, she learned that her only purpose in life is to be a full expression of herself. To love herself to the core of her being and to share that fearlessly in the world.

Moorjani emphasizes that no faith in anything was required for her healing. "Rather, I'd say that it was the complete suspension of all previously held beliefs, doctrines, and dogma that caused my body to heal itself," Moorjani explains. Strongly held ideas actually worked against her. Having concrete beliefs limits possibilities because they keep us locked into what we think we know, and our knowledge of this world is limited by our physical senses. As Moorjani stresses, being comfortable with uncertainty opens us up to possibilities. When we step into the realms of ambiguity, we are at our most powerful.

After a lifetime of confusion, the universe finally made sense to her. From a spiritual standpoint, she came to realize that God isn't a being, but a state of being. We are forever in that state of being, but we just don't realize it. Our moment on Earth is intricately woven into an infinite tapestry, our time here one more strand woven into that web. All the other strands represent our relationships, including every life we've touched.

Moorjani's experience embodies the truth that we all have the inner power and wisdom to overcome even life's most adverse situations. Yet, from conception onward, we are taught by authorities on Earth what to believe about matters physical and spiritual. As a result, some people have difficulty believing accounts of near-death experiences. Yet, viewed as part of a continuum – from the influence of the mind's conscious and subconscious beliefs on our perceptions of eating and other experiences, to our responses to placebos and nocebos that show that the body's response to "medicine" is not linear and simplistic, to evidence of people's ability to spontaneously die from a nightmare or heal from serious disease without conventional treatment – a near-death experience can be seen as yet another manifestation of the awareness of a body's inner wisdom. The knowing that we aren't separate from the rest of reality, but that as physicists and mystics have discovered, all things are fundamentally interconnected.

Some early Jewish mystics (*Merkabah* mystics) believed all of the vastness of divinity is contained within.[29] The universe can be conceived as a set of wooden Russian *matryoshka* dolls, with each doll having a smaller one inside of it. The entire visible universe is the outermost doll, and nested inside it are galaxies, solar systems, stars, planets – right down to the smallest doll, which is you. But inside of you is an even smaller doll that also has the biggest doll inside of it. And that, of course, is the notion of the *holon* discussed at the beginning of the book.

Quantum theory has abolished the notion of fundamentally disconnected objects; has introduced the notion of scientists as participators as opposed to objective, independent observers; and may even find it essential to include consciousness in its description of the world. Quantum physicists have come to view the universe as an interconnected web of relations whose parts can be defined only through their interconnections with the whole. The Tantric Buddhist Lama Anagarika Govinda summarizes this worldview:

> *The Buddhist does not believe in an independent or separately existing external world, into whose dynamic forces he could insert himself. The external world and his inner world are for him only two sides of the same fabric, in which the threads of all forces and of all events, of all forms of consciousness and of their objects, are woven into an inseparable net of endless, mutually conditioned relations.*[30]

Buddhists speak of emptiness, which should be understood in terms of the interdependent nature of reality. By virtue of their dependent origination, all things are devoid of independent existence. At the first level is the principle of cause and effect, whereby all things and events arise in dependence on a complex web of interrelated causes and conditions. At the second level is the mutual dependence that exists between parts and wholes. Without parts, there can be no whole; without a whole, the concept of parts makes no sense. At the third level, all phenomena can be understood to be dependently originated because when we analyze them, we find that, ultimately, they lack independent identity. In the end, we cannot separate any phenomena from the context of other phenomena. We can speak only in terms of ever-changing interrelationships.

If nothing changed, life would be mind-numbingly boring. If nature changed predictably, life would be uninteresting. "The real trouble with this world of ours is not that it is an unreasonable world," as G. K. Chesterton wrote in *Orthodoxy*, "nor even that it is a reasonable one. The commonest kind of trouble is that it is nearly reasonable, but not quite. Life is not an illogicality; yet it is a trap for logicians. It looks just a little more mathematical and regular than it is; its exactitude is obvious, but its inexactitude is hidden; its wildness lies in wait."

Our collective mythologies – the stories we've told ourselves for generations – thwart our attempts to embrace the ever-changing nature of life. Our rigid beliefs cause disconnection and discord in the world, including within our own bodies. On Earth, we are bound to a past we little know or understand, linked to a future we can little predict or control. We spend most of our time thinking about a past that's history and projecting into a future that's a mystery. In the process, we miss the experience of *now*. We manifest the "thou in me" by being who we are, experienced and expressed in each moment. And that's exactly what the rest of creation is doing.

# CHAPTER 15

# When Understanding Trumps Beliefs

As a graduate student learning from goats and blackbrush, I believed that, with enough effort, scientists would understand the intricacies of ecosystems, from soil and plants to herbivores and humans. But after a lifetime of such pursuits, I question the notion that arose in the seventeenth century: Human actions can be built on a foundation of independently verifiable scientific truths.

The challenge has to do with the complex and ever-changing web whose strands are inseparably linked in space and time. To fathom this whole is impossible. The totality is boundless because reality is composed of holons within holons ad infinitum. Every whole is always a part in space and time. The entire universe is part of the next instant's universe. The combinations of climate, soils, plants, animals, and people are endless.

This daunting challenge is illustrated with nutrition. In 2015, the Dietary Guidelines Advisory Committee (DGAC) issued a report with a striking recommendation: Eliminate dietary cholesterol as a nutrient of concern.[1] This change astonished the public but is consistent with scientific evidence reporting no appreciable relationship between dietary cholesterol and serum cholesterol or clinical cardiovascular events in general populations. A less obvious change was the absence of an upper limit on total fat consumption. The DGAC report neither listed total fat as a nutrient of concern nor proposed restricting its consumption. With this report, the DGAC reversed nearly four decades of nutrition policy that placed priority on reducing cholesterol and total fat consumption throughout the population. As with other scientific fields from physics to clinical medicine, scientists contend understanding of nutrition has advanced in recent decades.

Yet, despite the apparent increase in understanding, obesity and diet-related diseases continue to rise. In 1960, fewer than 13 percent of Americans were obese, and fewer than 1,100 scientific articles were published on obesity and diabetes. Today, 49 percent of Americans are obese and the United States ranks number one among thirty-five nations in childhood obesity. In 2013, 44,000 scientific articles were published on these topics. In total, over 600,000 articles have been published alleging to provide worthwhile information on diet-related disorders.

Science writer Gary Taubes is critical of nutrition as a science, as he notes:

> *The 600,000 articles – along with several tens of thousands of diet books – are the noise generated by a dysfunctional research establishment. Because the nutrition research community has failed to establish reliable, unambiguous knowledge about the environmental triggers of obesity and diabetes, it has opened the door to a diversity of opinions on the subject, of hypotheses about cause, cure and prevention, many of which cannot be refuted by the existing evidence. Everyone has a theory. The evidence doesn't exist to say unequivocally who's wrong.*[2]

Taubes suggests that if nutrition scientists could just do the right kind of research, they would obtain reliable knowledge. But here's another possibility: The evidence will never exist to say unequivocally who's right and who's wrong in any ecological, economic, or social scientific discipline. The complexity and dynamism of life is beyond our ability to analyze scientifically. There is no one-cause/one-solution for anything. With diet, outcomes emerge from concurrent integration of effects that span generations; manifest within the lifetime of each individual, beginning in utero and early in life; emerge daily from interactions among internal and external contingencies; and are unique for each individual. The best scientists can do is understand relationships and establish likelihoods for various outcomes. But no one knows with certainty how any individual will respond, or at the level of populations, how individuals will interact with biophysical environments to create the emerging conditions that ever surprise us.

We like to watch sports because we never know for sure how the game will end. Though few thought they'd win, the Denver Broncos beat the Carolina Panthers in Super Bowl 50. Even as the outcome was emerging in the first half, "experts" couldn't foresee what was happening. At half time, they listed all the reasons why the Broncos would be defeated in the second half. And that's just a football game with fifty-some players, their abilities all well known to experts who follow football. What happens when hundreds of thousands of variables interact in the process of transformation we call life? We can't fathom that. Is it any wonder we can't predict emergent properties of natural systems? We can't even predict who will win a football game.

## Challenges in Nutrition Research

Scientists face two challenges in conducting studies. First, we must ask questions that provide insights into the processes of nature. As simple as that may seem, it's difficult to do. That's why most studies are merely variations on known and overworked themes. Second, we have to control variables to establish cause-and-effect relationships. If researchers want to understand how saturated fat affects cardiovascular disease, for instance, they must control for intrinsic factors such as sex, age, and ethnicity as well as extrinsic factors including diet and lifestyle. Ideally, only the variable of interest – saturated fat – will differ for people in the treatment and control groups. If saturated fat diminishes cardiovascular disease for the "average" person in the treatment group relative to the "average" person in the control group, then the treatment is said to be effective. If a result can't be repeated, it doesn't pass muster.

In practice, no one can control all the intrinsic and extrinsic factors that affect the outcome of a study nor can researchers change just one nutritional variable at a time. Saturated fat, for example, can't be eliminated from the diet without also decreasing calories. To maintain consistent intake of calories, another food must replace saturated fat. Should unsaturated fats or carbohydrates replace saturated fats? Which ones? And that's varying just one of hundreds of variables – sex, age, ethnicity, other dietary factors, and lifestyle – that affect health.

Beyond these issues of design, how does a researcher ensure that people in two different treatment groups eat particular foods and exercise in particular ways every day during a study of several weeks, months, or years? They can't. And for obvious reasons, researchers can't use placebos or double-blind studies – where neither the researcher nor the subjects know who is in the treatment and control groups – to determine the effects of different diets on health. Even if they could, as we just discussed in chapter 14, that, too, affects the outcomes of studies.

In the end, researchers can control variables or have realistic context, but not both. To the degree researchers control variables, they can ascribe cause to particular treatments but *only within the specific and limited context of their study*. Conversely, to the degree that researchers study realistic contexts, they lose the ability to control variables and are unable to ascribe cause to any one factor. Beyond that, contexts change remorselessly and stochastically.

Most issues of importance – whether social, economic, or environmental – can neither be controlled nor replicated. They are too multifaceted and ever changing. Ecological studies of landscapes and prospective cohort studies of populations describe events in the "real world," but they, too, are limited to *associations* in which study contexts vary endlessly. As a result, most conclusions are based on two logical fallacies. One is the notion that if X occurred before Y, then X caused Y. The other is the perception that if X changes similarly to Y, then X and Y are causally linked. The fallacies lie in coming to a conclusion based on the order or pattern of events, rather than accounting for a multitude of other factors that affect a proposed association. Life is not without causes, but the events we observe are so multicausal, with strands linked in a web in time and space, that control of all the variables is impossible to implement or interpret.

Even if researchers could control all of the extrinsic and intrinsic factors in a study, *individuals* would still differ in their responses. Studies can provide only *probabilities of events for populations*: if you smoke, you are more likely to die of cancer than if you don't smoke; if you eat fruits and vegetables, you are less likely to die of cancer or cardiovascular disease than if you don't eat fruits and vegetables. But at the *individual level*, no one knows what will happen. Some people who smoke cigarettes never get cancer. Others who avoid eating fruits and vegetables live long lives devoid of cancer or heart disease. How each person behaves depends as much on what he or she believes as on what science has to say about outcomes. In the end, means (averages) are snapshots in time that reflect the questions we ask, how we design the study to control for intrinsic and extrinsic factors, and how we collect, analyze, and interpret the data. Means masquerade as truths after they steal the identities of individuals.

Not surprising, then, scientists don't agree on most matters in nutrition. For example, appetite, food intake, and weight gain are stimulated by foods high in refined carbohydrates. Refined carbohydrates quickly break down to glucose, which elevates insulin, a hormone that increases deposition of fat and inhibits mobilization of fatty acids in fat tissue. As we accumulate more fat, our metabolism slows and we are always hungry because our bodies are allocating more energy to fat cells than to muscle cells. Thus, some hypothesize obesity is a disorder of excess fat accumulation from eating refined carbohydrates.[3] Satiety and weight loss are promoted by foods that lower insulin and increase release of fatty acids, directing them to muscle cells. For many people, according to this hypothesis, maintaining weight loss is easier when people limit intake of high-glycemic foods, including refined carbohydrates.

Not all scientists agree with this hypothesis. As they point out, calorie for calorie, restricting intake of fat results in greater loss of body fat than does restricting intake of carbohydrates in people who are obese.[4] When people eat less carbohydrates, insulin declines, but that doesn't lead to release of fat from fat cells. Rather, with a shift to a diet low in refined carbohydrates, there is a decline in the number of calories burned from fat. Notably, fat oxidation on reduced-fat diets remains unchanged and results in a greater rate of fat loss compared to reduced-carbohydrate diets. They argue, these and other studies fail to show that cutting carbohydrates boosts burning of calories or fat loss more than does cutting fat from the diet.

Cinnamon research provides another example of this. Cinnamon has potential as an insulin substitute for people with type 2 diabetes. An analysis of research, though, concluded more evidence was needed to support use of cinnamon for type 1 or type 2 diabetes and that better trials – double-blind, placebo-controlled – are needed to determine if cinnamon works.[5] An updated meta-analysis found consumption of cinnamon is associated with a significant decrease in levels of fasting plasma glucose and other desirable effects, but no effect was found on hemoglobin A1c, a long-term measure of blood-glucose levels.[6] They advised caution in applying the results of this analysis to patient care because of the uncertainty of the dose and duration of use of cinnamon and the further need to identify "the ideal patient population."

These researchers did not address the following, and more salient, question: What happens to my insulin levels if I'm *aware* of this research and I *believe* cinnamon is beneficial? In that case, diet becomes one part of an *exposome* that contains all of the physical and social exposures I've experienced from conception throughout life as they affect form, function, and behavior.[7] Exposures affect disease and health. Exposures are entangled and dynamic, making their measurement and the analysis of their effects exceedingly challenging and, for all practical purposes, impossible.

## From Bias to Deception

We want to believe research that underlies decisions about nutrition and health is impartial and accurate. For most studies, though, results are more likely to be false than true.[8] Findings are more likely to be false when sample size is small; when treatment effects are small; as the number of tested relationships increases; when more teams are seeking significant findings in an active (hot topic) field; when financial backing comes from industry; and as the flexibility in experimental designs, definitions, outcomes, and analytical methods increases. In reality, the questions we ask, the way we design and conduct studies, and how we interpret the findings affect outcomes and conclusions, which often reflect prevailing bias in a discipline.

Some of the most basic techniques researchers use to study "free-ranging" humans are questionable. For example, scientists use memory-based dietary assessment methods (M-BMs), such as food frequency surveys and 24-hour interviews, based on the respondent's memory and ability to estimate, recall, and honestly report past dietary intake. Actual dietary intake, as opposed to reported memories of perceptions of dietary intake, are two quite different phenomena. Actual dietary intake and associated energy and nutrient consumption are observable, falsifiable, and quantitative phenomena. In contrast, memories of perceptions of dietary intake are abstract constructs that don't exist outside of the mind of the respondent.

The many errors associated with M-BM data are nonquantifiable due to fabricated memories, confabulations, omissions, and intentional misreporting.[9] Just try to recall which foods you ate, and how much of each item, even yesterday, let alone weeks, months, or years ago. And then add in your biases. Neither respondents nor researchers can quantify what percentage of the recalled foods and beverages consumed are completely false, grossly inaccurate, or somewhat congruent with actual consumption. Thus, neither researchers nor the participants know the validity of the reported food and beverage consumption.

No surprise, then, M-BM data have been shown to be physiologically implausible, incompatible with survival and life, and inadmissible as scientific evidence.[10] Diets that can't support human survival shouldn't be used as the basis for dietary recommendations. Yet, M-BM data are used to populate the National Evidence Library (NEL) and by the National Health and Nutrition Examination Survey (NHANES) to inform public health policy through the US Dietary Guidelines for Americans. In so doing, they exclude results and conclusions of highly cited and widely publicized research that questions the validity of these studies.[11]

Other times, bias occurs due to belief in the validity of a particular hypothesis. Eating oily fish is endorsed as part of a heart-healthy diet, a recommendation that stems from studies published a half-century ago by Danish scientist Dr. Hans Olaf Bang and colleagues.[12] They argued the low incidence of coronary artery disease (CAD) among Eskimos of Greenland was related to their diet, which was rich in whale and seal blubber. Their study is still widely cited by scientists and medical professionals who endorse eating oily fish or grass-fed meat and dairy and taking fish oil supplements such as omega-3 fatty acids to prevent cardiovascular disease.

As it turns out, Bang and colleagues didn't actually study the cardiovascular health of Eskimos. Rather, they studied the *dietary habits* of Eskimos and only speculated that the high intake of marine fats produced a protective effect on coronary arteries.[13] They relied on annual reports produced by the chief medical officer of Greenland to ascertain CAD deaths in the region. For several reasons, those reports were inadequate. Turns out, Eskimos have a similar incidence of CAD to non-Eskimos, and Eskimos have high rates of mortality due to cerebrovascular events such as strokes. Their life expectancy is nearly ten years less than Danish populations used for comparison and their mortality is twice that of non-Eskimo populations.

If Bang and colleagues knew then what is now known about Greenland's death records, they might have labeled the Eskimos' diet as possibly dangerous to health. Instead, based on limited information, they construed a hypothesis that eating marine fats prevents CAD and reduces atherosclerosis. Consumers now support a billion-dollar industry that produces and sells fish oil capsules based on a questionable hypothesis. More than 5,000 scientific papers have been published on the alleged benefits of omega-3s. Ironically, many well-designed studies show ambiguous or negative results regarding the cardioprotective properties of omega-3s and fish oil supplements, and yet these substances are still recommended as part of a heart-healthy diet.

People promote grass-fed based on evidence that grass-fed meat and dairy are lower in ratios of omega-6 to omega-3 fatty acids for livestock fed grass rather than grain.[14] While some nutritionists believe a healthy diet should have no more than one to four times more omega-6s than omega-3s, many people consume far higher ratios of omega-6s to omega-3s.[15] Some contend this imbalance explains the incidence of rheumatoid arthritis, heart disease, cancer, autoimmune, and neurodegenerative diseases, all thought to stem from inflammation.[16]

Historically, omega-6s were thought to be pro-inflammatory, but that was not evident in a review of inflammatory effects of linoleic acid, and some studies attribute lower inflammatory markers to omega-6s.[17] An analysis of twenty prospective cohort studies from ten countries shows the omega-6 linoleic acid has long-term benefits for preventing type 2 diabetes.[18] Conversely, omega-3s don't exhibit consistent benefits, though fish consumption reduces inflammatory responses in people with metabolic syndrome.[19] The current recommendation is to eat oily fish with a much richer array of compounds, rather than take supplemental omega-3s.[20]

The traditional diet-heart hypothesis, the epitome of strongly held beliefs, predicts that replacing saturated fat with vegetable oil rich in the omega-6 fatty acid linoleic acid will lower serum cholesterol, diminish deposition of cholesterol in arterial walls, slow progression of atherosclerosis, reduce coronary heart disease, and improve survival. Nearly fifty years ago, American physiologist Ancel Keys led two randomized controlled feeding trials – the Sydney Diet Heart Study and the Minnesota Coronary Experiment – to test this hypothesis, but he didn't fully report the results.[21] Recently disclosed findings from both studies add to the evidence that piecemeal publication hyped the benefits of replacing saturated fat with vegetable oils rich in linoleic acid.[22] Replacing saturated fat with linoleic acid reduced serum cholesterol, but that didn't reduce the risk of death from coronary heart disease or all-cause mortality. Indeed, rates of death increased from all causes, coronary heart disease, and cardiovascular disease when saturated fats were replaced with linoleic acid in Key's trials. Jointly, these findings suggest lack of fat in the diet, along with a great excess of linoleic acid in the diet, contributed to poor health.

In industrial nations, intake of the omega-6 fatty acid linoleic acid comes mainly from vegetable oils processed in ways that remove healthful components such as fiber, micronutrients, and many other phytochemicals present in unprocessed vegetables and seeds. Concentrated sources of linoleic acid are widely used as cooking oils and added to processed and packaged foods. If these sources of linoleic acid are considered as supplements, the average American who eats a highly processed Western diet ingests the equivalent of eleven 1-gram capsules of linoleic acid a day over and above intake from natural foods.[23]

Clinical investigator Christopher Ramsden at the National Institutes of Health and his colleagues conclude the history of the traditional diet-heart hypothesis suggests nutrition research would be improved by not overemphasizing intermediate biomarkers, by cautious interpretation of nonrandomized studies, and by timely and complete publication of all randomized controlled trials. The biochemically complex foods we eat are substrates that regulate numerous biochemical pathways critical in health and disease. Given their complexity and the limitations of research, understanding of biochemical and clinical effects of foods is rudimentary. Scientists and practitioners should be humble, highlight limitations of knowledge, and set a high bar for advising intakes beyond what can be provided by natural diets.

## Merchants of Doubt

In addition to using questionable methodologies and fomenting bias, corporations have shaped and skewed the literature, manufactured and magnified uncertainty, and influenced government policy to their advantage. That's why several scientific review papers conclude that much research is biased by funding source, which adds confusion to recommendations.[24]

That is illustrated in research on the role of sugar (50-50 mix of sucrose and high-fructose corn syrup) in the epidemic of metabolic diseases.[25] Studies that suggest added sugar doesn't adversely affect health share a commonality: They were funded by the sugar industry or conducted by investigators who have received consulting fees from industries with strong financial interests in maintaining high levels of sugar consumption.

Cristin Kearns, a faculty member at the University of California, San Francisco, and her colleagues have examined this facet of the history of the sugar industry. They found that as early as the mid-1950s, the Sugar Association, a trade group that claims to be the scientific voice of the United States sugar industry, recognized that if Americans embraced low-fat diets, then per-capita intake of sucrose would increase by more than one-third. In the mid-1960s, they began working with scientists to identify fat and cholesterol as primary causes of coronary heart disease (CHD) and cancer and to downplay evidence that sucrose was a risk factor in either disease.[26]

In 1965, the Sugar Research Foundation (the former name of the Sugar Association) funded a review published in the *New England Journal of*

*Medicine*, which singled out fat and cholesterol as the dietary causes of CHD and downplayed evidence that sucrose was a risk factor. The Sugar Research Foundation, whose funding and role was not disclosed, set the review's objectives, contributed articles for inclusion, and revised drafts. Industry-sponsored research in the 1960s and 1970s successfully cast doubt on the hazards of sucrose while promoting fat as the culprit in CHD. They didn't disclose evidence of harm from animal studies that would have fortified the case that the risk of CHD is greater for sucrose than starch and that sucrose is likely a carcinogen. They used tobacco-style tactics to dismiss troubling health claims against their products. Now, as obesity, diabetes, and heart disease skyrocket, doctors are treating the first generation of children suffering from fatty liver disease.

Two recent scientific review papers conclude that studies funded by the sugar industry are more likely to present a conclusion of no or lesser association between sugar-sweetened beverages and weight gain or obesity compared with those without such conflict of interest.[27] That bias is evident in scientific papers based on a ten-million-dollar study funded by the Corn Refiners Association, the largest randomized control study done on sugar consumption.[28] As Kimber Stanhope, faculty member at University of California, Davis, points out, their experiments *have not* been the gold standard for reasons that include unsuitable methodology, lack of controls, use of suboptimal statistical models to analyze the data, and lack of compliance monitoring. Rather than address key questions regarding public health, the studies suggest that the goal of industry-sponsored studies is to generate results that assure the public that the current level of sugar consumption is safe.

The tobacco and food industries both rely on marketing to lure children into buying their products, fashioning lifelong customers by creating habits and addictions that keep children and adults buying more. The tobacco and sugar industries spent massive amounts of money lobbying Congress, denying the science underlying the addictive effects of nicotine and sugar, all the time appearing to act socially responsible. The documentary *Merchants of Doubt* illustrates the playbook invented by the tobacco industry and now used by other corporations – pharmaceutical, chemical, agricultural, food, fossil fuel – to create doubt about everything from drugs, pesticides, and foods to fossil fuels and climate change. They do so by conducting phony science and passing it off as legitimate research; harassing scientists who speak out with results or views inconvenient for industry; creating uncertainty about conclusions where little or no doubt actually exists; buying credibility through alliances with academia or professional societies; and manipulating government officials or processes to improperly influence policy.

As a result of such tactics, the United States is now in the midst of an epistemological crisis. Many people are no longer certain what's true and what's not true or what to believe and what not to believe. If science has limits and scientists are frauds, if nothing the media reports is reliable, if there is no verifiable truth, then how does one differentiate between fake news and reality? At that point, there is no objective reality out there waiting to reveal its secrets, just what you believe – and the biased beliefs of others who disagree with you.

## Science and Creative Systems

As societies increase in complexity, they become more dependent on information and their members require higher levels of education. So we are taught that acquiring knowledge is worthwhile, and that's generally true. Understanding processes of nature can empower adaptive problem solving. Yet, other than those who fund research and education, few people appreciate that acquiring knowledge has costs and increasing knowledge is not always cost-effective. The knowledge developed early in the life of a scientific discipline is usually inexpensive to produce. As easier questions are resolved and knowledge becomes more complex and specialized, science moves to more costly research organizations, topics, and projects.

In an excellent review, Daniel Sarewitz of the Arizona State University School for the Future of Innovation and Society and codirector of the university's Consortium for Science, Policy, and Outcomes points out that scientists, fueled by fifty years of public investments, have published millions of articles in thousands of scientific journals covering an ever-expanding array of phenomena.[29] But much of this hypothetical knowledge is turning out to be contestable, unreliable, unusable, or flat-out wrong. Evidence of poor quality is coming from fields of health, biomedicine, and psychology, but the problems are likely as bad in other areas of research. Science that is supposed to yield clarity and solutions is in many instances leading instead to contradiction, controversy, and confusion. Recent examples include:

- From 75 to 90 percent of all basic and preclinical biomedical research is unreplicable.[30]

- Scientists could replicate only six of fifty-three "landmark" preclinical cancer studies.[31]
- Researchers were unable to reproduce any of the published positive findings in mice for more than 100 potential drugs for treating amyotrophic lateral sclerosis.[32]
- The cell line used in more than 1,000 breast cancer studies turned out to be a skin cell line, not a breast cell line.[33]
- Annual mammograms confer little benefit for women in their forties.[34]
- Nearly 150 clinical trials for therapies to block human inflammatory response showed that every therapy supposedly validated with mice failed in humans.[35]
- Up to 70 percent of positive findings in roughly 40,000 studies of functional magnetic resonance imaging (fMRI) to map human brain function could be untrue.[36]
- A meticulous effort to assess 100 psychology trials couldn't replicate 61 percent of the original findings.[37]

The goal of science since the seventeenth century has been to gain reliable knowledge so people can understand and better manage natural phenomena. Nearly fifty years ago, Alvin Weinberg observed that society would increasingly use science to understand complex problems, many of which had their roots in science and technology. He noted, insightfully, that such problems "hang on the answers to questions that can be asked of science and yet which cannot be answered by science." Weinberg – who ran the Oak Ridge National Laboratory (originally part of the Manhattan Project) and was a tireless advocate for nuclear energy – coined the name *trans-scientific* for processes that are contingent and ever in flux.[38]

Ecological, social, and economic systems are trans-scientific. Unlike atoms of hydrogen, for example, which are identical and behave in predictable ways, at least at the atomic level, living creatures and the phenomena they create are variable, imprecise, uncertain, and forever subject to interpretation and debate. Unlike the more precise laws of physics, trans-science is messy. It involves multidimensional relationships linked in space and time. Any study involves the specific conditions (contexts in time and space) under which any phenomena is studied and the choices researchers make about how to define and study them. Thus, the trans-scientific behavior of creative systems can't be predicted by the sorts of general laws that physicists and chemists have called upon historically. While we can certainly understand the principles and processes that underlie trans-scientific phenomena, the world is still unpredictable – even if we figure out the rules of the game. Physical and biological features are continually in flux, so we must learn to embrace evolving patterns and processes that create behavioral changes.

An overabundance of incomplete truths, each defended by authorities, ensues when scientist try to answer trans-scientific questions. What is the cause of the obesity crisis? Are genetically engineered crops safe? Does exposure to any of thousands of synthetic chemicals affect childhood development and harm human health? What will be the future costs of a warming climate to a particular region? Do tax cuts stimulate economies? At what point might the human population – which increased from 3 billion to 7.6 billion between 1960 and 2017 and is on its way to 9.7 billion in 2050 – collapse ecologically, economically, and socially? In short, any complex question one chooses to ask is subject to the limitations of trans-science.

If scientific research and political debates over such questions seem to drag on endlessly, surely one reason is that we have the wrong expectations of science. We believe scientific truth is unequivocal – there is one fact of the matter, which is why the light always goes on when I flip the switch. But trans-scientific questions reveal multiple truths, depending on what aspects of an issue a scientist decides to study, which part of the elephant is under consideration. There is no point in asking who is right or wrong. The best we can hope for is to understand why experts disagree, which often hinges on highly technical choices about how to carry out research and analyze and interpret data, influenced by biases that impact those choices.

Appreciating the interconnectedness and dynamism of trans-science complicates ethical issues of responsibility, acclaim, and blame. It is no longer possible to assign linear cause-and-effect relationships in a conventional sense. Each act is influenced by a multitude of factors linked in time and space so we can't assign cause. Who's responsible for the obesity crisis, climate change, the arms race, or social unrest? In embracing trans-science, we give up the assumption there is an individual, or an individual agent, responsible. This perspective suggests everyone shares responsibility for outcomes generated by a system – ecologically, economically, and socially. Life is a participatory sport: As we play the game, we change the playing field.

From this locus, researchers must accept that we can't separate ourselves from the systems we study or our thoughts about those systems. We are entangled participators, not observers. Quantum physicists accept this property of nature. In *The Tao of Physics*, Fritjof Capra shows that, like the mystic, the modern physicist sees a universe of inseparable, interacting, and ever-changing components with the observer as an integral part of these processes.

Sarewitz concludes his review article on saving science as follows:

> *Advancing according to its own logic, much of science has lost sight of the better world it is supposed to help create. Shielded from accountability to anything outside of itself, the freedom to explore and discover begins to seem like little more than a cover for indifference and irresponsibility. The tragic irony is that the stunted imagination of mainstream science is a consequence of the very autonomy that scientists insist is the key to their success. Only through direct engagement with the real world can science free itself to rediscover the path toward truth.*

Programs like Behavioral Education for Human, Animal, Vegetation, and Ecosystem Management (BEHAVE) show that innovative science can be linked with groundbreaking practices to enhance ecological, economic, and social well-being of communities and landscapes, as I discuss further in *Foraging Behaviors*. The research that laid the foundation for behavior-based management of landscapes inspired researchers in such diverse disciplines as animal behavior and welfare, ruminant and human nutrition, wildlife damage science and management, veterinary science, plant chemical ecology, landscape restoration ecology, and pasture and rangeland science and management. These efforts led to the formation in 2001 of the international BEHAVE network of scientists and managers dedicated to integrating understanding of behavioral principles and processes with local knowledge to enhance ecological, social, and economic values.

The interchange between scientists and managers nurtures innovation. Everyone is a student striving to understand creative systems and to inspire people to understand and use knowledge of behavior to create relationships that integrate different ideas about how to manage landscapes. Once people grasp behavioral principles, they can create practices that are innovative, inclusive, and self-transforming. Embracing the inevitability of change alters people's beliefs and practices from rigid, unyielding, and draining to fluid, malleable, and invigorating. In this process, researchers and managers view themselves as part of a larger biophysical system, cultivating the capacity for self-renewal while learning to embrace change and uncertainty. They are participating in understanding and managing creative systems.

## Strands in a Web Characterized by Change

I spent many days during my childhood roaming on Methodist Mountain, south of Salida in Colorado. I recall long climbs to the "boot," an area midway up the mountain where the lack of conifer trees fashioned a boot, covered with grasses and forbs. Yet, in all those years, I never made the trek to the top of the mountain. I climbed Methodist Mountain for the first time in September of 1995. I was in Salida to visit with my father and four sisters at my childhood home; my mother was in the final stages of a ten-year ordeal with Alzheimer's disease.

I didn't set out to climb the mountain that fall day. What began as an early morning run ended up taking me twelve miles to the top of the mountain. As I began my journey in Salida, all I could see was the town, but as I gained elevation, I realized Salida was nestled in a valley surrounded by immense mountains. The higher I climbed, the better my perspective of Salida in relation to the 14,000-foot peaks that surround the area. During my journey, the location of Salida never changed, but my perspective of Salida changed dramatically. When I reached the summit, my views were still changing, and I realized that the more I climbed, the more my perspectives changed as the landscape emerged. It never ends. I'll never reach the summit.

During the past forty years, our research group published more than 250 papers in scientific journals. Assuming each paper reported on five experiments, we conducted roughly 1,250 different studies. During that time, I became increasingly struck by four realizations.

First, we found significant differences among treatments in most of the studies we conducted. Why? It wasn't because we were so clever. It was because we knew enough about the plants and animals we were studying to design experiments – create contexts and conditions – that gave insights into facets of the process we were studying.

Second, the more we learned, the more variables we had to control. Why? Everything is linked in time and space, as we've discussed throughout this book. The more we learned about how systems cocreate, the more I came to appreciate that our goal wasn't to predict or control. Rather, the goal was to understand how systems continually create their way into now.

Third, I came to appreciate that the less I knew, the less I realized how little I knew, and the less likely I was to recognize my ignorance. The more I learned, the more I could see how very little I knew about anything and the interconnectedness of everything.

Finally, I came to realize there is no absolute truth in science. All theories and concepts are limited and approximate. Science is a quest to understand natural phenomena. But science cannot be perceived as true or final in any absolute sense. Science is a tentative organization of working hypotheses that, for the moment, best account for processes whose biophysical elements are tightly intertwined strands of a multidimensional web characterized by change.

Creative systems emerge from an endless array of interactions occurring as myriad agents in ecosystems interact. Our Western culture teaches us to think in linear, hierarchical ways about these relationships, but there is no one central force controlling outcomes, only untold numbers of agents, all interacting and transforming with one another and their local environments. Highly complex patterns emerge, ever transform, and disappear from the local interactions among all of the participants. Due to the complexity of these interactions, modifications we make to the system produce outcomes we can never anticipate or predict in advance. That means that we must continually transform with ever-changing environments, we must participate in creating.

We don't understand creative systems except in a general way. Whenever we think we understand them, we learn that we don't. We simply interact with them, and adjust our behavior accordingly based on any feedback we can gain. Even our most enlightened efforts have unintended outcomes, either because we don't know enough, or because the world responds to our actions in ways no one expects, or because we are unable to transform with the changes we help to create. Ironically, while we are quick to point out how people were mistaken in the past, we don't acknowledge that we are equally naive. We dismiss earlier errors as bad thinking by people less knowledgeable and then confidently embark on new blunders of our own.

In 1949, Harvard psychologists recruited two dozen undergraduate students for an experiment on perception.[39] The students were shown a deck of cards and asked to identify individual cards as they were flipped. Most of the cards were true to form, but a few of the cards were altered so the deck contained, for example, a red six of spades and a black four of hearts. The students tended to overlook the incongruities if the cards were flipped quickly, calling the red six of spades a six of hearts or a black four of hearts a four of spades. However, when the cards were flipped more slowly, students had difficulty making sense of what they were seeing. The experiment revealed how people process dissonant information. Our first impulse is to force the unfamiliar discordant information to fit into a familiar framework. At the point the incongruity becomes flagrant, a crisis ensues – what the psychologists christened the "'My God!' reaction" and what Thomas Kuhn called a paradigm shift.

Kuhn was an American physicist, historian, and philosopher of science whose book *The Structure of Scientific Revolutions* was influential in both academic and popular circles. Kuhn introduced the term *paradigm shift*, which has since become an English-language idiom. Letting go of old beliefs is hard work for anyone, scientists included, which is why the paradigm changes Kuhn describes are so monumental – on the rare occasions when they do occur. As John Kenneth Galbraith famously said, "Faced with the choice between changing one's mind and proving that there is no need to do so, almost everyone gets busy on the proof."[40]

The strength of science is the premise that no one can ever prove a theory. That, in theory, keeps disciplines humble in the face of uncertainty and mystery. Science isn't perfect, but it is the best way anyone has yet devised to gain some understanding of the cosmos. It isn't all that different from Churchill's assessment of democracy as the worst form of government – except for all those other forms that have been tried. But contemporary science is trapped in a self-destructive vortex. To escape, science will have to relinquish its protected ivory tower status and embrace both its limits and its accountability to the rest of humanity. Science for the sake of science accomplishes little of value. Science that tackles important issues of the day and enhances understanding of natural phenomena is of genuine value both to science and to society.

So, how can we understand and manage relationships among facets of poorly understood, ever-changing ecological, social, and economic systems in light of a future not knowable or necessarily predictable? We typically think the solution lies in more sophisticated science and mathematical models or better management. But I'm no longer convinced that's the case. Rather, I've come to believe we must integrate ecological, social, and economic values by creatively engaging with people who have different experiences.[41] In the arena of constant transformation, anything is possible if we engage with one another and the environments we inhabit in ways that nurture creativity. Creativity comes from transcending the boundaries we create. Suspending assumptions – speaking and listening from our hearts – liberates scientists and managers from the arbitrary boundaries of prevailing theories and best management practices.

In *Call of the Reed Warbler*, Charles Massy describes vital lessons learned by innovators in regenerative agriculture.

> *In the end, the story of their minds opening up came down to three key things: The first was they began to understand how the key landscape functions and the entire ecological system worked, and how all were indivisibly connected: that none could function in isolation. The second was they got out of the way to let nature repair, self-organize and regenerate these functions. And the third and vital factor was they had the humility to 'listen to their land', to then change but also continue to learn with that same openness.*

Learning can't occur when hubris, arrogance, narcissism, and the like reign. Confronted with these disorders, radical changes in thoughts and actions occur only through crushing blows to what the cognitive-rational-analytical mind thinks it knows. The humility that ensues from devastating and humbling defeat opens the door for dialog with nature. When that occurs locally, individuals change. When it occurs globally, societies change. Ironically, the first step on the road from hubris to humility back to hubris is thinking you've discovered the answer. The next step is thinking your revolutionary answer will save the planet: "What a good boy am I."

# PART V

# Fading into Mystery

## CHAPTER 16

# The Harmony of Nature

Sue and I walked nearly four miles along a snow-dusted trail on this November day, accompanied by the melodies of Frenchman Creek tumbling out of the mountains, over moss-covered rocks, and into the valley below. We began our ascent where ponderosa pines thrive. We continued on through forests of lodgepole pines to 10,500 feet. Farther still, we trekked through neighborhoods of spruces and firs, up to 12,000 feet, where no trees grow.

We finally arrived at an alpine meadow, a patchwork of golden brown grasses and forbs punctuated everywhere by the red-gray bark of willows. The roofs of the beaver chalets on Frenchman Creek were translucent, icy blue, with water pouring through the wooden walls of the ice castles. The meadow was encircled by steely granite peaks that jutted 14,000 feet into the deep blue sky. We dined on a peanut-and-jelly sandwich and an apple, as three mountain goats fed on grasses and forbs that carpeted slopes to the west.

I've read signs at the entrances to national parks and wilderness areas that attribute destruction of alpine habitats to foraging by these "invasive" wooly white beings of the peaks and crags. As I watch the three ghostly forms move silently across the slopes, I find it hard to believe they are the sole destroyers. They are generations removed from their ancestors, who were transplanted by humans into the mountains south of here. They didn't come of their own volition. I doubt they know they're invasive species destroying "pristine" alpine environments. How ironic, I think, that the most invasive species on the planet labels other species invasive. As we eat our sandwich and fruit, I'm consumed by thoughts about the ways we battle against the constancy and inevitability of transformation in matters ecological, economic, and social.

## Native Species, Untrammeled Wilderness

Until I enrolled in undergraduate studies in wildlife biology, I didn't discern native from invasive species or untrammeled from trammeled wildernesses. These are things I was taught and the connotation was clear: If an organism wasn't native or an environment wasn't untouched by our European ancestors, then it wasn't deemed to be in harmony with nature. On the ranch where I worked at the time, I knew I was surrounded by invasive species of plants such as alfalfa, orchardgrass, and barley and animals the likes of goats, sheep, and cattle, but I didn't realize the region's brown trout and mountain goats were invasive species, too.

My disbelief developed in earnest when, as a graduate student, I realized range ecology was based on discovering "pristine" or "relict" sites, and much of land management was built on attempts to return plant and animal communities to the prehistoric conditions that existed prior to European colonization. At that time, few were aware and no one acknowledged how much aboriginal peoples had modified plant and animal communities historically.

Since then, people such as agricultural ecologist and ethnobotanist Gary Paul Nabhan have described how four to twelve million aboriginal peoples "speaking two hundred languages variously burned, pruned, hunted, hacked, cleared, irrigated, and planted in an astonishing diversity of habitats for centuries."[1] When these newly arrived "colonists" moved from the Bering Strait into ice-free country, they played a role in Pleistocene extinctions, which amounted to a loss of 73 percent of North American mammals weighing 100 pounds or more. As E. C. Pielou writes in *After the Ice Age*, aboriginal peoples contributed to the extinctions that occurred from 12,000 to 9,000 years ago: Nearly fifty species of mammals went extinct due to changing climates and hunting by Clovis peoples. As in North America, most of Australia's large mammals went extinct after humans colonized the continent 50,000 to 45,000 years ago.[2]

Despite these dynamics, many scientific disciplines and management practices are grounded in how people imagine things used to be historically. We believe nature in the absence of humans is unspoiled and that through our actions, we've upset the balance of nature. This belief romanticizes nature in harmony, a notion that appeals to our desire for constancy. But the facts don't match our perceptions – climates, landscapes, and life change relentlessly, remorselessly, and stochastically.[3] These facts we little concede or embrace. "All conservation" wrote G. K. Chesterton, "is based upon the idea that if you leave things alone you leave them as they are. But you do not. If you leave a thing alone, you leave it to a torrent of change."[4]

In ecology, the term *succession* refers to changes in organisms that inhabit a landscape. Primary succession begins on new, unoccupied habitat, such as a lava flow or bare ground on a landslide, as pioneering plants and animals invade and then create conditions for other species to invade. The community gradually develops through increasing complexity until it becomes briefly stable – the so-called climax community. Secondary succession may be initiated by any number of disturbances including logging, grazing, fire, or blowdowns due to high winds.

While the term succession refers to plants and animals, it applies to any ecological interaction, including those with viruses, bacteria, and immune systems. The immune system protects against a range of agents, from viruses to parasitic worms, by distinguishing pathogens from the organism's own healthy tissues. Even unicellular bacteria have a basic immune system in the form of enzymes that protect them against bacteriophage infections. But pathogens evolve ways to avoid detection and neutralization by the immune system. In response, immune systems evolve still other defense mechanisms that recognize and neutralize pathogens – an unending game of attack and counterattack. Plants and herbivores, too, are involved in games of escalation. Plants produce herbicides to retard the growth of other plants and pesticides to resist attack by herbivores, which in turn develop ways to counter plant chemical defenses. People, too, enter into the game of succession in the many wars we declare – on predators, on insects, on weeds, on cancer, and on one another. All these wars escalate as creatures coevolve.

Succession has been at the heart of a thought-provoking issue in the Greater Yellowstone Ecosystem. Humans declared war on wolves early in the twentieth century to minimize predation on wild and domestic herbivores. When wolves were removed from Yellowstone early in the twentieth century, riparian areas changed due to excessive browsing on willows by elk that were no longer being hunted by wolves. Wolves were reintroduced into the northern range of Yellowstone in 1995. The growth of the wolf population from 1995 to 2010 coincided with a 70 percent decline in the number of elk.[5] With fewer elk munching on willows, many people thought willows would flourish and riparian areas would be restored. But that didn't happen.

Rather, riparian ecosystems in Yellowstone are now determined not by wolf predation on elk but by a lack of beavers.[6] Beavers and willows interact symbiotically in riparian ecosystems. By offering essential food and dam-building materials, willows create habitat for beavers. By making dams, beavers create conditions well suited to willows. Without willows and beaver dams, fast-flowing streams cut deep into terrain in many of Yellowstone's streams. The water table dropped below the reach of willow roots and willows disappeared.

There's irony in this tale, as ecologist Arthur Middleton wrote in an article for the *New York Times*:

> *This story – that wolves fixed a broken Yellowstone by killing and frightening elk – is one of ecology's most famous. It's the classic example of what's called a "trophic cascade" and has appeared in textbooks, on National Geographic centerfolds and in this newspaper. Americans may know this story better than any other from ecology, and its grip on our imagination is one of the field's proudest contributions to wildlife conservation. But there is a problem with the story: It's not true.*[7]

Reestablishing large carnivores isn't likely to fix the many environmental changes that have occurred due to past carnivore extirpation.[8] Changes in social and physical environments that transpired after the removal of large carnivores make systems resistant to restoration. No surprise, then, the hypothesis – that wolf restoration in Yellowstone National Park caused ecosystem restoration to prehistoric times – was eventually rejected in scientific studies. Yet subsequent literature and cinema ignored these findings and in its place repeated the story that Yellowstone had been restored following the reintroduction of wolves.[9]

We also attempt to get rid of "invasive species," not realizing that during the past 18,000 years many of the species we consider natives – from spruces, pines, and maples to mule deer, elk, and mountain sheep – are new arrivals. We pick points in the past we consider to be "pristine" and try to recreate those conditions by removing "invasive species" and adding species we consider were "native." How successful have we been? More than 50,000 invasive species of plants and animals now live in the United States, and the number is rising.[10] We spend more than $120 billion annually attempting to control and mitigate their damages, with little success.

We take solace in narratives about *Homo sapiens* putting nature back in harmony – like the story of wolves killing elk to save willows – yet nothing could be further from the truth. As contexts and contingencies change, the behavior of ecosystems change and no amount of intervention is going to return the system to the way things used to be. When humans declare war on any part of nature, even something as limited as making a change in the Yellowstone ecosystem, it's likely that the consequences will be other than what we hoped for or predicted.

If supposed states of harmony and balance don't exist, then attempts to describe untrammeled settings and restore nature to unspoiled states are pointless. We attempt in vain to reconstruct and interpret snapshots of times past, which we assume to be the way things were, and to predict futures that never will be. We are "photographing" moving targets: Things never were the way they were and they never will be again. Moreover, we humans are participating in cocreating nature's dynamism and unpredictability, which ultimately is impossible to control, no matter how much Cartesian reasoning and technological knowhow we attempt to assert upon it.

## Not So Silent Spring

In *Silent Spring*, Rachel Carson gives a nod to Charles Darwin: "If Darwin were alive today the insect world would delight and astound him with its impressive verification of his theories of the survival of the fittest." She points out that before 1945 only about a dozen species had developed resistance to pesticides, but by 1960, 137 species had become resistant to pesticides. She emphasized how quickly insects became resistant, in some cases within a year. Carson states: "Humbleness is in order, there is no excuse for scientific conceit here."

She devotes the final chapter of her book to what appeared at the time to be better ways to control insect pests biologically, as opposed to chemically, including research on using beneficial insects to predate on pest insects, including imported species; various techniques to sterilize males; pheromones to attract insects and poisons to then kill them; and biological control with spores of a bacterium, *Bacillus thuringiensis* (Bt).

Carson concludes with a warning about biocides and humans:

> *The "control of nature" is a phrase conceived in arrogance, born of the Neanderthal age of biology and philosophy, when it was supposed that nature exists for the convenience of man. The concepts and practices of applied entomology for the most part date from*

> *that Stone Age of science. It is our alarming misfortune that so primitive a science has armed itself with the most modern and terrible weapons, and that in turning them against the insects it has also turned them against the earth.*

She was unaware that the novel techniques she was promoting were no different from the Stone Age science that she was rebuking in Darwin's survival of the fittest. In November 2009, scientists found the pink bollworm had become resistant to first-generation Bt cotton in Gujarat, India, as well as in Australia, China, Spain, and the United States.[11] Monsanto countered with a second-generation cotton with multiple Bt proteins, but insects not affected by Bt toxins surged within a few years after farmers began to use Bt cotton.[12] In China, for instance, the mirids – often called leaf bugs or grass bugs – have in some cases eliminated benefits of Bt cotton. More pesticides are now needed to control emerging pests such as aphids, spider mites, and lygus bugs. Alternatively, organic farming practices can enhance pest control and match or outperform the abilities of conventional farming practices to limit pathogens and animal pest infestations.[13]

## War on Weeds, War on Truth

Weeds are nature's Band-Aids that prepare soil for a succession of perennial plants. According to tenets of conventional agriculture, weeds compete with crops for water, nutrients, and sunlight, which decreases crop production and causes billions of dollars in crop losses worldwide annually. Although alternatives to use of herbicides have been developed and verified, Monsanto has chosen to ignore them and focus its might on developing and marketing a patented package: GMO varieties of corn, soybeans, and other crops resistant to the company's signature herbicide Roundup, the most extensively used herbicide in the history of agriculture.[14]

Use of the GMO seed allows more use of Roundup to produce superior weed control – at least in theory. Some scientists argue herbicide-resistant GMO crops are less harmful than conventional crops, and they consider the study by PG Economics, a consulting firm in Dorchester, UK, to be one of the most extensive assessments of environmental impacts. That study contends GMO technology reduced spraying by 474 million kilograms globally from 1996 to 2011.[15] According to Graham Brookes, codirector of PG Economics and a coauthor of the industry-funded study, GMO technology delivered an 8.9 percent improvement to the environmental impact quotient – a measure that considers factors like toxicity to wildlife.

Conversely, other scientists maintain the use of herbicides increased by 174 million kilograms in the United States from 1996 to 2008, the first thirteen years of GMO technology.[16] This dramatic increase overwhelmed the decrease in use of insecticides attributable to GMO crops such as Bt corn and Bt cotton, making the overall chemical footprint of GMO crops distinctly negative. The high cost of skyrocketing herbicide use erodes farm-level returns and poses public health concerns. That's news to many people because marketing claims lead us to believe crops derived from biotechnology are reducing pesticide use and environmental hazards.

Use of other types of herbicides has also increased as weeds become resistant to glyphosate, the active ingredient in Roundup.[17] The number of glyphosate-resistant weeds has increased since 1996, with forty species globally.[18] While many of these weeds first appeared in cropping systems where glyphosate was used without a glyphosate-resistant crop, the most severe outbreaks have occurred where glyphosate-resistant crops have facilitated overuse of Roundup. The list includes many of the most challenging weeds, including Palmer amaranth (*Amaranthus palmeri*), common ragweed (*Ambrosia artemisiifolia*), horseweed (*Conyza canadensis*), and Johnson grass (*Sorghum halepense*), which now occur on millions of hectares.

As a solution to glyphosate-resistant weeds, companies developed second-generation GMO crops with "stacked resistance" to multiple herbicides, including cultivars of soybean, cotton, corn, and canola resistant to glyphosate, dicamba, *and* 2,4-D.[19] People in industry theorize planting cultivars with stacked traits will prevent evolution of herbicide-resistant weeds because weeds will have to become resistant to all three herbicides. They argue the likelihood of a mutation conferring resistance to a single herbicide is small, so the likelihood of resistance to two herbicides is even less. However, when an herbicide with a new mode of action is introduced into a region where weeds resistant to an older mode of action are common, the probability of selecting for multiple target-site resistance is not the product of two independent, low-probability mutations. Rather, the value is closer to the probability of finding a resistant mutation to the new mode of action within a population already widely resistant to the old mode of herbicide action.

No surprise, then, that several weeds have evolved cross-resistance, in which a metabolic adaptation allows them to degrade herbicides with several different modes of action.[20] As of January 2018, there were 487 unique cases of herbicide resistant weeds globally, with 253 species (147 dicots and 106 monocots). Weeds have evolved resistance to 23 of the 26 known herbicide sites of action and to 163 different herbicides. Herbicide-resistant weeds have been reported in ninety-two crops in seventy countries. One hundred and eight biotypes in thirty-eight weed species across twelve families are resistant to two or more herbicides. And 44 percent evolved since 2005.

Researchers predict herbicide use will rise from 1.5 kilograms/hectare in 2013 to more than 3.5 kilograms/hectare in 2025 in the United States for biological as well as financial reasons.[21] Crops with stacked resistance will increase weed resistance, further increasing use of herbicides. That's unlikely to change as seed and chemical companies garner vast revenues selling herbicides and transgenic seeds. Developing new crop varieties is lucrative business. Each herbicide-tolerant trait qualifies the patent holder for a technology fee premium.[22] Little wonder, then, research that concludes glyphosate is safe comes from industry or is carried out by researchers who have financial and professional conflicts of interest.

Several studies raise concerns about effects on liver and kidney function when animals eat GMO foods.[23] Typical ninety-day tests are inadequate to evaluate chronic toxicity. Studies must be longer and mandatory; sex hormones and reproduction must be assessed; studies must be multigenerational; and research must be done by labs independent of industry. Currently, no regulatory authority requires such studies for edible GMOs and formulated pesticides.

In a long-term study, French molecular biologist Gilles-Eric Séralini and colleagues fed rats GMO maize (NK603) and provided them water that contained glyphosate. Their two-year study demonstrated adverse effects on liver and kidney function, which was more severe in males than females. Females developed 3.25 times more mammary tumors, which led to premature deaths, compared with female rats in the control group. These ailments were associated with a 2.4-times increase in pituitary dysfunctions by the end of the trial. In 2012, they published these findings in the journal *Food and Chemical Toxicology*.[24]

In 2014, Elsevier, the publisher of the journal, retracted their paper, allegedly because the strain of laboratory rats (Sprague-Dawley) they used is prone to cancer and the number of animals (ten per treatment) was too small to assess risk of cancer. The editor-in-chief requested the retraction while acknowledging no misconduct or misinterpretation of data. This was a rare and extraordinary action. Séralini and colleagues republished their paper in *Environmental Sciences Europe*.[25] In another paper, they argue the decision to retract the paper can't be rationalized scientifically or ethically and they highlight conflicts of interest.[26]

Some contend their original paper was retracted due to "corporate terrorism" given the simultaneous placing of a Monsanto employee as an associate editor for the Biotechnology subsection of *Food and Chemical Toxicology*. As the *Institute of Science in Society* notes,

> *This is not just a blatant violation of publishing ethics, it means conspiring to remove from the public record results that could be of great importance for public health. Furthermore, it is an abuse of science and amounts to corporate terrorism on independent science and scientists. It strikes at the very heart of science and democracy, and the aspiration of scientists to work for the public good.*[27]

As journalist Carey Gillam highlights in *Whitewash*, Monsanto has ghost-written research papers and articles that appear to come from independent scientists, paid scientists outside the company to promote Monsanto's version of the science, and set up front groups and organizations that appear to be independent but are backed by Monsanto.

## War on Farmers Below and Above Ground

Conventional agriculture has waged chemical and physical warfare on huge swaths of Earth, destroying once diverse communities of animals and plants aboveground and belowground as well as in rivers, lakes, and oceans worldwide. Traditional agronomic practices degrade soil structure and function by plowing soil and sowing crops, harvesting them with heavy machines, plowing under residual stubble, and leaving soil bare until the following growing season. High-yielding crops require high inputs of fertilizer, water, and herbicides to be able to grow and compete with weeds.[28] At the rate we are diminishing soil health due to agricultural practices, we have roughly sixty years before we deplete the ability of soils to grow crops.

Unlike their native counterparts, crops such as corn, wheat, and soybeans are selected to allocate most of their resources to grain. They provide little litter or root exudates to nourish life in the soil. Without litter on the soil surface, organisms such as beetles, millipedes, earthworms, and fungi that process litter disappear.[29] Belowground, bacteria quickly decompose plowed-under stubble, respiring carbon dioxide and adding little to building humus – the more enduring form of organic matter. In *Under Ground*, author Yvonne Baskin writes: "Homeless, underfed, and out of work, an array of creatures from mycorrhizal fungi and fungal-feeding nematodes to pest-eating spiders drop out, leaving behind an impoverished, bacterial-dominated food web."

Traditionally, farmers and researchers at land-grant universities cooperated in generating new varieties of crops. They built on one another's successes to continually evolve varieties that thrived under local conditions. Not unlike what happens in natural landscapes, they were creating new varieties as local conditions changed. As exchanges of locally evolving varieties of seeds among farmers have been supplanted by patented seeds and herbicides sold by giant corporations, seed and pesticide industries have entered into the game of succession, with implications for resilience. Nowadays, corporate control of agriculture is constraining crops to a few transgenic plants that are relatively productive in a range of environments, rather than expanding diversity to include an array of plants valuable in diverse local environments. Increasing homogeneity in global food supplies has frightening implications for food security.[30]

Seed and chemical industries have consolidated their efforts to pressure farmers to use herbicide-resistant GMO crops. Once enough growers adopt resistant plants, remaining growers must follow suit to reduce risk of crop injury and yield loss by their neighbors' herbicide-resistant crops. Growers with concerns about the new technology struggle to find high-yielding varieties that don't contain transgenic herbicide-resistance traits. Short-term fixes provided by GMO technology discourage research and education in more ecological practices.

In *Saving Capitalism*, Robert Reich discusses how democracy has been corrupted by corporations through the revolving door between Washington, DC, and businesses. Monsanto's dominance, for example, grew out of a carefully crafted strategy to patent genetically modified seeds and create herbicides to kill weeds but not crops. This combination, which initially saved farmers time and money, came with a catch: Monsanto's engineered crops don't produce seeds of their own, so farmers become permanent customers of Monsanto. To further ensure control, Monsanto prohibits seed dealers from stocking competitors' seeds, and Monsanto has bought most of the remaining small companies. In less than fifteen years, most of America's commodity crop farmers have become dependent on Monsanto. The result is higher prices, far beyond the rise in cost of living. Between

1994 and 2011, the price of corn seed rose 259 percent and the average cost of planting an acre of soybeans increased 325 percent.

At every stage, Monsanto's growing economic power has enhanced its political power and ability to change laws to its advantage, further adding to its economic power. Monsanto has increased protection of its intellectual property in genetically engineered seeds, beginning with the Plant Variety Protection Act of 1970 and extending through a series of court cases. Monsanto also used its clout in Washington, DC, to fight moves in other nations to ban genetically modified seed. For many years, Monsanto effectively fought attempts in states and in Congress to require labeling of genetically engineered foods. Yet if the GMO/glyphosate combination is so environmentally friendly and safe for humans, why did they so strongly resist labeling?

To enforce and ensure dominance, Monsanto employs a bastion of lawyers. They've sued other companies for patent infringement and sued farmers who save seed for replanting. Their lawyers have prevented independent scientists from studying its seeds, arguing that would infringe on their patents. In 2012, Monsanto put an end to a two-year investigation by the antitrust division of the Justice Department into Monsanto's dominance of the seed industry. They use mandatory licensing agreements to require competitors to use whole lines of their products and to prevent customers from using competing products. Favorable court rulings, advantageous laws, and administrative decisions to forgo antitrust lawsuits or bring them against competitors extend these de facto standards to entire sectors of the economy.

Reich argues this can be reversed through grassroots efforts to rewrite legislation. Whether that's possible is debatable, given the revolving door between corporations and Washington, DC. As he points out, no one should expect this inevitable transition to occur smoothly. The moneyed interests have too much at stake. While they would be wise to support more equitable legislation – because they will do better with a smaller share of a faster-growing economy whose participants enjoy more of the gains and will be more secure in an inclusive society whose citizens feel they are being heard – they will nonetheless resist. The status quo is too comfortable, and the prospect of countervailing power too risky and unpredictable.

## Battling the Emperor of All Maladies

In 1971, President Nixon declared war on cancer. Since then, the federal government has spent over $200 billion for research on cancer, including money invested by public and private sectors and foundations. President Obama pledged $1 billion for a cancer moon shot focused on immunotherapies that activate the immune system to destroy cancers specific to individuals.

Ironically, a 2012 report in the journal *Nature* suggests scientists and drug companies haven't won the war on cancer because much of the research is unreliable.[31] A team of 100 researchers at Amgen Inc. and MD Anderson Cancer Center tried to replicate landmark discoveries for cancer. In forty-seven of the fifty-three studies, it was impossible. Folks often blame researchers who did the studies. No one mentions the dynamism and evolutionary dexterity of cancer cells may be "to blame."

While cancer-causing mutations are due to inherited and environmental factors, recent studies highlight the role of mutations that arise from errors as DNA replicates. Cancer Research UK announced in 2011 that 40 percent of cancers are due to environment and 60 percent aren't.[32] Based on health records from sixty-nine countries, preeminent cancer researchers C. Tomasetti and B. Vogelstein conclude that just one-third of the variation in cancer risk among tissues is due to inherited predispositions or environmental factors; two-thirds of cancer-causing genetic mutations arise when a healthy, dividing cell makes random mistakes as it copies its DNA.[33] That's why as the number of stem cell divisions in the lifetime of a given tissue increases, the lifetime risk of cancer in that tissue also increases. That's also why the older we get, the more likely we are to get cancer. Tomasetti and Vogelstein go to great lengths to point out, though, that doesn't mean two-thirds of cancers are beyond the reach of prevention as we've discussed.

In *The Emperor of All Maladies*, physician, oncologist, and author Siddhartha Mukherjee describes the complexities involved in understanding and treating cancer. Cancer genome sequencing validates 100 years of clinical observations: Every patient's cancer is unique because every cancer genome is unique. The evolution of normal cells into cancer cells is an amazing display of life mutating, evolving, and ultimately consuming itself and its host.

Mukherjee describes Germaine, a psychologist who fought cancer with all the cunning she and medicine could muster. As he recalls: "Germaine had fought cancer obsessively, cannily, desperately, fiercely, madly, brilliantly, and zealously – as if channeling all the fierce, inventive energy of generations of men and women who had fought cancer in the past and would fight it in the future. " He concludes that Germaine captured something essential about the war against cancer: "In that haunted last night, hanging on to her life by no more than a tenuous thread, summoning all her strength and dignity as she wheeled herself to the privacy of her bathroom, it was as if she had encapsulated the essence of a four-thousand-year-old war." To keep pace with this malady, humans must keep inventing and reinventing, learning and unlearning strategies.

## War to End All Wars

The wars we declare – on invasive species, weeds, diseases, terror, drugs, obesity, climate change, and one another – and the battles we fight are against anything that threatens constancy under the guise of saving the world. Invasive species are metaphors for the ever-changing nature of life and for our attempts to go back to the way things used to be. While we occasionally win a battle, invasive species forever win the wars. In the blink of a cosmic eye, herbicides to end weeds generated herbicide-resistant plants and antibiotics to end bacterial infections created antibiotic resistant superbugs. And the First World War – the war to end all wars – morphed into the Second World War, the Cold War, and the War on Terror.

Not unlike arms races among plants and herbivores, military arms races are archetypes for vastly diminishing returns on complexity.[34] Nations quickly match an adversary's advances in weaponries, armies, logistics, or intelligence, so investments in these areas produce no enduring advantage or sanctuary. Ever more money, resources, and personnel are spent on that most ephemeral of returns: military advantage. The costs of being a competitive state incessantly rise, while the returns on investment relentlessly decline in an all-consuming race to nowhere.[35]

Like human societies from antiquity, nations still invest heavily in defense, conflict, and war. The total spent for all the nations of the world in 2016 was equivalent to $1.686 trillion. The United States topped the list at $611 billion, more than the combined total for the next eight nations: China, Russia, Saudi Arabia, India, France, United Kingdom, Japan, and Germany.

In fiscal year 2018, the federal budget is $4.094 trillion, while estimated revenue is $3.654 trillion. That amounts to a $440 billion deficit for October 1, 2017, through September 30, 2018. Spending is in three categories: mandatory ($2.535 trillion), discretionary ($1.244 trillion), and interest on the national debt ($315 billion). We have been fortunate because interest rates have been low due to worldwide demand for Treasury notes. Now that the global economy is strengthening, Treasury yields are rising, and so will interest payments. Interest on the almost $20 trillion debt is already the fastest-growing federal expense.

Mandatory spending is $2.535 trillion. Social Security is by far the biggest expense, at $1.005 trillion. Medicare is next, at $582 billion, followed by Medicaid at $404 billion. Discretionary spending is $1.244 trillion. By far, the largest category of discretionary spending is the Pentagon and related military programs, budgeted at $824.7 billion. The biggest expense is the Department of Defense base budget, at $574.5 billion. Related Overseas Contingency Operations cost $76.6 billion. Military spending also includes $173.6 billion for defense-related departments, including Homeland Security, the State Department, and Veterans Affairs.

In *Myths to Live By*, Joseph Campbell makes a case for the inevitable necessity of allocating resources for defense on planet Earth:

> *It is for an obvious reason far easier to name examples of mythologies of war than mythologies of peace; for not only has conflict between groups been normal to human experience, but there is also the cruel fact to be recognized that killing is the precondition of all living whatsoever: Life lives on life, eats life, and would otherwise not exist. To some this terrible necessity is fundamentally unacceptable, and such people have, at times, brought forth mythologies of a way to perpetual peace. However, those have not been the people generally who have survived in what Darwin termed the universal struggle for existence. Rather, it has been those who have been reconciled to the nature of life on this earth. Plainly and simply: It has been the nations, tribes, and peoples bred to mythologies of war that have survived to communicate their life-supporting mythic lore to descendants.*[36]

Hunting peoples were warriors. They killed animals, and because meat supplies were limited, clashes arose among competing tribes. Compared with hunting and warring nomads of the northern plains, the largely settled village peoples of the tropics inhabited a vegetable environment, where plant food was their basic diet. So we might expect to find a more peaceful world there, with little requirement for war. But that wasn't the case: The prevailing belief throughout the tropical zones was based on observation that in the plant world new life arises from decay, from death comes life, and from the rotting of last year's growth new plants arise. Thus, the dominant mythological theme of many of these peoples supports the belief that through killing one increases life, and in those parts of the world arose the most horrible and grotesque rituals of human sacrifice, their inspiration arising from the notion that to activate life one kills.

"When we talk about settling the world's problems," Joseph Campbell concludes in *The Power of Myth*, "we're barking up the wrong tree. The world is perfect. It's a mess. It has always been a mess. We are not going to change it. Our job is to straighten out our own lives." Those sentiments are echoed in the Hindu view of life. "All talk of social progress," as Huston Smith writes in *The World's Religions* of the Hindu view of our visit to Earth, "of cleaning up the world, of creating the kingdom of heaven on earth – in short all dreams of utopia – are not just doomed to disappointment; they misjudge the world's purpose, which is not to rival paradise but to provide a training ground for the human spirit."[37]

Perhaps there is nothing to protect or preserve, just the illusion of the self, struggling for power, control, and permanence in a world of impermanence. Maybe only humans are naive enough to think the world needs to be saved and arrogant enough to think we're the ones to do it. Possibly, in our attempts to protect and preserve everything from ourselves to the planet, we miss the point. Maybe the point is to merely transform with loving kindness and compassion for all things – those we perceive as good *and* those we perceive as malevolent.

The greatest challenge we face in addressing any "critical" issue is crossing the divides that polarize and isolate us. The strange and wonderful irony is that now we must work together as a world community to transcend the boundaries we have created so we can address the really big issues – socially, economically, and ecologically. The airscapes, waterscapes, and landscapes of Earth are unified and influenced by the peoples of all nations around the globe.

In that sense, "love your enemies" has practical implications. By opening up to others, we increase the diversity of options upon which to act. Of all the challenges, love your enemies is the greatest, and we do that by declaring love, not war, on peoples and landscapes. While I take to heart the state of affairs on Earth, I can't help but wonder how the world would be if we spent all the money we allocate to wars on nurturing one another and mother Earth. I know that's an absurd thought, but we ought to give it a try before the sixth mass extinction takes us all out.

# CHAPTER 17

# Alice in Wonderland

The universe where Earth is located is approximately 13.8 billion years old. The Milky Way Galaxy where Earth resides – one of 100 to 200 billion galaxies in this universe – is roughly 13.6 billion years old, give or take 800 million years. The Milky Way is 100,000 light years in diameter, small compared to M87, an elliptical galaxy 980,000 light years in diameter, or Hercules A, which takes light 1.5 million years to cross.

Cosmologists are finding, much to their astonishment, that a massive black hole resides at the center of each galaxy, devouring anything that crosses its event horizon in a strangely synergistic way that benefits both the galaxy and the black hole. A black hole is formed when a massive amount of matter collapses into a tiny area. Think, for example, of a star ten times more massive than the Sun squeezed into a sphere roughly the diameter of New York City. The result is a gravitational field so strong nothing, not even light, can escape.

Light travels at 186,000 miles per second. Even at this amazing speed, the rays from the sun, now warming my body, require a little more than eight minutes to reach Earth. That means when we look at the sun, we don't see the sun as it is now. We see it as it was eight minutes ago. The farther away the object, the more we are looking into the past. If we lived 4.5 billion years from Earth and could travel far enough and fast enough toward Earth, we'd watch the history of Earth as a movie. Watching the movie unfold, we'd become aware of the changing nature of this planet. Rather than imperceptible changes during our fleeting tenure, we'd see continents crash together and pull apart; mountains rise and fall; vast deserts replace oceans; plant and animal species coming and going like the ever-changing colors and shapes of a kaleidoscope.

At 4.54 billion light years, we'd see the creation of this planet we call home. At 3.8 billion light years, we'd see life begin in the form of single-celled prokaryotic cyanobacteria. We'd witness striking changes in life on Earth ushered in during the five major extinction-regeneration events: the Ordovician–Silurian extinction 450 to 440 million years ago (Ma); the Late Devonian extinction 375 to 360 Ma; the Permian–Triassic extinction 252 Ma (Earth's largest extinction, which killed 57 percent of all families of organisms, 83 percent of all genera, and 90 to 96 percent of all species); the Triassic–Jurassic extinction 201.3 Ma; and most recently the Cretaceous–Paleogene extinction 66 Ma (the second largest extinction, which killed 27 percent of all families, 57 percent of all genera, and 60 to 70 percent of all species).

As we observed Earth from nearby, we'd see the more recent evolution that lead to a species called *Homo sapiens*. That drama has been playing out over the past ten million years, though anatomically modern *Homo sapiens* first appear in the fossil record in various regions across Africa 200,000 to 300,000 years ago. In the endless parade of species on Earth, *Homo sapiens* are unique in our ability to use science to fuel technology to massively impact Earth.

Nowhere is that more evident than in agriculture, though *Homo sapiens* didn't invent farming. Ants were the first farmers. They've been growing fungus for food in the rain forests of South America since the time of the Chicxulub meteor impact, which led to the Cretaceous–Paleogene extinction.[1] Ants grow fungi in climate-controlled chambers underground; they weed and water their gardens; they even use antibiotics to keep harmful bacteria away from their crops.

Only in the past 10,000 years did humans embark on an agricultural adventure, which now involves some 570 million farms worldwide. Over 90 percent of these farms are run by an individual or a family and rely primarily on family labor. Family farms produce about 80 percent of the world's food. Worldwide, agriculture employs more than 1.3 billion people, or close to 40 percent of the global workforce. In roughly fifty countries, agriculture employs half of the population, up to 75 percent in the poorer nations. In the United States, roughly 2.2 million farms cover 922 million acres. Agriculture is the world's largest provider of jobs.

The movie would reveal *Homo sapiens* flourishing during the warm, relatively stable climate of the Holocene. We discovered a source of energy – fossil fuels – that enabled our populations to grow exponentially, little different from bacteria provided with nutrients in a petri dish. In the process, we've declared wars on one another and on other species all over the planet to the extent that *Homo sapiens* are now participating in creating the sixth mass extinction-regeneration, dismantling the extant web of life at an accelerating rate. Except for the asteroid-caused extinction that killed the dinosaurs, the other four historical extinctions were all caused by climate change from greenhouse gasses. The Permian–Triassic episode began when carbon warmed the planet by 5°F, accelerated when that warming triggered the release of methane in the Arctic, and ended by killing nearly all life on Earth.

Through our use of fossil fuels, we are adding carbon to the atmosphere at a much faster rate than in the Permian-Triassic period. By most estimates, at least ten times more rapidly, and the rate is accelerating. Arctic regions are warming about twice as fast as other parts of Earth. The loss of frozen ground in the Arctic is in turn causing more warming. The rise in emissions has been so dramatic that arctic regions may be shifting from carbon sinks to carbon sources.

## Welcome to Wonderland

As a boy, I loved to disassemble watches and look at all the parts. I was mesmerized by the gears and springs, turning and spinning, the ruby jewels glistening. Even as a child, I knew I had to remember which pieces went where in order to put the watch back together again.

Nature doesn't seem to care about watch repair. If she cared, why has she gotten rid of nearly every biological part that ever existed? Paleontologists tell us 99.99 percent of the five billion species that once lived on Earth are extinct.[2] Nor has nature read Aldo Leopold's eloquent admonitions and precautions about intelligent tinkering: Keep all the cogs and wheels. Why does she ever alter the climates and habitats that sustain life? Why is she intent on disassembling and creating new watches? Maybe she's the ultimate tinkerer, ever creating.

All things in the universe are born, they enjoy their moment in the sun, and they pass away. Astronomers estimate the universe we inhabit was born 13.75 billion years ago. Our middle-aged sun and solar system were born 4.54 billion years ago. Cyanobacteria, some of the oldest forms of life on Earth, originated 3.5 billion to 2.8 billion years ago. Multicellular plants and animals began to appear in the oceans about one billion years ago. Plants and fungi began to colonize land, followed by arthropods and other animals around 530 million years ago. By comparison with these species, the 200,000 to 300,000 years *Homo sapiens* have been on Earth is a blink. And the kingdoms humans create last even less time. The Chinese Empire lasted

2,000 years, but no dynasty within it lasted more than 500 to 800 years. The Roman Empire passed away at roughly 1,500 years of age. The United States has existed a mere 240 years. The average lifetime of a Fortune 500 company is less than half that of a human being.

Earth's age seems long only in relation to a human lifetime. For perspective, if the age of Earth – 4.54 billion years – is scaled relative to the number of seconds in a year, 1 second is equal to 143 years. On that scale, a species that lives 10 million years is on Earth a mere 19.5 hours. Milankovitch Cycles, which alter climates at 100,000-year intervals, last only 11.7 minutes. The interglacial warming of 20,000 years we are experiencing, within the current ice age, will last 2.4 minutes. A human who lives 100 years is on Earth 0.7 seconds. Relative to Earth's life, we barely set foot on the world stage and we are snuffed out. Yet, for no apparent reason, we behave not as guests dining briefly at the table of life, but as permanent residents.

Like every individual, each species finally dies. A typical species goes extinct within ten million years of its first appearance, but some species survive with virtually no morphological changes for hundreds of millions of years. Mammalian species live about one million years, though some persist for ten million years. Given this average life span, the background extinction rate for mammals is expected to be approximately one species lost every 200 years. Of course, the occurrence of extinctions is uneven. Some centuries might see more than one species extinction, while several centuries might pass without the loss of any species. Yet, during the past 400 years, of the roughly 5,000 species of mammals now on Earth, 89 mammalian species have gone extinct, nearly 45 times the predicted rate, and another 169 mammal species are endangered.

Based on a sample of 27,600 terrestrial vertebrate species and a detailed analysis of 177 mammalian species, researchers recently showed that there is an extremely high dwindling of populations in vertebrates, even in common species of low concern.[3] Conservatively, almost 200 species of vertebrates have gone extinct in the last 100 years, a rate of 2 species per year. Compared with background or normal rates of extinction during the last 2 million years, it should have taken up to 10,000 years for these 200 species to go extinct. Extinctions of all species are now occurring at more than 1,000 times the background rate. One in five species now faces extinction, and some scientists estimate that up to half of existing plant and animal species may be extinct by 2100. From the perspective of a human lifetime, we can little appreciate the magnitude of the current rate of species extinctions.[4]

The mass of humans now on Earth, which is an order of magnitude greater than that of all wild mammals combined, is creating immense impacts on the global biomass of prominent taxa, including mammals, fish, and plants.[5] By creating long-term stressors and short-term shocks, humans are impacting land, water, and air that creatures require for life. Habitat loss, exploitation, competition from invasive species, pollution, and changing climates are causing appalling declines in the number, distribution, and size of populations of common and rare vertebrate species in the past few decades. Diminishing population sizes and ranges amount to a massive human-caused erosion of biodiversity. And the focus on iconic species such as polar bears has created a fallacy that Earth's biota is not imminently endangered, just slowly entering an episode of biodiversity loss. This view overlooks all of these current population declines.

The koalas of Australia are a revealing example. Herbivorous marsupials, koalas live in eucalyptus woodlands of Queensland, New South Wales, Victoria, and South Australia. They are arboreal (live in trees) and folivorous (eat leaves). Eucalyptus leaves make up most of their not-so-nutritious diet, so koalas are sedentary. They sleep as many as twenty hours a day. Koalas don't go into torpor, fly, or take shelter in hollows, so they lack ways to avoid weather extremes, which makes them particularly susceptible to the impacts of climate change.[6]

The distribution of koalas has shrunk by more than 50 percent since Europeans arrived in Australia in 1788. Their demise has come largely due to destruction of habitat, much of that fertile areas along watercourses, which provide refuge habitat. In 2000, Australia ranked fifth in the world in deforestation rates, having cleared 564,800 hectares. In rural areas, native forests were cut for wood and cleared for agriculture. In coastal areas, habitat has been lost to urbanization. Koalas can live in urban areas if enough trees are present, but urban populations have liabilities. Collisions with vehicles and attacks by dogs kill about 4,000 koalas annually. Increasingly erratic climate and high temperatures exacerbate loss of habitat. A heat wave and drought in 1979 to 1980, for example, killed 63 percent of the koala population in southwestern Queensland. Individuals that survived lived in good quality habitat along permanent watercourses. In suboptimal habitat away from permanent water, eucalyptus trees lost their leaves and the koalas were left with no food, water, or shelter. Some scientists are now exploring the use of fecal transplants to provide gut microbes to enable sensitive species such as koalas to eat plants they ordinarily are unable to consume in sub-optimal habitats.[7]

Another reason humans don't grasp the magnitude of extinctions is changes that occur over long spans of time are not perceptible from one generation of humans to the next. I fished the Madison River in Montana for the first time forty years ago. From Earthquake Lake to Ennis Lake, the Madison is one long riffle, tumbling over multicolored rocks, creating outstanding habitat for trout. Forty years ago, I was impressed by the number and size of trout I caught. So I was surprised one day when an old-timer told me he could no longer bear to fish the Madison – his memory was too long. What I thought were respectable trout – fourteen to eighteen inches or more in length – he considered inferior compared with the fish he'd caught fifty years ago.

When I fish the Madison now, I feel the sentiments he expressed forty years ago. I catch fewer trout and they are half the size of the fish I caught years ago. During the past century, the Madison River has become increasingly popular for anglers, especially guided float-fishing trips. An employee of a local shuttle service told me they run ninety shuttles each day during the height of the season, and they are just one of three shuttle companies. Little wonder the fish population has changed to fewer, more clever fish. The Madison River is a microcosm for what *Homo sapiens* have done to Earth without noticing because most of the destruction occurs gradually.

In western Germany, biologists, chemists, and electrical engineers, among others, documented decreases in abundance of flying insects over the past three decades.[8] In sixty-three nature preserves, equal to ninety-six location-year combinations, insect biomass declined by 76 percent, with midsummer declines as high as 82 percent. Loss of diversity and abundance causes downstream effects on the abundance of species that depend on insects, which play key roles in pollination, herbivory and detrivory, and nutrient cycling and as food for birds, mammals, and amphibians. An estimated 80 percent of plants depend on insects for pollination, while 60 percent of birds rely on insects as a source of food. The ecosystem services insects provide are estimated at $57 billion annually in the United States. Rachel Carson's *Silent Spring* has arrived.

Marine species are suffering, too. During 2015 to 2016, record temperatures triggered a pan-tropical episode of coral bleaching, the third global-scale event since mass bleaching was first documented in the 1980s.[9] Researchers were shocked at the severity of the latest major bleaching events, illustrated by aerial and underwater surveys of Australian reefs in 1998, 2002, and 2016. Water quality and fishing pressure had minimal effect on the unprecedented bleaching in 2016, suggesting that local protection of reefs affords no resistance to extreme heat. Nor did past exposure to bleaching in 1998 and 2002 lessen the severity of bleaching in 2016. Scientists conclude immediate global action to curb warming is essential to secure a future for coral reefs.

By analyzing fossil records, evolutionary biologist Leigh Van Valen showed that adaptation to past surroundings can adversely affect survival if environments change too quickly, a proposition he called the Red Queen Hypothesis.[10] To stave off demise, organisms must change at a rate equal with the rate of change. But eventually that doesn't happen. Sooner or later, biophysical environments will change more rapidly than do species. Then, like Alice and the Red Queen in Lewis Carroll's *Through the Looking Glass*, organisms end up "running" ever faster just to stay in the same place, and running in the same direction only makes matters worse.

In *Through the Looking Glass*, the Red Queen grabs Alice by the hand and off they go, running until Alice is gasping for air. "Faster! Faster!" the Red Queen cries. The wind nearly blows the hair off Alice's head, but the world stands still, as if the landscape below is running, too. "Faster!" the Red Queen cries, until Alice falls down, exhausted. She looks around and says, "But we've been under this same tree the whole time. Everything is just as it was." "Of course it is," the Red Queen replies. "It takes all the running you can do just to stay in the same place."

That's true for corporations, too. Organic food has never been so popular among American consumers, but, ironically, that's bad news for Whole Foods, the company that made organic a household name. In February 2017, Whole Foods reported its worst performance in a decade, revealing its sixth consecutive quarter of falling same-store sales and cutting its outlook for the year. The company is closing nine stores, the most it has ever closed at one time. In the recent past, Whole Foods projected it would grow its 470 locations to more than 1,200.

What happened? Four decades ago, organics were on the fringe. Today, they've become so thoroughly conventional that organic chains now face big-box competitors. Mass-market retailers accounted for 53 percent of organic food sales in 2015, according to the Organic Trade Association, while natural retailers were a little over 37 percent. Whole Foods is not the only store feeling the squeeze: Sprouts and Fresh Market, the second- and third-largest publicly traded organic stores, have also seen falling stock prices. In the process of exploiting existing niches, new niches are created by the players themselves, each competing to survive.

And so it goes: Energy from the sun is replaced by wood and coal are replaced by natural gas is replaced by nuclear power is replaced by who knows what? Handwritten letters are replaced by telephone calls are replaced by email and texts are replaced by Facebook, which will be replaced by who knows what? Travel by foot is replaced by horse and buggies are replaced by automobiles and airplanes are replaced by driverless vehicles are replaced who knows what? And the jobs that once supported these industries give way to new trades requiring different skills, and in many cases now, by robots with artificial intelligence that can learn and make far fewer mistakes than their human counterparts. And that's also true for indigenous groups and species, including *Homo sapiens*, who briefly come into and fade out of existence locally, regionally, nationally, and internationally: The age of dinosaurs gave way to the age of mammals gave way to the age of . . . existence in the process of creatively consuming itself.

As mathematical ecologist E. C. Pielou reminds us in *After the Ice Age*, ours is an unstable world. Plants, animals, and environments ever transform. She concludes extinctions of species are as inevitable as the deaths of individuals, and the roots of extinctions are as diverse as the causes of individual deaths. Calamitous predictions of global warming aside, the chilling past Pielou describes is likely a preview of upcoming warming and cooling events. And although we typically view death as an uninvited guest, every act of passing is also an act of creating. The mass extinction events have been followed by a proliferation of new species. By eliminating dominant groups, extinction events make way for new life – from individuals in communities to corporations in business-scapes to civilizations in human-scapes to species in landscapes.

Organisms are actors on stages they help create, contending in vain for roles they hope to sustain. All things – single cells, organisms, nations, dynasties, and species – enjoy a moment in the sun only to be ushered off by the next suite of participants in the game. Eventually, all things run out of creative energy: They are unable to cope with changes they helped to generate.[11]

## Civilizations Coming and Going in Wonderland

I like to visit a massive Douglas fir log that rests on a rise at the western edge of our woodlot. Its thick bark is gone. All that remains is the gray, brown, and reddish black fire-scarred timber where sap once flowed. The wood appears as gentle riffles, as if fashioned with an ax, though no ax ever touched this icon. In the log is a soil-filled cranny where some moss, a few grass plants, and a Douglas fir seedling have taken up residence. Has destiny doomed this seedling, growing in such an inauspicious dwelling? Will it put its roots into the soil below its current log home, still rock solid? Douglas fir trees live from 500 to 1,000 years. With time, this seedling, too, will pass, but how it will live and when it will die is a mystery only time will tell. I don't know, and I doubt if the seedling knows – or cares. Would it matter if we knew? So long as the sun rises and rain and snow fall, that will be enough to provide for yet another day.

Unlike Douglas fir trees, human civilizations typically live no more than 250 years or ten generations. During their life, they evolve through five stages William Ophuls – the pen name of Patrick Ophuls, an American political scientist, ecologist, and author – describes in *Immoderate Greatness*: the Ages of Pioneers, Commerce, Affluence, Intellect, and Decadence.

The Age of Pioneers is characterized by innovation, hard work, and a sense of shared purpose. This combination creates political stability and economic conditions favorable for the Age of Commerce with prowess in manufacturing, trade, and commercial ventures.

During the Age of Affluence, the society seems marvelous and invincible. Success breeds hubris manifest as arrogance, self-righteousness, overconfidence, and complacency. Beneath the surface, greed and selfishness overcome duty and service to country. A spoiled society is slowly rotting from within. Another symptom of rot is an increasing focus on welfare, with a corresponding loss of personal responsibility and independence.

Affluence sets the stage for the Age of Intellect. Some wealth in excess of need is used for pursuing knowledge. That produces important advances in many scientific fields. With time, however, the increase in intellectual output is accompanied by a decline in quality. The downside of this bourgeoning intellectual climate is excessive discussion, debate, and disagreement as public affairs drift from bad to worse amid unending disagreements that frustrate the power of action.

During the final stage – the Age of Decadence – moral decay, self-indulgence, corruption, and injustice reign. This age is characterized by narcissism, superficiality, aestheticism, hedonism, consumerism, materialism, nihilism, cynicism, pessimism, and fatalism – all manifest in religious and political fanaticism. Politicians are corrupt, and life is unjust as corporate and political elite accrue wealth and power at the expense of the citizenry, fostering fatal conflict of interests between the haves and the have-nots.

In the United States in the twenty-first century, we are leaving the Age of Intellect and entering the Age of Decadence. The threat is the concentration of power and resources in the political, corporate, and financial elite. They are rewriting the rules by which economies run.[12] Politicians serve the wealthy who put them in office, while they look down on common folks, whom they lie to regularly, believing them too unwitting to notice. In the United States we have increasing levels of social and economic inequity, combined with record levels of peacetime debt. At the time I am writing this, the only serious legislative achievement of a Republican-controlled Congress is a colossal transfer of wealth to the elite, funded by increasing the national debt by $1.5 trillion. We are experiencing the initial blows of climate change, and the current administration's other key achievement is wholesale removal of environmental regulations.

Representatives have deceived us about three economic practices: how markets have come to be organized differently now from a half century ago; why their current organization is failing to deliver the widely shared prosperity it did then; and what the basic rules of markets should be to distribute resources equitably. Legislation can encourage economies to create either virtuous or vicious cycles. Virtuous cycles occur when wages of workers increase. When that happens, workers buy more, productivity increases, companies hire workers, economies expand, workers get better educated, tax revenues increase, and governments invest more. Conversely, vicious cycles occur when wages stagnate. When that happens, workers buy less, companies downsize, tax revenues decrease, governments cut programs, the workforce becomes less educated, unemployment rises, and deficits grow. Enticed into believing the falsehood that work is indispensable to civilizations and lured by promises of rising prosperity, American workers have allowed themselves to be exploited for generations, only to reach this present moment when robotics may end the need for the wealthy to employ any human servants at all.

Those who attempt to govern a civilization in the process of decay are engaged in a war they can't win. The previous succession of ecological, social, and economic changes renders the society increasingly dysfunctional and ungovernable. Humans rightly concentrate on the affairs of the moment, yet the vagary and unpredictability of the future renders decisions increasingly useless. In early stages of a civilization, when affairs are relatively simple and morale is high, a united public thinks and acts with prudence, foresight, and respect for future generations. In latter stages, when affairs are complex and morale is low, selfishness trumps sacrifice, the interests of the masses and the elites diverge, and even the elites divide into warring factions.

Now more than at any point in modern history, both the Republican and Democratic parties have become so divided within themselves that historians, political analysts, and party members themselves say that these relatively long-lived political groups face the possibility of splintering. The rise of antiestablishment figures shows the desire for change in both the Republican and Democratic parties. And though the governing elites may appreciate the dysfunction, reform requires major sacrifice on their part, so they instead fight to preserve their privilege and power. As increasingly inept and corrupt leaders fail to confront critical issues, even capable and honest leaders have no viable way forward in the face of deteriorating ecological, economic, and social conditions.

Previous investments in social and physical infrastructures increasingly polarize and limit choices. Now, for instance, Republicans are ideologically against efforts to reduce carbon emissions linked with climate change, because they view that as regulatory overreach. Financially, they are hesitant to act because they receive the vast majority of their campaign contributions from oil, gas, coal, and other energy industries and from states that produce and consume the greatest amounts of fossil fuels. Politically, they are reluctant to act because they receive overwhelming support from states that produce oil and consume large quantities of coal-fired electricity for manufacturing. Democrats, on the other hand, depend primarily on the states mostly along the east and west coasts that produce little energy and have transitioned toward a low-carbon – information-age and service-based – economy. For each party, the economic interests of their core states reinforce their ideological inclinations. Thus, the infrastructures we've created in our mindscapes and across our landscapes become impossible to change.

In principle, regeneration is possible at any stage, but in practice that doesn't occur. Though these fateful patterns have been described repeatedly over centuries by many authors for numerous civilizations, humans are unable to appreciate what's happening during their own tenure. Until the brute forces of physical and moral decay impose themselves, people living in the various succeeding generations imagine themselves to be on a manifest march to greatness. People living in the Age of Pioneers – characterized by honor, shared purpose, and devotion to a strict moral code – can't imagine their actions are setting the stage for decay. Nor do people living during the Ages of Commerce or Affluence foresee the ecological exhaustion they are creating or the social degradation they are generating through welfare states that don't account for their long-term political, economic, sociological, or psychological consequences. Nor do those living in the Age of Intellect fathom that a society in which diverse beliefs are encouraged will foster division instead of consensus and render society ideologically polarized and intractable.

The evolution of human civilizations is characterized by seemingly inexorable evolution to ever-greater complexity, specialization, and socio-political control. Those changes are accompanied by processing greater quantities of energy and information; forming ever-larger settlements; and evolving more complex technologies. Cities become "black holes" for resources such as soil, water, and food "drained" from rural landscapes, which puts city and country inhabitants at odds over whose resources they are and for what uses. Biophysical limits are increasingly imposed by exponential growth of populations, ecological exhaustion of natural resources, and excessive complexity of human societies. Practical failures escalate as ever more energy is transformed into matter of little useful value. These changes all accelerate decay.

In *The Collapse of Complex Societies*, American anthropologist and historian Joseph Tainter argues that, while environmental degradation, diseases, and wars appear to cause societal collapse, the underlying reason is diminishing returns on investments in biophysical complexity and ensuing inability to solve ever more difficult problems.[13] As easy solutions for complicated issues are exhausted, problem solving moves to greater complexity, higher costs, and lower returns. Simply maintaining the attained complexity in infrastructure, regulation, and expertise consumes ever more resources. Disputes over how to allocate increasingly limited resources to maintain social services and infrastructures comes to dominate discussions from coffee shops to capitals. These challenges enable politicians who espouse seemingly simple solutions to complex problems to take control. Each increment of complexity appears to be a rational response to a challenge, but the society, running ever faster just to stay in the same place, eventually collapses.

We are now at the end of the 11,500-year-old Holocene Epoch, a time when we morphed from hunter-gatherers who lived in small clans into industrial-scale farmers and food producers. Industrial-scale ranching, farming, and feedlots are the extended family and clan subdivisions of labor that allowed small settlements to grow into cities as we developed technologies of mass production and distribution, and flexible economies based on money.[14] Hard-working humans today may think of a subsistence lifestyle as a constant struggle to survive, but the opposite is true. The !Kung San people of the poorly productive Kalahari Desert work just two and a half days per week to obtain all the food they need. The Kapauku Papuans of New Guinea work about two hours a day at agricultural tasks and ninety minutes a day fishing. They spend the rest of the day dancing, wrestling, or resting. That's also true of farm families in Volokolamsk, Russia: They value leisure more than marginal return on extra labor. Tainter summarizes the situation as follows: "The natives and peasants produce less than they might for the simple reason that increasing production yields diminishing returns on labor."

At one time, we thought human dominion over nature would lead to shared prosperity and leisure. If we created machines to do our work, we would have more time to spend with our families and communities, enhancing our minds and spirits. That promise never came to pass. Now the automation of labor is increasing more rapidly than ever, carrying once middle- and working-class people into unemployment and despair. The great promise of mechanical mastery of nature had generated swelling inequality, and for many, impoverishment and new forms of servitude. Given what "primitive" peoples once had, the irony couldn't be greater.

## One Minute till Midnight

The First Law of Thermodynamics states that energy can neither be created nor destroyed. The Second Law of Thermodynamics states that entropy increases as energy changes forms: Every transformation incurs a loss of useful energy. Inefficiencies characterized by a loss of useful energy (entropy) reaches an apex with industrial agriculture. People accelerate entropy by "mining" soil to produce food. The by-products include erosion, compaction, leaching, and losses of organic matter, water, and nutrients. Historically, nutrients in foods people ate were recycled into soil by much smaller aboriginal populations – estimates for North America range from 2.1 million to 18 million – than the 323 million people now living in the United States. While options exist for managing manure, nowadays nutrients in urine and feces are excreted into latrines and sewers and dispersed to rivers, lakes, and oceans never to return, except as pollution.[15] These conversions represent huge losses of useful energy through entropy.

Industrial agriculture turns petroleum into calories at a ratio of about 10 calories of fossil fuel energy for 1 calorie of food energy. Prior to industrial agriculture, that ratio was roughly 2.3 calories of human energy per 1 calorie of food energy. The proportion of global agricultural dry biomass consumed as food amounts to just 9 percent for energy and 7.6 percent for protein.[16] The greatest losses are due to livestock production. The largest absolute losses of biomass occur prior to harvesting crops, with sizable losses after harvesting – 36.9 percent of energy and 50.1 percent of protein lost after harvest and prior to human consumption. If food eaten by people in excess of nutritional need is included as an inefficiency, 53.2 percent of energy and 42.3 percent of protein are lost. Overeating adds at least as much as food waste to losses in the food system.

Like bacteria in a petri dish, humans discovered a source of energy that enabled agriculture and stimulated populations to grow exponentially. Our petri dish is Earth and our source of energy is fossil fuel. Between 1960 and 2017, the global population grew more – from 3 billion to 7.6 billion – than in the past 2 million years. The current world population of 7.6 billion is expected to reach 8.5 billion by 2030, 9.7 billion in 2050, and stabilize at 11.2 billion people by 2100. Life expectancy has also increased more now than in the past two million years. In 1950, life expectancy was forty-seven years. In 2011, it was seventy years. As we grow in numbers and longevity, we massively transform landscapes, waterscapes, and airscapes.

One of the greatest shortcomings of the human race is our inability to comprehend the exponential function. To illustrate the elusiveness of exponential growth, as Ophuls does, imagine it is 11 p.m. and we put a single bacterium that divides once a minute in a petri dish that will be full in 1 hour. When will the dish be half full? It will be half full at 1 minute to midnight. Though "only" half full, by 11:59 pm the outcome is inescapable, as there's simply far too much momentum in the system. The only questions are when and how the system will collapse. And we gain nothing, even if we make the dish four times larger. By one minute past midnight, the bacteria will fill half of the greatly expanded dish. By 12:02 am, they will fill the dish. Exponential population growth and use of natural resources are insidious and explosive.

Some look to nature as a guide, but human systems don't emulate natural systems. In theory, a human economy can mimic natural economies, which are highly efficient in thermodynamic terms because the flow of solar energy is not only consumed but also used to build a rich and diverse capital stock stored in webs that link soil, plants, and animals. But that would involve a radical transformation of civilization. Societies would have to be far more intimately coupled with one another and the landscapes we inhabit, and individuals would have to tolerate strong checks on human wants, as was the case in aboriginal societies. Even if that were acceptable, it is unlikely humans have the managerial capacity to sustain such a system. As Ophuls argues, we'd have to renounce greatness in favor of simplicity, frugality, and fraternity.

## Extending Our Stay in Wonderland

Near the end of his life, Joseph Campbell spoke of the need for a myth appropriate to our times:

> *You can't predict what a myth is going to be any more than you can predict what you're going to dream tonight. Myths and dreams come from the same place. They come from realizations of some kind that have then to find expression in symbolic form. And the only myth that is going to be worth thinking about in the immediate future is one that is talking about the planet, not the city, not these people, but the planet, and everybody on it. That's my main thought for what the future myth is going to be. And what it will have to deal with will be exactly what all myths have dealt with – the maturation of the individual, from dependency through adulthood, through maturity, and then to the exit; and then how to relate to this society and how to relate this society to the world of nature and the cosmos. That's what the myths have all talked about, and what this one's got to talk about. But the society that it's got to talk about is the society of the planet. And until that gets going, you don't have anything.*[17]

Roughly 2,500 years ago, Confucius devised ways for people to address the challenges of civilizations.[18] To do so, he taught that societies must become intricately coupled with one another and with the natural landscapes they inhabit; cultures must tolerate strong checks on human wants; and nations must develop the managerial capacity to sustain such a system.

Confucius's proposal was based on the need to merge tradition and change. Confucius understood that when tradition is no longer adequate to hold society together, humans face the gravest crises. Only by linking what people know and are accustomed to with the unknown will a new vision be largely accepted. At the same time, people must take an unflinchingly realistic account of developments that are rendering old ways untenable.

Confucius's proposal is based on deliberate tradition, which demands careful attention to moderate the individualism that can erode traditions dedicated to self-cultivation and secular ethics. The Confucian goal is to become more fully human, a person who is ever trying to become better. The self is the center of this endeavor, but not in the sense of self-centered Western individualism. Rather, Confucius saw the individual as a node, not an entity unto itself. The self is a place where lives converge.

Confucianism is grounded in a heart-mind interface (*hsin*) that expands in concentric circles. They begin with one's self and spread from there to include successively one's family; one's face-to-face community, including the natural environment; one's nation; and finally all of humanity. Confucius understood that in shifting the center of life from the self to the family, one transcends selfishness and egoism. The shift from family to community gets us beyond nepotism. The move from community to nation overcomes parochialism. And the move from nation to embrace all of humanity counters ethnocentrism and chauvinistic nationalism.

The vision of Confucius thus moves beyond secular humanism, which advocates man as the measure of all things, to include all things in the cosmos. The outwardly expanding circles of ethical consideration begin with an individual's concern for self, which is expanded to include the individual's family, tribe, region, nation, and the entire human race. Earth and heaven are a continuum, an unbroken procession in which death is no more than a promotion to a more honorable state. The two realms are mutually associated and in constant touch.

Under a succession of dynasties, the Chinese Empire lasted for more than 2,000 years. As Huston Smith points out in *The World's Religions*, if we multiply the number of years the Chinese Empire lasted (more than 2,000) by the number of people it embraced (roughly one-third of the human population), it emerges as the most impressive social institution ever devised by human beings.

Confucianism and Taoism are complementary, rotating around one another like *yin* and *yang*. Confucianism stresses social responsibility; Taoism praises spontaneity and naturalness. Confucianism wanders within society; Taoism wanders beyond. The classical shades of Confucianism are thus balanced by the romantic hues of Taoism. The *Tao Te Ching* (*The Way and the Power*) is a testament to humanity's at-one-ness with nature and the universe.[19] Nonaction by those in harmony with the *Tao* is natural and uncontrived. Water is an apt analogy.

Water molds to landscapes, nourishing all, as the *Tao Te Ching* states: "The supreme good is like water, which nourishes all things without trying to. It is content with the low places that people disdain. Thus it is like the *Tao*." Water holds a power unknown to rigid things. In a stream, it flows among the hard edges of stones, turning them into smooth pebbles. "Nothing in the world is as soft and yielding as water. Yet for dissolving the hard and inflexible, nothing can surpass it. The soft overcomes the hard; the gentle overcomes the rigid. Everyone knows this is true, but few can put it into practice." A person who embodies these traits – infinitely supple, yet consummately strong – can act without strain and achieve results without violence, coercion, or pressure. "A leader is best when people barely know that he exists. Of a good leader, who talks little, when his work is done, his aim fulfilled, they will say, 'We did this ourselves.'"

Our relationship with nature was the crux of Aldo Leopold's land ethic. Humans are members of natural communities: What we do to nature, we do to ourselves. Leopold feared we wouldn't come to this realization before things "natural, wild, and free" were gone, run over by the technological progress of humanity. Only then, when it is too late, will we appreciate the vastness of what has been lost and the cost to be paid by humanity and all other species. As he wrote, "Now we face the question whether a still higher 'standard of living' is worth its cost in things natural, wild, and free. For us of the minority, the opportunity to see geese is more important than television, and the chance to find a pasque-flower is a right as inalienable as free speech."

Leopold concludes:

> *Conservation is getting nowhere because it is incompatible with our Abrahamic concept of land. We abuse land because we regard it as a commodity belonging to us. When we see land as a community to which we belong, we may begin to use it with love and respect. There is no other way for land to survive the impact of mechanized man, nor for us to reap from it the esthetic harvest is it capable, under science, of contributing to culture.*

Humans now espouse two broad paths for evolving with environments. Along one path, people use science to develop technologies that attempt to control outcomes ecologically, socially, and economically. These "new-technologies" people promote biotechnology as the way to feed the people of the planet. The green revolution has morphed into the gene revolution, where transgenic technologies are promoted as the way to feed a hungry world. Multinational corporations attempt to control every phase of production from patented seeds to pesticides to manufactured "food" products with little regard for developing a dialog with the landscapes we inhabit. The basis for this approach involves producing and consuming more resources, which as we discussed, raises complexity, increases costs, and decreases economic efficiency.

Though seldom acknowledged in the argument for the need to feed a future population of 10 billion people, the problem isn't inadequate food production. Agriculture already produces enough to feed 11 billion people and could feed twice that many. Globally, farmers and ranchers produce 17 percent more calories per person now than thirty years ago, despite a 70 percent increase in population. This is enough to provide everyone in the world with at least 2,720 calories each day.[20] The challenge is income inequality and a lack of food distribution. In 2015, 795 million people in the world did not have enough food to lead a healthy, active life. Most of these people live in developing countries, where 13 percent of the population is undernourished. Policies of the World Bank and other organizations have exacerbated malnutrition in many parts of the world. They have encouraged large-scale industrial agriculture, which caused small farmers, who formerly produced food locally, to move to cities where they have no work.

Along another path, people in the "mimic-nature" group attempt to empower farmers, ranchers, and communities to develop relationships with landscapes. In essence, this is a move toward a simpler mode of life, more of a subsistence lifestyle that minimizes costs by working with nature. Revolutions are occurring in agriculture at a grassroots level in the United States and around the world, fueled by a growing awareness of the unsustainability of industrial agriculture.

"Mimic-nature" advocates value science for understanding processes of adaptation and creation.[21] They avoid costly technological fixes that make soil, plants, animals, and people dependent on GMOs and fossil-fuel–based fertilizers, herbicides, and insecticides to grow and protect plants; antibiotics and anthelmintics to maintain the health of herbivores; or pharmaceuticals to sustain health. They emphasize "small, local, and biodiverse" in evolving creatively with landscapes. They manage holistically and work with nature to cut costs. The goal of the "mimic-nature" movement is to reduce harmful agricultural impacts.

As activists and writers such as Courtney White emphasize, the regenerative toolbox is overflowing with practical, profitable, and effective practices.[22] People are protecting soils with plants that increase organic matter, enhance water infiltration into soil, and increase fertility. To do so, they use perennial crops and forages in rotations; diversify annual cropping systems to include legumes; use cover crops along with row crops; change plow tillage to no-till cropping. To minimize adverse effects of fertilizers on life in soils, they use less nitrogen fertilizer; fertilize with legumes and controlled-release, nano-enhanced fertilizers with nitrification inhibitors; apply biotic fertilizer formulations that feed soil microbes and mycorrhiza and reduce nitrogen and phosphorus runoff; and use precision agriculture to moderate the rate and timing of application of agrochemicals and water. Farmers are also integrating animals – goats, sheep, cattle, hogs, and chickens – back into cropping systems to enhance soil organic matter and nutrients. On extensive landscapes that can't be farmed, ranchers are converting marginal and degraded croplands into pastures and forests, restoring wetlands, and managing grazing more intensively. People are also using permaculture to grow food and imitate nature in form and function.

Enabling local adaptation of communities to climate change will be critical for reducing poverty and increasing food security.[23] In the developing world, 10 to 15 percent of one billion hectares are farmed using traditional methods. Roughly 475 million people cultivate food in small farms. Local production of food has many benefits, including promoting food diversity and crop resilience, reducing carbon footprints during production and transport, boosting local economies, providing healthier foods for people, and enhancing ecosystem services. Local food production also increases a sense of self-reliance and embeddedness of provisioning services, resulting in tighter social connectivity among individuals within communities.

Ecologist and farmer Gary Kleppel discusses these issues in *The Emergent Agriculture*, an insightful discussion about the future of regenerative agriculture. He argues that industrial food production is incompatible with the realities of nature, science, and ethics. He makes the case for locally based food systems that promote bonds between farmers and consumers, remain stable in the face of economic and environmental uncertainty, and acknowledge the stewardship and skills of people who grow food. Studies from various fields suggest the solution to food supply worldwide is small, organic agro-ecological farms that supply food in their regions.

Diet links human health with environmental health. Nowadays, agriculture contributes more than 25 percent of all greenhouse gases, pollutes fresh and marine waters with agrochemicals, and uses nearly half of the ice-free land area on the planet as cropland or pastureland. Worldwide increases in population and demand for food are bringing about the clearing of tropical forests, savannas, and grasslands, threatening many species with extinction. The ongoing global shift to diets high in processed foods, refined sugars, refined fats, oils, and meats has contributed to 2.1 billion people becoming overweight or obese. These changes are associated with increased incidence of chronic noncommunicable diseases, especially type 2 diabetes, coronary heart disease, and some cancers. If unimpeded, these dietary trends will be a major contributor to an estimated 80 percent increase in global agricultural greenhouse gas emissions due to food production and land clearing by 2050.

Relative to animal-based foods, plant-based foods have lower greenhouse gas emissions. Greenhouse gas emissions from ruminants (beef and lamb) are about 250 times those of legumes, per gram of protein. Eggs, dairy, non-trawling seafood, traditional (non-recirculating) aquaculture, poultry, and pork all have much lower emissions per gram of protein than ruminant meats. Nevertheless, as discussed in chapter 8, livestock production can increase food security, dietary quality, and provide multiple environmental benefits when ruminant meat and dairy animals are sustainably grazed on lands unsuitable for cropping and fed crop residues.[24]

Diet affects environmental health, as illustrated when researchers compared vegetarian, pescetarian, Mediterranean, and omnivorous diets.[25] A vegetarian diet consists of grains, vegetables, fruits, sugars, oils, eggs, and dairy, and generally not more than one serving per month of meat or seafood. A pescetarian diet is a vegetarian diet that includes seafood. A Mediterranean diet is rich in vegetables, fruit, and seafood and includes grains, sugars, oils, eggs, dairy, and moderate amounts of poultry, pork, lamb, and beef. Omnivorous diets, such as the 2009 global-average diet and the income-dependent 2050 diet, include all food groups. Relative to conventional omnivorous diets, across the three alternative diets, incidence rates of type 2 diabetes were reduced by 16 to 41 percent, cancer was reduced by 7 to 13 percent, relative mortality rates from coronary heart disease were reduced by 20 to 26 percent, and overall mortality rates for all causes combined were reduced by up to 18 percent.

Changes towards healthier diets in turn reduce greenhouse gas emissions. Global population is projected to increase to 9.7 billion people by 2050. When combined with the projected 32 percent increase in per capita emissions from worldwide dietary shifts, the net effect is an estimated 80 percent increase in greenhouse gas emissions from food production. In contrast, there would be no net increase in food production emissions if by 2050 the worldwide diet has become, on average, a Mediterranean, pescetarian, or vegetarian diet.

If widely adopted, any of the alternative diets could reduce agricultural greenhouse gas emissions, reduce land clearing and ensuing species extinctions, and help prevent diet-related non-communicable diseases. Because diet links human and environmental health, the dietary transition is one of the great challenges facing humankind. The dietary choices that individuals make are influenced by culture, nutritional knowledge, price, availability, taste, and convenience, all of which must be considered if the dietary transition that is taking place is to be counteracted. The evaluation and implementation of dietary solutions to the tightly linked diet-environment-health trilemma is a global challenge – and opportunity – of great environmental and public health significance. Meaningful solutions will require the combined efforts of nutritionists, agriculturists, public health professionals, educators, policy makers, and food industries.

In *Call of the Reed Warbler*, radical farmer Charles Massy describes how ecological agriculture can restore Earth and human health. He highlights how indigenous peoples created enduring relationships with harsh Australia landscapes over tens of thousands of years through their nurturing and discerning approaches to "Caring for Country" ecologically and spiritually. For the Aboriginal Australian, *Country* is a nourishing place that gives and receives life. Country nourishes body, mind, and spirit. A healthy Country is one in which all elements nourish one another. In this way, life becomes meaningful, and art, music, dance, philosophy, religion, and ritual and daily activity celebrate and promote life. Every living thing is participating in creating a living system. Each Country is surrounded by other unique Countries, and the relationships among the Countries ensure they create constellations of alliance networks.

Massy contrasts the Aboriginal way with how post-1788 land was regarded as an exploitable resource and indigenous people as nonpeople. The post-1788 period in Australia was a clash of the so-called "primitive" culture of balance, holism, and longevity pitted against a "civilized" mechanistic-thinking culture imbued with ideas of conquest and short-term gain. The nonindigenous immigrants from Europe lacked physical or spiritual linkages with the land they conquered. The outcome was inevitable. Both Australia's indigenous peoples and their many diverse and interconnected Countries were the losers in the short term, with devastating results for both the land and the Aboriginal and European peoples in the long term.

Massy contends humans morphed from an *Organic* mind into a *Mechanical* mind, which is now evolving into an *Emergent* mind – a change in consciousness that embraces self-organizing processes. The Emergent mind has a deep biophilic capacity to see humans in harmony and accord with Earth and its systems and with all other creatures, not dominant over them. This is more than functional ecology and landscape literacy because it also includes social elements such as food safety and quality, public health that arises from environmental health, and broader issues that have to do with power and control, justice and equity for all inhabitants of Earth. Massy concludes, we can heal Earth but only by transforming ourselves and our place within Country: The reed warbler "is calling us in a poignant, heartfelt cry for all creation – a metaphor for us humans to once more become the enablers, the nurturers, the lovers of Earth."

His conclusion is in line with the assertions of Hans Rosling (along with Ola Rosling and Anna Rosling Rönnlund) in *Factfulness*. They present strong evidence that the peoples of the world are in much better condition than we typically think. On many fronts – from poverty, education, and life expectancy to disasters, diseases, and future population numbers – the human race is transforming for the better economically and socially. They contend that in most cases we worry about issues that have improved a great deal during the past century. On the other hand, they caution that we should be worried into action about climate change and extreme poverty, as well as the possibilities of a global pandemic, financial collapse, and World War III. These issues will require determined and unified efforts by all of the nations of the world.

Historically, the consequences of a failed civilization were catastrophic for a particular local society and its inhabitants, but they weren't fatal to *Homo sapiens* as a species. We now live in an interdependent, global civilization, and the consequences of decay of civilizations are different. The destinies of all peoples across the globe are linked ecologically, economically, and socially. What will become of *Homo sapiens*? Can today's societies become more intricately coupled with one another and the landscapes we inhabit? Can we learn the value of curbing human wants? Do people have the managerial capacity to sustain such a system? The answers to these questions are a mystery only time will reveal. By participating, we are creating.

# CHAPTER 18

# The Mystery of Being

When I was a child, I never thought about what I was going be when I grew up. I never spent any time pondering how my life would unfold. As I look back now, I can see how events that at the time seemed accidental and of little meaning were key features in creating the tale.

I went to a Catholic grade school, and in the eighth grade I had a most caring middle-aged teacher named Jerry who, while studying to become a priest, had a nervous breakdown. My cousin Bobby and I befriended Jerry. We eased his loneliness and enjoyed his companionship. We also liked that he'd let us drive the parish jeep all over the mountains. Jerry went on to become a priest, and I remained friends with Father Jerry until he died many years later.

When I finished high school, I went to Colorado State University to study wildlife biology, not because I thought it would lead to a job, but because I was fascinated by wild fish, birds, and mammals. The summer before I entered college, my friend Bernie asked if I'd like to earn extra money, in addition to what I was making working at a greenhouse. He said we'd make eight cents a bale – four cents each – hauling hay at DeLuca's Ranch. We hauled hay in the evenings and on weekends. I loved it so much that I worked on the ranch throughout college.

In college, I learned about soil, plants, animals, and ecology. I also learned that I didn't want a career as a wildlife biologist, so I managed the DeLuca Ranch for two years after I graduated from college. During those years, I began to think about graduate school, though I honestly don't know why that interested me. I had no knowledge or experience of research. I applied to graduate schools at a couple of universities, not really knowing what I was doing, and I wasn't accepted. At that point, Father Jerry suggested I apply to Utah State University, for no reason other than that he knew the pastor at the Catholic Church in Logan, Utah.

I decided to follow Father Jerry's suggestion, and I applied to the Wildlife Department, where no one was interested in me. They did, however, pass my application to the Range Science Department, where a new professor – Phil Urness – was looking for graduate students. That June, Phil sent me a letter indicating he'd take me as a student, though he couldn't provide any financial assistance. My wife and I would have to come on our own money. We were delighted that I was finally accepted, even if we did have to pay our own way. We knew in our hearts that it was the right path. The forty years that flowed out of that decision have been filled with many adventures, none of which either of us anticipated. We simply followed our hearts.

I've often reflected on my encounters with Father Jerry, my hay-hauling friend Bernie, and the ranch. Had I not hauled hay that summer at DeLuca's, I would never have acquired an interest in ranching. Had I not met Father Jerry, I would never have gone to graduate school in range science at Utah State University. Of course, we all have relationships involving people like Bernie, Father Jerry, and Henry DeLuca. To what degree are these encounters predestined? Or are they evolving as we participate? Or perhaps both things are happening simultaneously.

German philosopher Arthur Schopenhauer points out that after a person reaches an advanced age and looks back over their life, it can seem to have had an order and plan, as if composed by a novelist.[1] Events that seemed accidental and of little importance turn out to be indispensable factors in the composition of a consistent plot. Who composed the plot? He suggests that just as our dreams are composed by an aspect of our self of which our consciousness is unaware, so, too, is our whole life. And just as people whom we will have met, apparently by chance, became agents in creating our life, so, too, we will have served unknowingly as an agent giving meaning to the lives of others. It is as if our lives are the features of the dream of a single dreamer in which all the dream characters are dreaming, too. Everything links to everything else, moved by the will to life and being, which is the universal will in nature.

## Mysteries of the Visit: Fate and Destiny

In mythology, *fate* and *destiny* both refer to the belief in an order to the cosmos. Though the words are often used interchangeably, they have different connotations. On the one hand, *fate* implies a predetermined course of events that is inevitable and thus unavoidable. In classical European mythology, three goddesses bestow the fates of mortals – Moirai in Greek mythology, Parcae in Roman mythology, and Norns in Norse mythology. They determine the events of the world through their mystic spinning of threads that characterize the lives of individuals. Greek legends teach the tragic futility of trying to out-maneuver an inexorable fate. On the other hand, *destiny* implies a probabilistic future with a sense of intention in which an individual willfully participates in creating the order of events that fashion outcomes projected into the future.

Like fate, *determinism* is the belief that for every event there are complex circumstances that preclude other possible developments. *Causal determinism* is the belief in an unbroken series of occurrences that extend back to the origin of the universe. Causal determinists believe nothing is uncaused or self-caused: All happenings are inextricably linked in a web of causality that extends through time and space. According to Newton (classical physics) and Einstein (relativistic physics), if we can prestate the particles, forces, laws, and initial and boundary conditions, then we can compute the consequences for the full (finite) range of possibilities.

Prior to the seventeenth century, the principal worldview was one of a spiritual, organic, living universe, which was deeply mysterious and in many ways unknowable and frighteningly unpredictable. That view changed during the seventeenth century to one in which nature, though complex, was thought to be both knowable and predictable scientifically. The machine became the model and the clock the metaphor for this worldview. The belief was that if we could understand the parts of "the machine," and how the parts interacted, we could predict outcomes in time and space: The sum of the parts would equal the whole.

Pierre-Simon, Marquis de Laplace, the French mathematician and astronomer whose work was pivotal in developing mathematical astronomy and statistics, summarized this causal deterministic view:

> *An intellect which at a given instant knew all the forces acting in nature, and the position of all things of which the world consists — supposing the said intellect were vast enough to subject these data to analyses — would embrace in the same formula the motions of the greatest bodies in the universe and those of the slightest atoms; nothing would be uncertain for it, and the future, like the past, would be present to its eyes.*

In contemporary cinema, characters in movies are archetypes for these worldviews. In the late twentieth-century film *Forrest Gump*, the character of Lieutenant Dan is an archetype for the interplay among fate, determinism, and destiny: He thought his fate was to die on the battlefield like his ancestors, but that wasn't his destiny. When Forrest saved him from his fate and ushered in his destiny, he was angry and wretched until he abandoned his deterministic view of life and became more like Forrest and his mother, who were archetypes for destiny – participating in creating their lives. Mrs. Gump's motto was "Life's like a box of chocolates, Forrest. You never know what you're going to get." Forrest lived spontaneously, in the moment, without expectations. He embraced the many twists and turns life delivered and he excelled at everything he did – from football to Ping-Pong to fishing – simply by being Forrest Gump.

The feather is a metaphor for the dynamism of life, destiny, and chance co-occurring. The feather darts and glides unpredictably, moved by the winds of chance, each twist and turn linking past, present, and future into one reality. Forrest merges these worldviews when, in tears at the grave of his lifelong love, Jenny, he says he isn't sure if mama was right or if Lieutenant Dan was right. Do we each have a destiny or are we all just floating around, chancelike, on a breeze? Forrest concludes, maybe it's both happening at the same time. And he's in good company.

Nobel laureate Ilya Prigogine and philosopher Isabelle Stengers conclude likewise in *Order Out of Chaos*:

> *Self-organizing processes in far-from-equilibrium conditions correspond to a delicate interplay between chance and necessity, between fluctuations and deterministic laws. We expect that near a bifurcation, fluctuations or random elements would play an important role, while between bifurcations the deterministic aspects would come to dominate.*

Change and chance collaborate to make ecological, economic, and social systems evolve in ways that aren't knowable or predictable. Change and chance create challenges, opportunities, and creativity in life, but they come at a price: The cost of admission is endless transformation. The choices organisms make are influenced by the history of the actors, their needs, and chance.

These notions notwithstanding, Albert Einstein didn't believe in "chance." Indeed, he often commented, "God doesn't play dice with the universe." To Einstein, if we just knew enough about every facet of nature, all events would be as understandable and predictable as the movements of billiard balls on a pool table. Einstein understood the implications: Everything anything will ever do is quite predictable, so no creature, human or otherwise, ever makes a choice. We live under the illusion that we have free will and make choices.[2]

The notion of free will and its nature and definition have long been debated in philosophy because free will has religious, ethical, and scientific implications. In the religious realm, free will implies individual will and choice coexist with an omnipotent deity. In ethics, free will has implications for whether individuals can be held morally accountable for their actions. In science, free will suggests different pathways for predicting and modifying human behavior.

According to one view of quantum physics, *chance* and *uncertainty* are built into the fabric of the universe. The Copenhagen interpretation, one of the most commonly taught interpretations of quantum physics, was devised by Niels Bohr, Werner Heisenberg, and others from 1924 to 1927. They argued quantum mechanics doesn't yield *a deterministic description* of reality but provides only probabilities of observing, or measuring, various aspects of energy quanta, entities that fit neither the classical idea of particles nor the traditional view of waves. According to their interpretation, the act of measuring causes probabilities to instantly assume one of countless possible values. This feature is known as the collapse of the wave function. According to quantum mechanics, nature is knowable, but not predictable, even if we know the rules, because *chance* influences the behavior of elementary particles that make up the universe.

Another discovery of quantum physics is that solid objects of classical physics dissolve into wave-like patterns of probabilities at subatomic levels. These patterns aren't *probabilities of things*, but *probabilities of relationships*. Nature isn't "isolated building blocks" but rather appears as a web of relationships. Theoretical physicist John Archibald Wheeler wrote:

> *Nothing is more important about the quantum principle than this, that it destroys the concept of the world as "sitting out there," with the observer safely separated from it by a 20 centimeter slab of plate glass. Even to observe so miniscule an object as an electron, he must shatter the glass. He must reach in. He must install his chosen measuring equipment. . . . Moreover, the measurement changes the state of the electron. The universe will never afterwards be the same. To describe what has happened, one has to cross out that old word "observer" and put in its place the new word "participator." In some strange sense the universe is a participatory universe.*[3]

As business consultant Margaret Wheatley concludes in *Leadership and the New Science*,

> *The new physics cogently explains there is no objective reality out there waiting to reveal its secrets. There are no recipes or formulae, no checklists or advice that describes "reality." There is only what we create through our engagement with others and with events. Nothing really transfers; everything is always new and different and unique to each of us. In this realm, there is a new kind of freedom, where it is more rewarding to explore than to reach conclusions, more satisfying to wonder than to know, and more exciting to search than to stay put.*

## Mysteries in Physics, Evolutionary Biology, and Mathematics

As a child, I'd lie in bed at night pondering the origins of existence. How, I wondered, can anything *be*? If *nothing* were, what would nothing be? Over the years, I have come to realize that existence and its origins are deep mysteries – neither scientists nor preachers have answers to these questions. At some point, science and religion both become matters of faith.

The limit for physicists is Planck time, $10^{-43}$ seconds before the big bang of creation of this universe. The equations of science can't describe what occurred before then. As cosmologist John Barrow writes in *The Origins of the Universe*, "The Planck time of $10^{-43}$ seconds is significant, because when we reach this extraordinarily early time the size of the visible universe becomes smaller than its quantum wavelength and is thus enshrouded by quantum uncertainty. When quantum uncertainty overtakes everything, we don't know the positions of anything, and we can't even determine the geometry of space. This is when Einstein's theory of gravitation breaks down." In his book entitled *Now*, Richard Muller makes the case that it isn't any one physics theory – quantum or relativistic – that is incomplete; physics itself is incomplete.

Some scientists theorized, based on inflationary cosmology and string theory, that our universe is unique, but not alone; ours is one of many universes. Physicists who developed the string theory during the late 1960s hoped its rigid mathematical framework would yield a unique set of conclusive and testable predictions about the universe. As the years passed, however, detailed analyses of the theory's equations revealed copious solutions. The tally of possible universes now stands at $10^{500}$, a number so vast it defies comprehension. Scientists believe the laws and physical properties of each universe are unique. If so, then we now know the answer to Einstein's inquiry about the creation of the universe: "What really interests me is whether God had any choice in creating the world." The answer is yes, God had $10^{500}$ possible choices.[4]

Beyond physics, the limit of knowledge in evolutionary biology is the origin of life. No one has proved how something as complex as life began. As medical doctor and theoretical biologist Stuart Kauffman points out in *At Home in the Universe*, the simplest free-living cells are called *pleuromona*, a bacterium that is the least complex organism known to be alive. "All free-living cells have at least the minimum molecular diversity of pleuromona. . . . Why is there this minimal complexity? Why can't a system simpler than pleuromona be alive?"

No one knows for certain. It's also debatable whether even pleuromona meets the definition of simple, except in the relative sense. The number of genes in pleuromona is estimated at a few hundred to a thousand, compared with the 4,288 genes in *E. coli*. And still, the information content in even a single enzyme of a "simple" organism such as pleuromona or *E. coli* is so vast it can't be generated merely by chance, even over vast spans of time. Though many have tried, no one has shown how something as complex as life began.

In their calculations of the probability of life emerging by chance, astronomer Fred Hoyle and mathematician-astrobiologist Chandra Wickramasinghe assume the first living cell was much simpler than today's bacteria. Their calculation for the likelihood of even a simple enzyme arising at the right time and place is 1 in 1,020. As there are thousands of enzymes with diverse functions in the simplest living cell, they calculate that about 2,000 enzymes are needed with each one performing a specific task to form a single bacterium like *E. coli*. They calculate the odds at 1 in $10^{40,000}$ for all these enzymes to form in one place at one time to produce a single bacterium. They conclude life could not have appeared by random processes even if the entire universe consisted of primeval soup. The information content of even the simplest organism can't be generated by chance. They conclude in *Evolution from Space*: "The chances that life just occurred are about as unlikely as a typhoon blowing through a junkyard and constructing a Boeing 747."

Dean Overman, scholar of information theory, physics, and religion, examines theories for the origins of life in *A Case Against Accident and Self-Organization*. He concludes,

> *Life cannot be explained by an appeal to accident; and if life transcends the laws of physics and chemistry, then the origin of life will never be demonstrated by an adequate self-organization scenario, but remain an intractable or indeterminate problem such as that represented by Gödel's Incompleteness Theorem . . . If life transcends the laws of physics and chemistry, the cause of life is more than a thing. At this point, we enter into metaphysics . . .*

Just as there are limits to what we can know in physics and biology, so, too, with mathematics. Logician and mathematician Kurt Gödel set out to show anything can be proved mathematically, but his incompleteness theorems showed the limits of what is knowable. In *A Madman Dreams of Turing Machines*, mathematician and author Janna Levin describes the life and times of Gödel, depicting his fascination with the liar's paradox, an ambiguity illustrated in this statement: "This sentence is false." As we study this sentence, we see it can't be true – for then, as it declares, it is false; nor can it be false – for then, it is true.

Gödel's first theorem shows that no consistent system of axioms whose theorems can be listed by an "effective procedure" – such as a computer program or any algorithm – can prove all truths about the relations of the natural numbers. For such a system, there will always be statements about natural numbers that are true, but that are unprovable within the system. His second theorem, an extension of the first, shows that such a system cannot demonstrate its own consistency. Thus, many things we know to be true can never be proved mathematically.

This brings to mind something Einstein said, near the end of his life, about his beliefs. Einstein spent the first ten years of his career in a patent office trying to imagine what it would be like to be a beam of light; out of his imagination came the special theory of relativity, a theory that changed our views of the universe. He spent the next ten years developing the general theory of relativity, which to this day is one of the pillars upon which physics is built. So, you might think his statement of belief would include something about physics, but it doesn't.

Rather, he said:

*I believe in the brotherhood of man and the uniqueness of the individual. But if you ask me to prove what I believe, I can't. You know them to be true but you could spend a whole lifetime without being able to prove them. The mind can proceed only so far upon what it knows and can prove. There comes a point where the mind takes a higher plane of knowledge, but can never prove how it got there. All the great discoveries have involved such a leap.*

As for creativity, that was simple enough:

*The important thing is not to stop questioning. Curiosity has its own reason for existence. One cannot help but be in awe when [one] contemplates the mysteries of eternity, of life, of the marvelous structure of reality. It is enough if one tries merely to comprehend a little of this mystery each day. Never lose a holy curiosity.*[5]

## Mysteries in Mythology

The agnostic astronomer Robert Jastrow suggests the big bang of creation began under conditions that may make it impossible now or ever to determine how this universe came into being. In his book *God and the Astronomers*, he concludes: "For the scientist who has lived by his faith in the power of reason, the story ends like a bad dream. He has scaled the mountains of ignorance; he is about to conquer the highest peak, and as he pulls himself over the final rock, he is greeted by a band of theologians who have been sitting there for centuries."

The spiritual journey begins with personal experience of the mysteries and wonders of being.[6] Our words, thoughts, and beliefs are simply metaphors for the mystery of being, which transcends all thought. Theology and religion codify thoughts about symbols that emanate from personal experience and organize them in different forms culturally across time and space. Institutionalized dogma often transports the journey away from the individual to formulas for how to reach God, prescribed to ease the uncertainty of life. The fear of hell is added to make sure the faithful follow the "right" path. The spiritual journey becomes a dogma of fear to control the masses. In so doing, religion sacrifices uncertainty, wonder, and mystery for rigid beliefs. When personal inspiration gives way to concretization, the spiritual journey is lost.

Spiritual teachings are often concretized against the wishes of their architects. The Buddha embraced Hindu insights into the need for each person to discern their own path to God. He protested against how the original intent of that message became mired down in orthodoxy and dogma. Ironically, his revolt against rites, speculation, grace, and the supernatural ultimately did not succeed, and Buddhism transformed Buddha – who was an atheist as far as a personal God was concerned – into a God himself. In like manner, Martin Luther protested rigid, authoritarian versions of Jesus's teachings by the Catholic Church. Protestants believe life is too fluid to allow God's Word to be enclosed in doctrine or institutions. American journalist Lincoln Steffens wrote a parable about a man who climbed to the top of a mountain and grasped hold of the Truth. Satan, suspecting mischief from this upstart, had directed one of his underlings to tail him. But when the demon reported with alarm the man's success – that he had seized hold of the Truth – Satan was untroubled. "Don't worry," he yawned. "I'll tempt him to institutionalize it."[7]

Just as no "average" diet exists for nourishing individuals, no "average" religion exists for sustaining the multitudes. By some estimates, the world has roughly 4,200 religions. The primal religions were about the experiences of life and the mysteries and wonders of existence. The more recent historical religions – Hinduism, Buddhism, Confucianism, Taoism, Judaism, Christianity, and Islam – of the past 4,000 years are but the tip of the religious iceberg, preceded by hundreds of thousands of years of rich primal religions. The lack of agreement among and within religions highlights the transcendent mystery that underlies the origins of spirituality. While most contemporary Western thinking is based on belief in a personal God, most Eastern and much "primitive" thinking view the Gods as impersonal. They are manifestations of the energies of the universe inflected in bewildering forms including life.

Though the world's religions are undeniably diverse and often superficially opposed to one another, they do exhibit some shared aims. The perennial philosophy views each of the world's religious traditions as sharing universal truth, posited as mystical insofar as it views one purpose of human life as a recognition of union with the transcendent, which can be achieved by undertaking programs of physical and mental "purification" or "improvement" such as fasting and meditating. According to philosopher Aldous Huxley, who popularized the idea of a perennial philosophy, the perennial philosophy is expressed succinctly in the Sanskrit formula, *tat tvam asi* ("That thou art") – the Atman, or immanent eternal Self, is one with Brahman, the Absolute Principle of all existence – and the last end of every human being is to discover that for one's self.[8] This universal truth, Huxley contends, lying at the heart of each religion, has been discovered in each epoch by saints, sages, prophets, mystics, theologians, and common folk.

In *Misquoting Jesus*, New Testament scholar Bart Ehrman describes his quest as an evangelical teenager to read the authentic "Words of God." That quest took him first to the Moody Bible Institute, then to Wheaton College, and finally to the Princeton Theological Seminary. If the full meaning of scriptures could be grasped only by reading them in Hebrew and Greek, then Ehrman intended to learn those languages. No obstacle was too tough in his quest. The more he studied, however, the more he was confronted by a conundrum: If God had indeed written the words, "he" could have preserved them intact. But "he" didn't. We have only error-ridden copies of the Bible, and the vast majority of these are centuries removed from the originals and different from them, evidently, in thousands of ways. During the next thirty years, Ehrman came to view the Bible as a very human document. Interpreting biblical texts based on beliefs is a very human endeavor, as Ehrman came to appreciate. People interpret the Bible in thousands of ways. One need only reflect on all the different Christian denominations, filled with people of faith who base their views of how the church should be organized and function on the Bible, yet all of them – Baptists, Pentecostals, Roman Catholics, Appalachian snake handlers, Greek Orthodox, and so on – come up with radically different interpretations.

The challenge isn't acquiring beliefs. One need only be conceived, born, and reared in a particular culture. The challenge is accepting people have diverse beliefs, all incomplete and imperfect. As long as we teach dogmatically, and believe without questioning, we will never appreciate that humans wrote the world's sacred scriptures to inspire people who visit Earth. They were written at different times and in different ways – the ideas are culturally inflected to speak to different peoples at different times and places, reflecting the fact that no peoples or individuals experience Earth in the same way as cultural knowledge and norms transform.

The difficulty comes when cultures elaborate and then concretize their beliefs in space and time. In *The Power of Myth*, Joseph Campbell warned of the danger of concretizing myths, which are metaphors for deep mysteries, not meant to be taken literally. But that's what we do. As he noted:

> *Jesus ascended to heaven. The denotation would seem to be that somebody ascended to the sky. That's literally what is being said.*

*But if that were really the meaning of the message, then we have to throw it away, because there would have been no such place for Jesus literally to go. We know that Jesus could not have ascended to heaven because there is no physical heaven anywhere in the universe. Even ascending at the speed of light, Jesus would still be in the galaxy. Astronomy and physics have simply eliminated that as a literal, physical possibility. But if you read "Jesus ascended to heaven" in terms of its metaphoric connotation, you see that he has gone inward — not into outer space but into inward space, to the place from which all being comes, into the consciousness that is the source of all things, the kingdom of heaven within. The images are outward, but their reflection is inward. The point is that we should ascend with him by going inward. It is a metaphor of returning to the source, alpha and omega, of leaving the fixation on the body behind and going to the body's dynamic source.*[9]

Historically, religions were a powerful way to unite peoples in tribes and nations. Today, though, we are a world community and religions are often a source of division among the many tribes that now inhabit the nations of Earth. The interconnectedness of the peoples of Earth now necessitates that we see our myths as metaphors for deep cosmic mysteries. I embrace what the Dalai Lama said in a conversation with Desmond Tutu:

*Just to pray or rely on religious faith is not sufficient. It will remain a source of inspiration, but in terms of seven billion human beings, it's not sufficient. No matter how excellent, no religion can be universal. So we have to find another way to promote these values. I think the only way really is, as we have said, through education. Education is universal. We must teach people, especially our youth, the source of happiness and satisfaction. We must teach them that the ultimate source of happiness is within themselves. Not machine. Not technology. Not money. Not power.*[10]

From infancy we train children what and what not to think. We rarely challenge them to make the inward journey to discover and develop their own beliefs about life in the here and now and the hereafter. Our thoughts as adults merely reflect what we've been taught. Any belief system that doesn't encourage us to explore the mystery and wonder of being is being taught by preachers, not teachers, and it is about dogma, not enlightenment. True teachers empower us to explore the mysteries and wonders of life. Preachers tell us what and how to believe. Institutions are filled primarily with preachers, not teachers. While our adventure on Earth is personal, many relinquish the quest to authorities based on assurances for our well-being in the here and now and the hereafter. We do so based on fear, which abolishes inspiration. So our creative imaginations gradually dissolve. We marvel when we observe genius in matters physical and divine inspiration in matters spiritual, never realizing the genius and spiritual inspiration that reside in each of us. We don't appreciate the wisdom of our own bodies.

Jesus was a teacher. Unlike the dogma that has been built upon his teachings, he didn't tell people what or what not to believe. He invited them to see things in a different way based on engaging their hearts. If listeners were to accept his invitation, the place where they were being summoned would have to seem real to them. My teachings are true, he said in effect, not because they come from me, or even from God through me, but because your own hearts attest to their truth. The unconditional love emanating from Jesus liberated people in three ways, all related to the here and now: He freed them from fear (of death); he freed them from guilt (you can do no wrong); and he freed them from self (love your enemies and bless those who curse you). In so doing, he gave them, respectively, *infinite being*, *infinite consciousness*, and *infinite bliss*. According to Hindu thinking, the chief attributes of God are *sat*, *chit*, *ananda*: infinite being, infinite consciousness, infinite bliss. We possesses these attributes, but the infinite is down in the darkest, most profound crypt of our being, in the forgotten well house, the deep reservoir.

Like religion, science uses metaphorical language. As Richard Lewontin highlights in *The Triple Helix*:

> *It is not possible to do the work of science without using a language that is filled with metaphors. Virtually the entire body of modern science is an attempt to explain phenomena that cannot be experienced directly by human beings, by reference to process and processes that we cannot perceive directly because they are too small, like molecules, or too vast, like the entire known universe, or the result of forces that our senses cannot detect, like electromagnetism, or the outcome of extensively complex interactions, like the coming into being of an individual organism from its conception as a fertilized egg.*

These explanations, which are meant to appeal to the understanding of the world that we have gained through experience, must unavoidably involve the use of metaphorical language. And so, physicists speak of "waves" and "particles" even though there is no medium in which those "waves" move and no solidity to those "particles." And biologists speak of genes as "blueprints" and DNA as "information."[11]

Forms change endlessly – from soil, plants, herbivores, and humans to planets, stars, and galaxies – yet they are one: Energy manifests as matter in myriad ways. Einstein's equation describes the relationship between energy and matter ($e = mc^2$). The laws of thermodynamics tell us energy can neither be created nor destroyed, it merely changes forms. But neither Einstein's equation nor the laws of thermodynamics nor the words of theologians can tell us the source of the energy from which all things come and back into which they go, the energy that is the ground of being, beyond names and forms. Ultimately, the mysteries of life are unfathomable.

And so with mythology, as with physics, biology, and mathematics, our attempts to link back to the phenomenal source leads to silence and mystery: the silence from which all things come and back into which they go, the mystery that transcends all thought. Perhaps the Artist could have painted a crystal clear picture of "our purpose on Earth." Instead, we dine with change and dance with uncertainty as the Artist's painting fades into mystery. That has given rise to a fantastic array of beliefs. Sadly, we come to view our beliefs as uncontestable truths to be defended – even to the death – in order to maintain our religious and political sway over others who believe differently from we do. We have sealed our hearts against the right of each human to follow an individually valid spiritual journey.

## Ever-Changing Verses in the Language of I Am

In the geologic book of the life of planet Earth, the impact of the Chicxulub crater in the Gulf of Mexico 66 million years ago is recorded as a thin layer of sediment found throughout the world in marine and terrestrial rocks.[12] This sediment is high in iridium, rare in Earth's crust but abundant in asteroids. The massive explosion filled the sky with debris that blocked the sun's rays for decades, causing global temperatures to plummet 50°F. Three-quarters of the species on Earth went extinct. The reign of the dinosaurs was over and the rise of the mammals had begun. Had this massive asteroid entered the atmosphere just 30 seconds earlier or later, it would have landed harmlessly in the deep waters of the Atlantic or Pacific Ocean, rather than in the shallow coastal waters where the underlying sediments were filled with the gypsum that created the clouds that darkened the Earth. Most plants and animals would have survived. The age of mammals, and the evolution of *Homo sapiens*, might never have occurred.

What we deem to be preordained has elements of chance: the genes we inherit; whether or not we get cancer; who lives or dies when unlikely events, such as asteroids, strike; meeting the people who end up influencing our lives; a presidential election decided by only 77,000 out of 136 million votes – cast in just three states: Pennsylvania (44,292 votes), Wisconsin (22,748 votes), and Michigan (10,704 votes). Humans attempt to explain events retroactively by creating patterns where none exist.[13] We can't integrate manifold levels of causation, across time and space. We don't like uncertainty, and we certainly don't stop to ponder the role of chance in our lives. We want to know why things happen. We want to feel in control of our surroundings. Yet, complexity and chance render our abilities to foresee how the future will unfold meager at best.

The myth of each civilization is built upon the stories people tell of their manifest destinies. We like to think of *Homo sapiens* as unique in time and space – our species, our race, our tribe, our nation, ourselves, myself. Indeed, that's true. But this anthropocentric view ignores a broader awareness: The same is true for the 500 billion species and unfathomable number of individuals who have visited Earth during the past 4.54 billion years. As for our perception of transcendent uniqueness, science has falsified the assertions that underlie that tired old claim, right down to the belief that consciousness resides in humans and humans alone. Perhaps, if some cosmic stargazers watch flickers of Earth from space sometime henceforth, they might write our epitaph as one uncontested claim: *Homo sapiens* were far too clever and arrogant for their own good. They consumed the planet, which in turn, consumed them.

*I am; without beginning or end; unfathomable.*
*I can neither be created nor destroyed; the first law of thermodynamics; indisputable.*
*I change forms endlessly: the second law of thermodynamics; irrefutable.*
*A universe, a species, a life; ever-changing verses in the language of I am; fleeting.*
*Origins of I am; beyond all categories of thought; incomprehensible.*

## DINING ON EARTH

# A Visitor's Reflections

There's a rhythm in temperate regions that draws creatures into its cadence. The song begins in spring, when plants elevate their shoots to greet the sun, whose presence is increasingly felt warming the landscape. Wild and domestic mammals, whose young are no longer bathed in the warmth of mother's womb, are now bathed in the rays of the sun. The tempo picks up in summer as plants fashion leaves, flowers, and fruits, and parents educate offspring in the ways of the world. The beat slows in autumn, a time for taking stock of what just transpired, for relishing the fall harvest, enjoying the colors as deciduous trees of green turn to reds, salmons, yellows, and oranges. Winter is time for re-creation, as snow blankets the landscape from the foothills to the tops of the high peaks. New life – already in the wombs of mother Earth and the plants and animals she nurtures – is awaiting rebirth to planet Earth in spring.

At sunset this evening, Sue and I strolled to the top of a small hill. To the south, the steep, jagged peaks of the Sangre de Cristo Mountain Range rise to over 14,000 feet. To the north, the smooth, rounded peaks of the Mosquito Range rise to similar elevations. In between are large valleys, parks, and rolling hills. As I gazed at the spectacle, I imagined how these grass-covered parks would've looked with bison families dispersed about. Native American artifacts are scattered all over these haunts. How many people have passed this way, enjoyed the view, and vanished? And where has this piece of ground I stand on traveled over the millennia? The supercontinent Pangaea existed during the late Paleozoic and early Mesozoic eras. It assembled from earlier continental units roughly 300 million years ago, and it began to break apart about 175 million years ago. The ground where I stand has traveled a great distance since then.

My friends Henry and Rose DeLuca had been married sixty-one years when Rose died during surgery in March of 1985. Her warmth and love left an ache in Henry's heart and hunger pangs in his gut. She was a marvelous cook, and he couldn't boil water. Out of sympathy, one of his brothers, Dominic, came to live with Henry after Rose died. It was ironic he returned to the ranch after so many years. The original ranch had been subdivided into four ranches with the passing of their father, Joe, who homesteaded the properties in the late 1800s. Henry often remarked – with a mix of distain and satisfaction – that all of his brothers "sold out, ended up working in town for wages, and didn't have a pot to piss in or a window to throw it out."

Henry was part of these landscapes. The last thing he wanted was for his ranch to be subdivided. Nevertheless, despite his best efforts to prevent that happening, not long after his death in November 1986, the ranch was subdivided among the three children he loved. Nor would he have approved of the many subdivisions put together in the vicinity of the ranch. The uninhabited landscapes he once knew now support more high-end mansions than cows.

None of the newcomers would have a clue that Henry was ever part of that landscape. Long gone are his cattle, sheep, goats, hogs, chickens, and geese. Nor is there any sign of the sawmill that helped him through the Great Depression and provided extra income during his life. Not a trace of Henry, who ranched in the 1900s, nor of his father, Joe, the immigrant from Italy who homesteaded the land in the 1800s, nor for that matter of the Ute Indians and their ancestors who lived on the land for some thirty-five centuries before the DeLucas. Except for mountains named after Ute Chiefs Shavano and Ouray and Ouray's wife, Chipeta, not a trace of the Utes remains.

Henry DeLuca was born in 1900. He died in 1986. I met Henry in 1969 when I was a senior in high school. I was eighteen years old. He was sixty-nine years old. I worked for him on the ranch in Colorado from 1969 to 1973 and I ran the ranch from 1973 to 1975. Every day, without fail, during those years he'd lament with a heavy Italian accent "I ache all over," and then he'd describe the parts of his body hurting on that particular day. At the time, I listened to him, and I certainly believed him, but his sentiments meant little to me as a spry young teenager.

We're born in pain, most of us will die in pain, and along the way we get sick. Aging, too, has emotional and physical pains. As Henry would've anticipated, I've experienced my share of physical pain. Like Henry, the older I get, the less the physical pains go away. I've also experienced, and sadly inflicted, my share of emotional pain. And though I would have chosen to avoid the physically and emotionally painful times, I've experienced three trials that have transformed me during my visit to Earth. I view those trials now as reminders. I'd become an intergalactic traveler with amnesia, and the trials reminded me why I came to visit Earth.

## Awakened by Depression

My first trial began in the late 1980s when I became agonizingly familiar with the despair of depression. Having never been unhappy a day in my life, my newfound melancholy came as a complete surprise. I'd always been so engaged and excited about every facet of life. Now, for the life of me, I couldn't figure out where I was, this totally unfamiliar mindscape of depression, how I got there, or how I could escape. As the years went by, I came to believe I'd never escape from this prison. The bottom had fallen away, and all I could think was "so this is all there is for the rest of my life. And if so, how can I, and why should I, go on?" All the things I used to live to do were dead in me, not a whisper, not a spark of emotion. I had quickly climbed the ladder of academic success and then fallen into a crevasse so deep that I had to look up to see a worm.

During that time, I began to read about the mythologies of peoples across time and from around the world, the wisdom traditions for planet Earth. I was captivated reading books and watching videos by Joseph Campbell. He spoke about mythology in ways that connected deeply with me. I came to appreciate the wisdom in each of the traditions – the primal religions, Hinduism, Buddhism, Confucianism, Taoism, Christianity, Islam, and Judaism – and the relationships among the wisdom traditions. They all resonated deeply within me.

I recall from my childhood only one meaningful exchange about religion. I asked my dear mother why only Catholics were going to heaven. I listened as she provided a historical account for why the Catholic faith was the only true religion. I didn't argue, but I didn't believe her. My rationale was simple: What about all the other children who belong to different Christian faiths? What about Hindus, Buddhists, Taoists, Judaists, and Muslim children? If each believed their way was the only way, it seemed to me everybody was going to hell.

I also began to read how revolutions in quantum and relativistic physics transformed the ways physicists viewed the world. I was stunned to realize the Newtonian worldview I learned in high school and college was only one small part of the worlds physicists began exploring in the early 1900s, from subatomic particles to galaxies. I was amazed to learn how the worlds physicists were exploring were deeply aligned with mystical views, as Fritjof

Capra described so eloquently in *The Tao of Physics*. All of this new knowledge fit so well with deep mysteries I'd pondered since I was a child. I knew these worlds, but nobody had ever talked about them.

I couldn't help but wonder why no one ever discussed such matters inside or outside of the classroom, inside or outside of churches. Didn't other people wonder about these mysteries? If so, why didn't they discuss them, and if not, why not? I was dismayed that no one even hinted at, let alone discussed in depth, any of these topics in all my years from grade school through graduate school. I became keenly aware I'd been a prisoner of what I'd been taught, confined to a cell fashioned by the culture in which I was raised. Until my depression, I was too busy amassing all the cultural information to explore the insights now being revealed to me.

Midway through my five years of depression, our family moved to Armidale, New South Wales, in Australia. We spent a year there on my sabbatical. A short while after I got to the shores of Australia, I was amazed to realize what I missed the most about home. I didn't miss all the accolades for the groundbreaking research and accomplishments of our group. Rather, I missed the people. I realized that loving relationships were creating the innovative research.

I experienced depression for five years, from the late 1980s, when the bottom fell out, into the early 1990s when the haze lifted and the world came into focus. I had been a cosmic voyager with amnesia, and I emerged seeing through transformed eyes. I'd been confusing what I was doing – following a career as an academic – with what I'd come to learn. Being a professor, doing research and teaching, had become ends in themselves, not means to an end.

After depression, I never taught a course the same way. I came to view the subject matter as a vehicle to relate the details of the material to life in the broadest sense – from personal relationships to the dynamism and uncertainty, wonders and mysteries of our moment on Earth. Teaching for me became both affective (in the spirit of love) and reflective (intellectual).

During the 1990s, students from different disciplines at Utah State – from environmental studies, wildlife biology, and fisheries biology to range science, animal science, and agronomy – were required to take a course I taught titled "Managing Dynamic Ecological Systems." Students majoring in range science, animal science, and agronomy were more production-oriented, and not nearly as concerned about environmental well-being as students pursuing programs in environmental studies, fisheries biology, and wildlife biology. At that time, a strong social movement was underway to remove all livestock from grazing on federally administered lands. Emotions ran high. To say the least, they all had very different views about how natural resources should be managed. The first class meetings of the semester weren't all that different from the town-hall meeting to discuss repealing the Affordable Care Act.

Late in the first week of one semester, a student named Wally finally came to class. He later told me he was dreading taking the course, and on that day, when I asked him to introduce himself to the class, his anger was palpable as he raged about having spent spring break hiking in "cow-shit canyon," a.k.a. Grand Staircase-Escalante National Monument. He hated cows being there. A little later that same day, an animal science student remarked that people had to eat meat to be healthy. To that, Wally stood up, ripped his shirt off, and said, "The hell you do."

After class that day, the teaching assistant, Cody, told me, "I looked to see what you were going to do, and you never skipped a beat." I was used to strong sentiments, and by then I had learned that the class was a wonderful opportunity to turn blame into understanding by creating an atmosphere where each person was comfortable speaking from the heart, without fear of recrimination from anyone. The classes were about dialog in the true sense of the word, a free flow of meaningful thoughts and sentiments about issues of local, regional, national, and international importance regarding how to manage natural resources for ecological, economic, and social integrity of the landscapes that nurture us all. As we explored those relationships, we touched on every conceivable theme from world mythologies and politics to local communities and families and natural resources. I was ever amazed at the collective wisdom of the students.

We spoke in meaningful ways of challenges, opportunities, living in an evolutionary spirit, and transcending the boundaries we create. Class discussions were now facilitated by students, called upon at random to lead discussions. Students had to come to class prepared. The courses were participatory: no more "I lecture, you take notes, I test to see if you've mastered what I told you." I told students, "I don't want you to think like I think. I don't care *what* you think. I just want you to think." I came to see myself as a vehicle for facilitating dialog, not as a source of knowledge. The class was rigorous, with many assignments. No more you-tell-me-what-I-told-you style of exam. For the take-home exams, students had to review materials we'd covered and then write about what the content meant for them personally and professionally.

Early in the semester, especially for the first exam, some of the personal comments were aggressive and antagonistic. Some students clearly felt uncomfortable with the format of the course, with opening up to others, and with thinking for themselves. But as the course progressed, amazing transformations occurred. We laughed, we cried, we spoke from our hearts.

I worked for many years to learn how to teach, but I never really got there until the end of my career. With time, I came to realize I had to transform. I had to transcend my own imaginary fears and boundaries. I had to learn to become open and vulnerable in front of a group, to speak from my heart without preconceived notions. That, in turn, enabled the group to do likewise.

One former student, Christy Mack, captured some of the sentiments we experienced:

> *The most important subject ever discussed in my opinion is how an individual's beliefs and traditions from throughout their life can affect their beliefs in natural resource management. It amazes me to listen to people and realize that there are many times that a person cannot even begin to comprehend other's beliefs because their mind-set is so totally different from the other person's. More than once, ideas that have seemed supported on all sides and written in stone have been totally dynamited by a single person's explanation of a different viewpoint. To get a person to question those things they have taken for granted all their life is an amazing thing. Listening to the other side and trying to understand their reasoning is a challenge I love to face. Sometimes both sides have such good reasoning I cannot tell you that there is a correct belief. People's belief systems have been my favorite topic of this class.*

And Wally? He ended up managing a ranch in southern Utah.

In *Embraced by the Light*, author Betty Eadie tells of her near-death experience (NDE). During her NDE, she learned our experiences on Earth are by intent and design. Although we don't consciously realize it, we planned our lives prior to our sojourn to Earth. We are all at different levels of spiritual development, and we come to Earth in the stations that best suit our spiritual needs. Our experiences transform us and people we encounter "apparently by mere chance" during our visit. Transformation is an essential part of our experience of Earth, and the people we like the least are the ones who can help us to grow the most, spiritually. I've often wondered how many of the people we "despise" come into our lives to teach us to love. We don't know enough to begin to judge people correctly here on

Earth. When we judge others for their faults or limitations, we are exhibiting similar shortcoming in ourselves.

Eadie's experiences align with beliefs about reincarnation.[1] Reincarnation tempers justice (credits/debits for past deeds) with mercy (chance to repay debits). In *Many Lives, Many Masters* Brian Weiss – American psychiatrist, hypnotherapist, and author who specializes in past life regression – sums up his beliefs: "We are immortal; we are eternal. We never die; we merely transform to a heightened state of consciousness, no longer needing a physical body. We are always loved. We are never alone, and we can never be harmed, not at this level." He concludes, paraphrasing the mystic Teilhard de Chardin: "We are not human beings having a spiritual experience. We are spiritual beings having a human experience."

I've often wondered where the souls of humans who believe in reincarnation will go when no living humans are left on Earth. Cosmologists estimate this universe has 200 billion galaxies, each with 400 billion stars, and who knows how many planets – and this may be just one of who-knows-how-many universes. Time has no beginning, no end. While I embrace the notion that we evolve spiritually by coming to Earth, my oh-so-mortal self wonders: Eternity is a long "time," what do the Gods do for eternity? I'm fascinated by the thought that Earth is merely one of untold places where the Gods come to play – they just play rough here. What we call life on Earth is merely one manifestation of universal consciousness. Perhaps we transform endlessly, energy manifest as ever-changing forms of matter in countless galaxies and universes.

Eben Alexander is a neurosurgeon who was skeptical of the accounts of the NDEs of his patients. He knew every scientific explanation why NDEs aren't real. And then he had an NDE of his own that couldn't be explained neurologically. As he describes in his book, *Proof of Heaven*, Alexander learned there isn't one universe but many – in fact, more than he could conceive. He saw the wealth of life throughout countless universes, including some whose intelligence was advanced far beyond that of humanity. He also saw countless higher dimensions, but like a visit to Earth as a human being, the only way to know these dimensions is to enter and experience them directly. Like physicists who try to explain complex mathematical relationships in words, Alexander notes that conveying that knowledge now "is rather like being a chimpanzee, becoming a human for a single day to experience all of the wonders of human knowledge, and then returning to one's chimp friends and trying to tell them what it was like." He also emphasizes that most of what we have to say about God and the higher spiritual worlds involves bringing them down to our level on Earth, rather than elevating our perceptions up to theirs.

We often speak of two dimensions – the here and now and the hereafter. For me, everywhere is a dimension of heaven. For me, there is only endless transformation. There is no birth, no death; no coming, no going; no being, no nonbeing. I am. I don't doubt that as I behold in awe the indescribable beauty manifest in each particle of nature. The Kingdom of God is not to be awaited in the field of time, but to be realized in the here and now of the Wisdom Body. This message is affirmed in the last lines of the Gnostic Gospel according to Thomas, when Christ's disciples asked him, "When will the Kingdom come?" He replied, "It will not come by expectation; they will not say 'See here,' or 'See there.' But the Kingdom of the Father is spread upon the Earth and men do not see it." We are in heaven. We just need to wake up.

## My Gift of Cancer

Ironically, events that occurred during the depression phase of my life inadvertently set up my next awakening. In order to afford our family sabbatical in Australia, we sold a house we'd built in Providence, Utah. Though I wasn't the contractor, I'd worked hard on the construction, and I'd done all the work to finish the basement. I had a lot of skin in the game. When it came time to sell our house, the market was depressed and we ended up selling the place for little more than we had put into it. We still had six months before we were leaving for Australia, so we rented a small, much more modest house than the one we built in Providence. All of this shocked me: selling the house I'd worked so hard to finish and moving into a "dump."

But after a couple of weeks in the "dump," I began to realize I didn't own the house in Providence – it owned me. The huge monthly payments for the "castle" vanished into small payments for the "dump." My family was together and we were warm and cozy during cold winter months. That's what mattered most. In Australia, we moved into a house that was even smaller and more primitive than the "dump." We had a cast-iron stove like my grandmother and great aunts used for cooking and heating their homes. When we returned to the United States, we built a modest home on 2.5 acres, raised big gardens, lambs, and calves. We owned one car, and I rode a bike to work and back for the next ten years. I was never again enamored with all the trappings of things. I was intently focused on paying off the debt on our house and retiring by the year 2000. But life had other designs for me, and I didn't retire until 2009. My intense focus, and the stress of working so hard so I could retire, caused my gift of cancer.

Betty Eadie writes in *Embraced by the Light*,

> *To my surprise, I saw that most of us had selected the illnesses we would suffer, and for some, the illness that would end our lives. Sometimes healing does not come immediately, or at all, because of our need for growth . . . sometimes it takes what we would consider negative experience to help develop our spirits.*

My second transformation began on the day before Thanksgiving in 1999. I knew it wouldn't be good news when the nurse called earlier in the week to tell me they needed to discuss the results of my colonoscopy. I knew they weren't calling me into the office to tell me how great my GI tract looked. But I never thought cancer.

So, there I sat, stunned, looking at pictures. I'd seen cancers, but I'd never "owned" an inflamed, bright red-and-white one. I listened as the doctor told me the seriousness of the cancer. He said he hoped that no cells had yet metastasized and gotten into my lymph system. He didn't have to say "it's all over but the crying if that's happened." I knew that from losing a good friend and mentor to colon cancer. Sadly, tellingly, as he spoke, I was thinking, "I just don't have time for cancer, not now, not ever. I have too much going on now." Through the mirage of words and images, I heard him say surgery. I remember thinking I couldn't do that until March 2000, at the earliest. Fortunately, I said nothing that would have revealed the utter foolishness and ignorance of the "human-doing" I'd become. I needed another transformation and cancer was it.

As the reality of cancer sank into my being, I felt a sense of relief. I'd have to stop working nonstop; no way could I keep going at this pace without a change of perspective and behavior. It was an amazing time – so many different feelings, but no dread at all, just excitement and a sense that this was something that needed to happen to me. The experience was surreal. The proximity of death magnified and intensified the experience of life. In retrospect, cancer changed how I experienced life and prepared me for the upcoming decade of my life.

I remember the day of the surgery, walking down the hall to the operating room with the feeling you have when you know you must go, but everything in you wants to run away. And then the surgery, having "my" body cut open was in one way the ultimate invasion of privacy, and in another way a total opening up of "my" self. I remember the intensity and magnitude of the pains after surgery – long, slow, painful walks down the halls, long agonizing nights. I recall the hustle and bustle during the day as nurses tended to my needs. I recall the peace and quiet at night, the Christmas lights shining exquisitely and peacefully into my room from the hallway.

After the surgery, I was overcome with deep peace, impossible to describe, a peace that transcended understanding, a feeling of unconditional love that surrounds and sustains us all. That combination of peace and love created a profound feeling that "everything is going to be alright." I was experiencing a sense of nirvana that comes from shutting down the endless prattle in my brain and opening up to the experience of peace and love that come from being. I still attain that deep peace in nature.

I experienced what Jill Bolte Taylor, a Harvard-trained brain scientist, described in *My Stroke of Insight*, which we discussed in chapter 5. When what she referred to as her "left hemisphere" shut down, her "right hemisphere," which experiences the present moment and our at-one-ness with all things, took over. As she weakened to the point that she couldn't walk, talk, read, write, or recall any of her life, she alternated between the euphoria of her intuitive and kinesthetic brain, in which she felt a sense of complete well-being and peace, and the logical, sequential brain that knew she was having a stroke. Her stroke was a revelation: Quieting our brain unearths feelings of well-being from the experience of our interconnectedness with all things, sensations that are often sidelined by our endless focus on the details of life.

When I returned to work in January following my surgery, a friend and colleague named Mary Lou told me she was sorry to hear I had cancer. She told me she, too, had cancer. She said the doctors were reasonably certain they "got it all." In my case, they were right. In her case, they weren't. Mary Lou died a year later. I've gone to many touching funerals, but Mary Lou's was the most poignant of any I've attended. During her service, many people spoke of the warm friendships they shared with her. Four people in particular evoked deep emotions.

Her son spoke of the remarkable love of a mother for a child and a child for his mother. One of her close friends then described peaceful mornings near the end of her life sitting in the backyard, holding hands, quietly absorbing the beauty and wonder of nature. Her husband's brother told a parable. In the tale, a man saw the butterfly struggling to emerge from its cocoon. Wanting to help the butterfly, he cut the cocoon unaware that the struggle was an indispensable step in the development of the butterfly:

Challenges create opportunities to transform. Finally, her husband spoke briefly. He began by thanking everyone for coming. He then talked about their experience of cancer: how his beautiful wife had been ravaged by tumors that ultimately commandeered her entire body, including her brain. He concluded by describing the emotional experiences of cancer: from dread and anxiety to anger, resentment, and frustration to hopes and dashed hopes. In the end, only one emotion remained . . . and that was love.

During my bout with cancer, I recall overwhelming love from people all over the world. I often thought it was worth getting cancer to feel so utterly enveloped in love. I'd never felt such unconditional love and compassion from so many people simultaneously.

In *The Power of Myth*, Joseph Campbell writes,

> *For, according to the Indian view, our separateness from each other in space and time here on earth – our multitude – is but a secondary, deluding aspect of the truth, which is that in essence we are of one being, one ground; and we know and experience the truth – going out of ourselves, outside the limits of ourselves – in the rapture of love.*[2]

In *Dying to Be Me*, Anita Moorjani writes of the unconditional love she experienced during her near NDE:

> *In my NDE state, I realized that the entire universe is composed of unconditional love, and I'm an expression of this. Every atom, molecule, quark, and tetraquark is made of love. I can be nothing else, because this is my essence and the nature of the universe. . . . I also realized that within that infinite, nonjudgmental realm, there's actually no need to forgive myself or anybody else. We're all perfect, exquisite children of the universe, and we exist out of pure love. Unconditional love is our birthright, not judgment or condemnation, and there's nothing we need to do to earn it. . . . The need to forgive is born out of seeing things as good and bad, but when there's no judgment, there's nothing for us to pardon.*

Love transcends duality. Anita no longer thought in terms of good/bad, right/wrong. During her NDE, she was not judged. Instead, she experienced unconditional love for herself and everyone else. After her NDE, she had only compassion for criminals, terrorists, and their victims. She understood people who commit these acts are confused, frustrated, in pain, and filled with self-hatred. Self-actualized, happy individuals never engage in such behaviors. People who love themselves share their love unconditionally. Perpetrators are the victims of their own self-hatred and pain. In a very real sense, we create heaven (love) or hell (lack of love) on Earth by the ways we behave. We make choices. Nothing theoretical here; we observe the outcomes every day we live. As Joseph Campbell often said, "Heaven and hell and all the Gods are in us."

Those experiences are Buddhist notions of wisdom and compassion. Cultivating wisdom means gaining insight into the interdependent nature of all things, culminating in an awareness of "emptiness," that is to say nothing arises independent of manifold causes and conditions and nothing lasts. Nurturing compassion means opening our hearts and minds to the struggles of all beings. Unifying these skills is manifest in loving kindness and compassion for all beings.

Eben Alexander echoes these views in *Proof of Heaven*. During an NDE, he received the same message:

> *You are loved and cherished. You have nothing to fear. There is nothing you can do wrong. If I had to boil this entire message down to one sentence, it would run this way: You are loved. And if I had to boil it down further, to just one word, it would (of course) be, simply: Love. Love is, without a doubt, the basis of everything. Not some abstract, hard-to-fathom kind of love but the day-to-day kind that everyone knows – the kind of love we feel when we look at our spouse and our children, or even our animals.*

Unconditional love is the core of everything that ever will exist, and no proper understanding of who and what we are can be achieved by anyone who does not know it, nor embodies it in all of their actions.

I've often thought life is the stage upon which the challenge to learn to love one another plays out, no matter one's vocation. We don't come to Earth to be farmers or ranchers; lawyers, doctors, or professors; plumbers or carpenters. We come to Earth to learn to love. The courage to love our enemies is the source of creativity in the world. For as paradoxical as it seems, creativity comes from unions of pairs of opposites, from dying to one's self only to be born anew. When we lose the ability to love, we lose hope. When we lose hope, we lose the ability to imagine new relationships. In so doing, we lose faith in our power to participate in creating them. In the end, there are but three things that matter, and the greatest of these is love.

## The Poignancy of Letting Go

My third transformation began ten years after my gift of cancer, with my retirement after thirty-five years of working at Utah State University. Though I knew that I was more than my career, the exceedingly gratifying relationships, accolades, and securities that were part of the job were hard to give up. I still loved teaching, research, and the stimulation of being in a university environment. The difference between this transformation and the previous two was that I knew that it was time to go, though by far the easier path would seem to have been to stay. From depression and cancer, I'd learned the value of transformation. My heart was telling me it was time to let go – of the rewards and certainties of my job with its strong allure of being on top – and move into the unknown of the backwoods of Colorado. I had more to learn.

I remember the melancholy feelings of cleaning out "my" office, filling a dumpster with memories, thinking that's a career, gone in a blink, into the trash. The ache of this trial was a vivid reminder of what I knew: Everything that's transitory is but an illusion, and everything is transitory. Impermanence wasn't causing my pain; longing for things to be permanent was the source of my suffering. With time, as I transitioned from the stimulation of university life into the forest for quiet reflection, I relaxed into the peace, beauty, wonders, and mysteries of Earth.

Prior to this transformation, as a student, I'd been fixated with identifying all the plants and animals that lived where I now lived. As a professor, I'd been obsessed with understanding how form, function, and behavior emerged from ever-changing interrelationships among soil and plants, herbivores, and humans. My years in science awakened me to the beauty, wonders, and mysteries of life, but my interest in studying nature dropped off in the backwoods of Colorado. I also found my interest in doing, achieving, and succeeding gradually vanished.

No longer caught up in the hustle and bustle of academic life, I became enchanted with the perfection of the displays of creation and the mysteries of existence, like the child I was years ago. I once again delighted in simply observing the exquisiteness of plants and animals, appreciating their marvelous abilities to survive in extremes of drought and temperature and live year-round, especially in far-below-zero extremes of snow-covered landscapes during long winters. I've often wondered what I could learn if I were to experience – for just one day – being a pasque flower, a ponderosa pine, a mountain bluebird, a rabbit, a red-tailed hawk, a badger, a bobcat, or a lynx. How differently would I view life after any of those experiences?

In *Thus Spake Zarathustra*, Friedrich Nietzsche describes three transformations. They begin as a child, symbolized by a camel, learning how to live in a society. This is a time for the obedience and instructions a child needs to live in a society in a particular landscape. The camel, fully loaded, struggles to its feet and walks into the desert, where it is transformed into a lion – the stronger the lion, the heavier the load. The task of the lion is to kill a dragon, and the name of the dragon is "thou shalt." On every scale of the dragon is a "thou shalt" that must be thrown off in order to come to a realization of one's own nature. When the dragon is killed – all the "thou shalts" overcome – the lion is transformed into a child, now living out of its true nature.

In the end, my trials were vehicles of transformation. The first taught me that people and relationships, not accomplishments, are what matters: All good things come from loving relationships. Love is the creative energy of the universe. The second trial taught me that I was a visitor on Earth experiencing a magnificent dimension of heaven. I had nothing to fear, and I should let the anxieties go. When I did, I experienced deep peace that came from shutting down the chatter that was cluttering my mind, preventing me from appreciating the beauties, wonders, and mysteries of life. The second trial also taught me compassion: We are family with all of existence. My third trial taught me that all things, including this thing I call "my" self are illusory. Our attempts to cling to them bring only suffering and discontent. Everything that's transitory is but an illusion, and everything in the Earthly realm is transitory. When I look back now, my seven decades on Earth seem like a dream, the dream of life from which I'm wakening.

## Tat Tvam Asi – This Is You

It's autumn once again and the glacial cirque is aglow with bright greens, yellows, oranges, and salmon reds of alpine huckleberry plants that blanket the spires that surround Island Lake atop the North Fork drainage. The azure sky is reflecting as colors morph from turquoise to emerald green in the lake. The soft breeze creates sparkling diamonds rippling across the water. As the gentle wind calms to a whisper, the cirque becomes a mirror dotted with the circles cutthroat trout make as they feed on the surface. The scene is stunningly placid. At moments like this, the consciousness of every cell in my body knows I am not separate from this sacred place. I am this creation and the experience is awe-inspiring. I feel autumn morphing into winter.

Though I'd like nothing better than to spend the autumns that remain in this meditation, I'm ready for winter. I know of no peace or silence as tranquil as snowflakes falling gently on a winter's night, the ringing glow of countless snowflakes tumbling gently to Earth. Am I among the last of the "snowflakes" who will have the opportunity to tumble to Earth?

I reflect on the words of Nobel laureate Erwin Schrödinger as he beckons us to imagine a setting such as this.[3] He points out that the granite peaks were here for thousands of years before I came to Earth. In a short time, my body will no longer exist, yet the cirque, peaks, and sky will continue, unchanged, for millennia. He asks what has called me so suddenly out of nothingness to experience this spectacle. For thousands of years, others sat here and gazed with awe and longing in their hearts. Like me, they were begotten of man and born of woman. He then asks: "Was this person someone else? Was it not you yourself? What is this Self of yours? What was the necessary condition for making the thing conceived this time into you, just you and not someone else? What clearly intelligible scientific meaning can this 'someone else' really have?"

He asks: If my mother had cohabitated with another man and had a son by him, and my father had done likewise, would I have come to be? Or was I living in them, and in my father's father thousands of years ago? Even if this is so, why am I not my sister, why is my sister not me, why am I not one of my cousins? What justifies me in discovering this difference – the difference between "me" and someone else – when objectively what is there is the same?

Pondering in this manner, we may see, in a flash, the profound rightness of the basic conviction in Vedanta (one of the schools of Hinduism): It is not possible that I came into being from nothingness at a given moment not so long ago. Rather this experience is eternal and unchangeable, one with all creation. But not in the sense that I am a part, a piece, of an eternal, infinite being, an aspect or modification of it. For we then have the same mysterious question: Which part, which aspect am I? What, objectively, differentiates me from others?

No, I am "all in all" – one with the transcendent. My life isn't merely a part. Rather, my life is the whole, though this whole is not so organized to be grasped in a glance. This is what Brahmins express in their mystic formula:

*Tat tvam asi: This is you.* "I am in the east and in the west, I am below and above, I am this whole world." This is Jesus's insight in the Gnostic Gospel of Thomas: "Cleave a piece of wood, I am there." This is the metaphysical insight, which can be known only when the masks of the "I am's" dissolve. Our apparent separateness is an illusion. We identify with our country, our ethnicity, our religion, our politics, our job, and so on – the "I am's." But that's a trap, inflicted upon us momentarily in the here and now. If we had been born at a different time or in a different place, the "I am's" would be completely different for each of us. When we transcend the "I am's" we come to the realization of "I am." That realization can occur in either of two ways: detaching from the finite self or connecting with all existence. Whichever way, when the experience of transcendence occurs, the fears, desires, and social duties of being on Earth fade into at-one-ness with all existence: *Tat tvam asi.*

I'm captivated to think I am dust of the Big Bang, the cataclysmic event cosmologists tell us created this universe some fifteen billion years ago. The cosmic atoms that have traveled the universe have created me for an instant too brief to measure on a cosmic clock of that magnitude. I am them and they are me, together on a cosmic journey. We are one. I am composed of some fifty trillion cells, each a universe with a life of its own, too short to measure by my clock.

My cells are organized into societies of organ systems, all conscious. They've all had a vote in what I did every minute of every day of my life. They are my life as I am their life. They meet my needs as I meet their needs. Their votes were tallied, without my knowledge or consent, within my body, which coordinated their behavior with the social and physical landscapes I momentarily called home. We are one, though I didn't always stop to ponder that.

I'm also a strand in the fabric of society, a fact I ponder more than I consider cell number 1.375 billion in my pancreas. I'm fascinated to think the society I inhabit is merely a strand in the ongoing journey of man. That journey began some 50,000 years ago when our most recent descendants left Africa, amidst global climate change and increasing aridity, to seek more favorable environs elsewhere, and ultimately they came to inhabit landscapes round the globe.

When I was a child, I dreaded death. I was afraid I'd die before I'd done whatever it was I'd come to do on Earth, perhaps to experience lessons I was supposed to learn. I didn't realize as a youth I had nothing to fear. Now, well into my seventh decade, I've had an "unforeseen" adventure on Earth. I'm no longer afraid of "death." I'm prepared to transform into whatever other dimensions of universal at-one-ness await me when the time comes.

I'm ready for the scavengers and decomposers to transform this carcass, which I've had the opportunity to inhabit, into soil, leaves of bluebunch wheatgrass, pasque flowers, mountain mahogany, antelope, coyote, raven, and eagle. But I will miss being human on Earth, with all the delights and dilemmas, sorrows and pains, qualms and uncertainties, wonders and deep mysteries the journey entails.

In awe and with humility, I give thanks for my moment on Earth – with all its horrors, beauties, wonders, and deep mysteries. I'm grateful for family and friends who accompanied me on the journey. I thank the plants and animals that grace this planet; that daily gave their lives to sustain my life through meals prepared with love. Bless this planet and its inhabitants for the greater good, whatever that may be, and however limited our ability to fathom that mystery.

# ACKNOWLEDGMENTS

I wrote the first draft of this book in 1999, the year I was diagnosed with cancer, then I set the book aside for the next ten years. I began to work on the manuscript again in 2008 when Sue and I moved from Utah to the backwoods of Colorado. Along the way, many people participated.

I could not have written this book without their help and support. In the beginning were colleagues at Utah State University, from mentors and professors the likes of Phil Urness, John Malechek, Dave Balph, and Carl Cheney to technicians and graduate students. Of note, Beth Burritt and Roger Banner worked with me throughout my career, and Juan Villalba was there during my last two decades at Utah State. Their friendship, collaboration, and wide-ranging conversations were invaluable. I also collaborated in thought and action with outstanding graduate students, whose research is cited throughout this book. Creativity flowed from warm professional and personal relationships with more than seventy-five graduate and postdoctoral students, visiting scientists, and colleagues during the past forty-five years.

Many people read and made useful comments on chapters of my book manuscript at various stages in this process, and some suggested additional useful reading materials. My thanks go to Anne Adams, Greg Brennan, Carl Cheney, Rich Cincotta, Natalie Gibson, Grady Grissom, Lorraine Munguia, Mark Schatzker, Mike Huffman, Richard Morris, Sue Rahilly, Becky Richards, Margaret Stojaks, Rachel Treasure, Juan Villalba, Nancy Warner, and Sandra Wise.

Other people provided words of encouragement throughout the process. I appreciate the suggestions on the manuscript offered by Cindi Anderson, Sandi Atwood, David Findley, Pablo and Mindy Gregorini, Douglas Hayes, Jessie Provenza, Stan Provenza, and Courtney White.

Finally, I want to thank Margo Baldwin and Chelsea Green Publishing for the opportunity to bring this project to fruition. Fern Marshall Bradley helped me translate this book from science speak into something considerably more fathomable and palatable for a general audience. Her in-depth and thoughtful questions, comments, and suggestions were invaluable. Thanks, too, to Nancy Bailey whose help as copy editor was invaluable in further honing the message, and to Angela Boyle for her careful proofreading of the manuscript. I also wish to thank Sarah Kovach for her capable and calm project management, Melissa Jacobson for the book's elegant design and layout, and production director Patricia Stone for her guidance throughout the process. I am deeply impressed with the professionalism and dedication of the entire staff at Chelsea Green.

In every way my partner of nearly fifty years, Sue, cocreated this story with me.

# NOTES

## Point of Departure: Transforming

1. William E. Holston, "The Diet of the Mountain Men," *California Historical Society Quarterly* 42 (1963): 301–9.
2. Malcolm Hollick, "Self-Organizing Systems and Environmental Management," *Environmental Management* 17 (1993): 621–28.
3. Camilo Mora et al., "How Many Species Are There on Earth and in the Ocean?" *PLoS Biology* 9, no. 8 (2011): e1001127.
4. David M. Raup, "Biological Extinction in Earth History," *Science* 231 (1986): 1528–33; David M. Raup, "The Role of Extinction in Evolution," *Proceedings of the National Academy of Sciences* 91 (1994): 6758–63.
5. Elizabeth Kolbert, *The Sixth Extinction: An Unnatural History* (New York: Henry Holt and Company, 2014).

## Chapter 1: Goats, Rats, and Clara's Kids

1. Frederick D. Provenza and Juan J. Villalba, "Foraging in Domestic Vertebrates: Linking the Internal and External Milieux," in *Feeding in Domestic Vertebrates: From Structure to Behavior*," ed. V. L. Bels (Oxfordshire: CABI Publication, 2006), 210–40.
2. M. Abrams et al., "Self-Selection of Salt Solutions and Water by Normal and Hypertensive Rats," *American Journal of Physiology* 156 (1949): 233–47.
3. Curt P. Richter and J. F. Eckert, "Increased Calcium Appetite of Parathyroidectomized Rats," *Endocrinology* 21 (1937): 50–54; Curt Richter and J. F. Eckert, "Mineral Appetite of Parathyroidectomized Rats," *American Journal of Medical Science* 198 (1939): 9–16.
4. S. Hao et al., "Uncharged tRNA and Sensing of Amino Acid Deficiency in Mammalian Piriform Cortex," *Science* 307 (2005): 1776–78.
5. B. J. Tepper and R. B. Kanarek, "Dietary Self-Selection Patterns of Rats with Mild Diabetes," *The Journal of Nutrition* 115 (1985): 699–709.
6. F. Q. Nuttall and M. C. Gannon, "Metabolic Response of People with Type 2 Diabetes to a High Protein Diet," *Nutrition & Metabolism* (2004): 6.
7. Curt Richter, "Total Self-Regulatory Functions in Animals and Human Beings," *Harvey Lecture Series* 38 (1943): 63–103.
8. Clara M. Davis, "Self Selection of Diet by Newly Weaned Infants," *American Journal Disabled Children* 36 (1928): 651–79.
9. Stephan Strauss, "Clara M. Davis and the Wisdom of Letting Children Choose Their Own Diets," *Canadian Medical Association Journal* 175 (2006): 1199–201.
10. Clara M. Davis, "Results of the Self-Selection of Diets by Young Children," *Canadian Medical Association Journal* 41 (1939): 259.
11. Bennett G. Galef, "A Contrarian View of the Wisdom of the Body as It Relates to Dietary Self-Selection," *Psychological Review* 98 (1991): 218–23.

12. Caitlin Dow, "Personalized Nutrition: Do You Need a Just-for-Me Diet?" *Nutrition Action Healthletter*, May 2018.

## Chapter 2: Challenges for Guests

1. Ian Heap, "The International Survey of Herbicide Resistant Weeds," http://www.weedscience.org.
2. Barbara McClintock, "The Significance of Responses of the Genome to Challenge," *Science* 226 (1984): 792-801.
3. Suzanne W. Simard and Daniel M. Durall, "Mycorrhizal Networks: A Review of Their Extent, Function, and Importance," *Canadian Journal of Botany* 82 (2004): 1140–65.
4. Oren Shelef et al., "Tri-Party Underground Symbiosis between a Weevil, Bacteria and a Desert Plant," *PLoS ONE* 8, no. 11 (2013): e76588.
5. Ted C. J. Turlings et al., "An Elicitor in Caterpillar Oral Secretions That Induces Corn Seedlings to Emit Chemical Signals Attractive to Parasitic Wasps," *Journal of Chemical Ecology* 19 (1993): 411–25.
6. Christopher J. Frost et al., "Plant Defense Priming Against Herbivores: Getting Ready for a Different Battle," *Plant Physiology* 146 (2008): 818–24.
7. S. A. Mousavi et al., "Glutamate Receptor-Like Genes Mediate Leaf-to-Leaf Wound Signalling," *Nature* 500 (2013): 422–26.
8. Gregg A. Howe and Georg Jander, "Plant Immunity to Insect Herbivores," *Annual Review of Plant Biology* 59 (2008): 41–66.
9. B. S. Meldrum, "Glutamate as a Neurotransmitter in the Brain: Review of Physiology and Pathology," *Journal of Nutrition* 130 (2000): 1007S–1015S.
10. Daniel A. Herms and William J. Mattson, "The Dilemma of Plants: To Grow or Defend," *Quarterly Review of Biology* 67 (1992): 283–335.
11. John P. Bryant and Peggy J. Kuropat, "Selection of Winter Forage by Subarctic Browsing Vertebrates: The Role of Plant Chemistry," *Annual Review of Ecology and Systematics* 11 (1980): 261–85.
12. Jonathan Gershenzon, "The Cost of Plant Chemical Defense Against Herbivory: A Biochemical Perspective," in *Insect-Plant Interactions*, Vol. V, ed. Elizabeth A. Berneys (Boca Raton: CRC Press, 1994), 105–73.
13. John P. Bryant et al., "Interactions Between Woody Plants and Browsing Mammals Mediated by Secondary Metabolites," *Annual Review of Ecology and Systematics* 22 (1991): 431–46.
14. May Berenbaum, "Patterns of Furanocoumarin Distribution and Insect Herbivory in the Umbelliferae: Plant Chemistry and Community Structure," *Ecology* 62 (1981): 1254–66.
15. Nancy Stamp, "Out of the Quagmire of Plant Defense Hypotheses," *The Quarterly Review of Biology* 78 (2003): 23–55.
16. Thomas P. Clausen et al., "Chemical Model for Short-Term Induction in Quaking Aspen (*Populus tremuloides*) Foliage against Herbivores," *Journal of Chemical Ecology* 15 (1989): 2335–46.
17. Ian T. Baldwin and Michael J. Karb, "Plasticity in Allocation of Nicotine to Reproductive Parts in *Nicotiana attenuata*," *Journal of Chemical Ecology* 21 (1995): 897–909.
18. G. Vourc'h et al., "Defensive Adaptations of *Thuja plicata* to Ungulate Browsing: A Comparative Study between Mainland and Island Populations," *Oecologia* 126 (2001): 84–93.
19. Frederick D. Provenza and David F. Balph, "Applicability of Five Diet-Selection Models to Various Foraging Challenges Ruminants Encounter," in *Behavioural Mechanisms of Food Selection*, ed. Roger N. Hughes (Berlin: Springer-Verlag, 1990), 423–60.
20. Anne-Marie Mayer, "Historical Changes in the Mineral Content of Fruits and Vegetables," *British Food Journal* 99 (1997): 207–11; D. R. Davis et al., "Changes in USDA Food Composition Data for 43 Garden Crops, 1950 to 1999," *The Journal of the American College of Nutrition* 23 (2004): 669–82; Brian Halweil, "Still No Free Lunch: Nutrient Levels in U.S. Food Supply Eroded by Pursuit of High Yields," *The Organic Center Critical Issues Report* (2007); D. R. Davis, "Declining Fruit and Vegetable Nutrient Composition: What Is the Evidence?" *Horticultural Science* 44 (2009): 15–19.

21. I. Loladze, "Rising Atmospheric $CO_2$ and Human Nutrition: Toward Globally Imbalanced Plant Stoichiometry?" *Trends in Ecology and Evolution* 17 (2002): 457–61.
22. D. R. Taub et al., "Effects of Elevated $CO_2$ on the Protein Concentration of Food Crops: A Meta-Analysis," *Global Change Biology* 14 (2008): 565–75; S. S. Myers et al., "Increasing $CO_2$ Threatens Human Nutrition," *Nature* 510 (2014): 139–42.
23. M. Dehghan et al., on behalf of the Prospective Urban Rural Epidemiology (PURE) study investigators, "Associations of Fats and Carbohydrate Intake with Cardiovascular Disease and Mortality in 18 Countries from Five Continents (PURE): A Prospective Cohort Study," *The Lancet* 390 (2017): 2050–62, doi: 10.1016/S0140-6736(17)32252-3.
24. David J. Augustine et al., "Elevated $CO_2$ Induces Substantial and Persistent Declines in Forage Quality Irrespective of Warming in Mixed Grass Prairie," *Ecological Applications* 28, no. 3 (2018): 721–35.
25. L. H. Ziska et al., "Rising Atmospheric $CO_2$ Is Reducing the Protein Concentration of a Floral Pollen Source Essential for North American Bees," *Proceedings of the Royal Society B* 283 (2016): 20160414.
26. John P. Bryant et al., "Carbon/Nutrient Balance of Boreal Plants in Relation to Vertebrate Herbivory," *Oikos* 40 (1983): 357–68; Phillis D. Coley et al., "Resource Availability and Plant Antiherbivore Defense," *Science* 230 (1985): 895–99; Jennifer R. Reeve et al., "Organic Farming, Soil Health, and Food Quality: Considering Possible Link," *Advances in Agronomy* 137 (2016): 319–67.
27. S. Anzman-Frasca et al., "Repeated Exposure and Associative Conditioning Promote Preschool Children's Liking of Vegetables," *Appetite* 58 (2012): 543–53; S. Bouhlal et al., "'Just a Pinch of Salt.' An Experimental Comparison of the Effect of Repeated Exposure and Flavor-Flavor Learning with Salt or Spice on Vegetable Acceptance in Toddlers," *Appetite* 83 (2014): 209–17; Antti Knaapila et al., "Pleasantness, Familiarity, and Identification of Spice Odors Are Interrelated and Enhanced by Consumption of Herbs and Food Neophilia," *Appetite* 109 (2017): 190–200.
28. A. D. Blatt et al., "Hidden Vegetables: An Effective Strategy to Reduce Energy Intake and Increase Vegetable Intake in Adults," *American Journal of Clinical Nutrition* 93 (2011): 756–63; M. K. Spill et al., "Hiding Vegetables to Reduce Energy Density: An Effective Strategy to Increase Children's Vegetable Intake and Reduce Energy Intake," *American Journal of Clinical Nutrition* 94 (2011): 735–41; J. S. Meengs et al., "Vegetable Variety: An Effective Strategy to Increase Vegetable Intake in Adults," *Journal of the Academy of Nutrition and Dietetics* 112 (2012): 1211–15.
29. Juliana R. Fritts et al., "Herbs and Spices Increase Liking and Preference for Vegetables among Rural High School Students," *Food Quality and Preference* 68 (2018): 125–34.
30. R. C. Havermans, "Increasing Children's Liking and Intake of Vegetables through Experiential Learning," in *Bioactive Foods in Promoting Health*, eds. R. R. Watson and V. R. Preedy (Oxford: Academic Press, 2009), 273–83; P. Pliner and C. Stallberg-White, "'Pass the Ketchup, Please': Familiar Flavors Increase Children's Willingness to Taste Novel Foods," *Appetite* 34 (2000): 95–103.
31. K. Brandt et al., "Agroecosystem Management and Nutritional Quality of Plant Foods: The Case of Organic Fruits and Vegetables," *Critical Reviews in Plant Sciences* 30 (2011): 177–97; M. Barański et al., "Higher Antioxidant and Lower Cadmium Concentrations and Lower Incidence of Pesticide Residues in Organically Grown Crops: A Systematic Literature Review and Meta-Analyses," *British Journal of Nutrition* 112 (2014): 794–811.
32. Linda M. Bartoshuk and Harry J. Klee, "Better Fruits and Vegetables through Sensory Analysis," *Current Biology* 23 (2013): R374–R378; Harry J. Klee and D. M. Tieman, "Genetic Challenges of Flavor Improvement in Tomato," *Trends in Genetics* 29 (2013): 257–62.
33. M. L. Schwieterman et al., "Strawberry Flavor: Diverse Chemical Compositions, a Seasonal Influence, and Effects on Sensory Perception," *PLoS ONE* 9, no. 2 (2014): e88446, doi:10.1371/journal.pone.0088446; D. Tieman et al., "The Chemical Interactions Underlying Tomato Flavor Preferences," *Current Biology* 22 (2012): 1035–39.

34. A. E. Oltman et al., "Consumer Attitudes and Preferences for Fresh Market Tomatoes," *Journal of Food Science* 79 (2014): S2091–97.
35. Gary Taubes, *Good Calories Bad Calories: Challenging the Conventional Wisdom on Diet, Weight Control, and Disease* (New York: Alfred A. Knopf, 2007).
36. A. Jacobs and M. Richtel, "How Big Business Got Brazil Hooked on Junk Food," *New York Times*, September 16, 2017.
37. K. Z. Guyton et al., on behalf of the International Agency for Research on Cancer Monograph Working Group, IARC, "Carcinogenicity of Tetrachlorvinphos, Parathion, Malathion, Diazinon, and Glyphosate," *The Lancet Oncology* 16 (2015): 490–91; John P. Myers et al., "Concerns Over Use of Glyphosate-Based Herbicides and Risks Associated with Exposures: A Consensus Statement," *Environmental Health* 15 (2016): 19.
38. "Eat the Peach, Not the Pesticide: A Shopper's Guide," *Consumer Reports*, May 19, 2015.
39. Barański et al., "Higher Antioxidant and Lower Cadmium Concentrations".
40. Brigitta Kurenbach et al., "Sublethal Exposure to Commercial Formulations of the Herbicides Dicamba, 2,4-Dichlorophenoxyacetic Acid, and Glyphosate Cause Changes in Antibiotic Susceptibility in *Escherichia coli* and *Salmonella enterica* serovar Typhimurium," *mBio* 6, no. 2 (2015): e00009–15, doi:10.1128/mBio.00009-15; H. C. Steinrücken and N. Amrhein, "The Herbicide Glyphosate Is a Potent Inhibitor of 5-Enolpyruvyl-Shikimic Acid-3-Phosphate Synthase," *Biochemical and Biophysical Research Communications* 94 (1980): 1207–12; E. Schönbrunn et al., "Interaction of the Herbicide Glyphosate with Its Target Enzyme 5-Enolpyruvylshikimate 3-Phosphate Synthase in Atomic Detail," *Proceedings of the National Academy of Sciences* 98 (2001): 1376–80.
41. David S. Ludwig, "Technology, Diet, and the Burden of Chronic Disease," *Journal of the American Medical Association* 305 (2011): 1352–53.
42. D. F. Lancy, *The Anthropology of Childhood: Cherubs, Chattel, Changelings* (Cambridge: Cambridge University Press, 2008).

## Chapter 3: No Two Alike

1. Frederick D. Provenza, "What Does It Mean to Be Locally Adapted and Who Cares Anyway?" *Journal of Animal Science* 86 (2008): E271–84.
2. Kevin D Welch et al., "The Effect of 7, 8-Methylenedioxylycoctonine-Type Diterpenoid Alkaloids on the Toxicity of Tall Larkspur (*Delphinium* spp.) in Cattle," *Journal of Animal Science* 90 (2012): 2394–401; Benedict T. Green et al., "Mitigation of Larkspur Poisoning on Rangelands through the Selection of Cattle," *Rangelands* 36 (2014): 10–15.
3. Frederick D. Provenza et al., "Conditioned Flavor Aversion: A Mechanism for Goats to Avoid Condensed Tannins in Blackbrush," *American Naturalist* 136 (1990): 810–28.
4. Lindsey L. Scott and Frederick D. Provenza, "Variation in Food Selection among Lambs: Effects of Basal Diet and Foods Offered in a Meal," *Journal of Animal Science* 77 (1999): 2391–97.
5. Michel Meuret and Frederick D. Provenza, "When Art and Science Meet: Integrating Knowledge of French Herders with Science of Foraging Behavior," *Rangeland Ecology and Management* 68 (2015): 1–17.
6. C. Hollenbeck and G. M. Reaven, "Variations in Insulin-Stimulated Glucose Uptake in Healthy Individuals with Normal Glucose Tolerance," *The Journal of Clinical Endocrinology and Metabolism* 64 (1987): 1169–73.
7. David S. Ludwig and Mark I. Friedman, "Increasing Adiposity: Consequence or Cause of Overeating?" *Journal of the American Medical Association* 311 (2014): 2167–68
8. Edward Archer, "In Defense of Sugar: A Critique of Diet-Centrism," *Progress in Cardiovascular Diseases* (2018): 10–19.
9. D. Zeevi et al., "Personalized Nutrition by Prediction of Glycemic Responses," *Cell* 163 (2015): 1079–94.

10. M. D. Jensen et al., "2013 AHA/ACC/TOS Guideline for the Management of Overweight and Obesity in Adults: A Report of the American College of Cardiology/American Heart Association Task Force on Practice Guidelines and The Obesity Society," *Circulation* 29 (2014): S102–38.
11. Dow, "Do You Need a Just-for-Me Diet?"
12. Raup, "The Role of Extinction in Evolution."
13. M. V. Novotny et al., "Biochemical Individuality Reflected in Chromatographic, Electrophoretic and Mass-Spectrometric Profiles," *Journal of Chromatography. B, Analytical Technologies in the Biomedical and Life Science* 866 (2008): 26–47.
14. D. S. Wishart et al., "HMDB: The Human Metabolome Database," *Nucleic Acids Research* 35 (2007): D521–26.
15. E. Jablonka and G. Raz, "Transgenerational Epigenetic Inheritance: Prevalence, Mechanisms, and Implications for the Study of Heredity and Evolution," *Quarterly Review of Biology* 84 (2009): 131–76; E. Heard and R. A. Martienssen, "Transgenerational Epigenetic Inheritance: Myths and Mechanisms," *Cell* 157 (2014): 95–109.
16. R. J. Schmitz et al., "Transgenerational Epigenetic Instability Is a Source of Novel Methylation Variants," *Science* 334 (2011): 369–73; D. Weigel and V. Colot, "Epialleles in Plant Evolution," *Genome Biology* 13 (2012): 249; D. C. Baulcombe and C. Dean, "Epigenetic Regulation in Plant Responses to the Environment," *Cold Spring Harbor Perspectives in Biology* 6 (2014): a019471.
17. Ian C. G. Weaver et al., "Epigenetic Programming by Maternal Behavior," *Nature Neuroscience* 7 (2004): 847–54.
18. C. Faulk and D. C. Dolinoy, "Timing Is Everything: The When and How of Environmentally Induced Changes in the Epigenome of Animals," *Epigenetics* 6 (2011): 791–97.
19. M. Pembrey et al., "Human Transgenerational Responses to Early-Life Experience: Potential Impact on Development, Health and Biomedical Research," *Journal of Medical Genetics* 51 (2014): 563–72.
20. R. C. Painter et al., "Prenatal Exposure to the Dutch Famine and Disease in Later Life: An Overview," *Reproductive Toxicology* 20 (2005): 345–52; R. C. Painter et al., "Transgenerational Effects of Prenatal Exposure to the Dutch Famine on Neonatal Adiposity and Health in Later Life," *BJOG: An International Journal of Obstetrics and Gynecology* 115 (2008): 1243–49.
21. B. T. Heijmans et al., "Persistent Epigenetic Differences Associated with Prenatal Exposure to Famine in Humans," *Proceedings of the National Academy of Sciences* 105 (2008): 17046–49.
22. J. G. Kral et al., "Large Maternal Weight Loss from Obesity Surgery Prevents Transmission of Obesity to Children Who Were Followed for 2 To 18 Years," *Pediatrics* 118 (2006): 1644–49; R. J. Loos, "Recent Progress in the Genetics of Common Obesity," *British Journal of Clinical Pharmacology* 68 (2009): 811–29; F. Guénard et al., "Differential Methylation in Glucoregulatory Genes of Offspring Born before vs. after Maternal Gastrointestinal Bypass Surgery," *Proceedings of the National Academy of Sciences* 110 (2013): 11439–44.
23. Jessica Smith et al., "Effects of Maternal Surgical Weight Loss in Mothers on Intergenerational Transmission of Obesity," *Journal of Clinical Endocrinology and Metabolism* 94 (2009): 4275–83; B. E. Levin, "The Obesity Epidemic: Metabolic Imprinting on Genetically Susceptible Neural Circuits," *Obesity Research* 8 (2000): 342–47; B. E. Levin and E. Govek, "Gestational Obesity Accentuates Obesity in Obesity-Prone Progeny," *American Journal of Physiology* 275 (1998): R1374–79; P. D. Taylor and L. Poston, "Developmental Programming of Obesity in Mammals," *Experimental Physiology* 92 (2007): 287–98; P. Iozzo et al., "Developmental Origins of Healthy and Unhealthy Ageing: The Role of Maternal Obesity – Introduction to DORIAN," *Obesity Facts* 7 (2014): 130–51.
24. C. A. Cooney et al., "Maternal Methyl Supplements in Mice Affect Epigenetic Variation and DNA Methylation of Offspring," *Journal of Nutrition* 132 (2002): 2393S–2400S; R. A. Waterland and R. L. Jirtle, "Transposable Elements: Targets for Early Nutritional Effects on Epigenetic Gene Regulation," *Molecular and Cellular Biology* 23 (2003): 5293–300; D. C. Dolinoy and

R. L. Jirtle, "Environmental Epigenomics in Human Health and Disease," *Environmental and Molecular Mutagenesis* 49 (2008): 4–8; D. C. Dolinoy et al., "Maternal Nutrient Supplementation Counteracts Bisphenol A-Induced DNA Hypomethylation in Early Development," *Proceedings of the National Academy of Sciences* 104 (2007): 13056–61.

25. W. Vanden Berghe, "Epigenetic Impact of Dietary Polyphenols in Cancer Chemoprevention: Lifelong Remodeling of Our Epigenomes," *Pharmacological Research* 65 (2012): 565–76.
26. The Human Microbiome Project Consortium, "Structure, Function and Diversity of the Healthy Human Microbiome," *Nature* 486 (2012): 207–14.
27. E. M. Quigley, "Gut Bacteria in Health and Disease," *Gastroenterology and Hepatology* 9 (2013): 560–69.
28. R. E. Hungate, *The Rumen and Its Microbes* (New York: Academic Press, 1966); M. D. Dearing et al., "The Influence of Plant Secondary Metabolites on the Nutritional Ecology of Herbivorous Terrestrial Vertebrates," *Annual Review of Ecology, Evolution, and Systematics* 36 (2005): 169–89.
29. M. Arumugam et al., "Enterotypes of the Human Gut Microbiome," *Nature* 473 (2011): 174–80.
30. E. C. Martens et al., "Recognition and Degradation of Plant Cell Wall Polysaccharides by Two Human Gut Symbionts," *PLoS Biology* 9 (2011): e1001221.
31. S. L. Schnorr et al., "Gut Microbiome of the Hadza Hunter-Gatherers," *Nature Communications* 5 (2014): 3654.
32. E. Patin and L. Quintana-Murci, "Demeter's Legacy: Rapid Changes to Our Genome Imposed by Diet," *Trends in Ecology and Evolution* 23 (2008): 56–59; G. H. Perry et al., "Diet and the Evolution of Human Amylase Gene Copy Number Variation," *Nature Genetics* 39 (2007): 1256–60; K. Hardy et al., "The Importance of Dietary Carbohydrate in Human Evolution," *The Quarterly Review of Biology* 90 (2015): 251–68.
33. J. C. Brand-Miller and S. Colagiuri, "Evolutionary Aspects of Diet and Insulin Resistance," *World Review of Nutrition and Dietetics* 84 (1999): 74–105; L. Cordain et al., "Macronutrient Estimations in Hunter-Gatherer Diets," *American Journal of Clinical Nutrition* 72 (2000): 1589–90; L. Cordain et al., "Plant-Animal Subsistence Ratios and Macronutrient Energy Estimations in Worldwide Hunter-Gatherer Diets," *American Journal of Clinical Nutrition* 71 (2000): 682–92; A. Ströhle and A. Hahn, "Diets of Modern Hunter-Gatherers Vary Substantially in Their Carbohydrate Content Depending on Ecoenvironments: Results from an Ethnographic Analysis," *Nutrition Research* 31 (2011): 429–35; K. Milton, "The Critical Role Played by Animal Source Foods in Human (Homo) Evolution," *Journal of Nutrition* 133 (2003): 3886S–92; A. W. Barclay et al., "Glycemic Index, Glycemic Load, and Chronic Disease Risk – A Meta-Analysis of Observational Studies," *American Journal of Clinical Nutrition* 87 (2008): 627–37.
34. H. M. Sinclair, "The Diet of Canadian Indians and Eskimos," *Proceedings of the Nutrition Society* 12 (1953): 69–82; H. V. Kuhnleini and R. Soueida, "Use and Nutrient Composition of Traditional Baffin Inuit Foods," *Journal of Food Composition and Analysis* 5 (1992): 112–26; J. D. Speth, "Early Hominid Hunting and Scavenging: The Role of Meat as an Energy Source," *Journal of Human Evolution* 18 (1989): 329–43; J. D. Speth, "Middle Paleolithic Subsistence in the Near East," *Before Farming* 2 (2012): 1–45.
35. Teicholz, *The Big Fat Surprise*, Chapter 7.
36. Anthony King, "Antibiotic Resistance Will Kill 300 Million People by 2050," *Scientific American*, December 16, 2014.
37. Terence S. Crofts et al., "Shared Strategies for β-lactam Catabolism in the Soil Microbiome," *Nature Chemical Biology* (2018): 556–64.
38. Erwin Schrödinger, *What Is Life? And Mind and Matter* (Cambridge: Cambridge University Press, 1944).
39. Frederick D. Provenza et al., "Complex Creative Systems: Principles, Processes, and Practices of Transformation," *Rangelands* 35 (2013): 6–13.

## Chapter 4: More Than a Matter of Taste


1. Provenza et al., "Conditioned Flavor Aversion."
2. Chris H. Titus et al., "Supplemental Polyethylene Glycol Influences Preferences of Goats Browsing Blackbrush," *Journal of Range Management* 54 (2001): 161–65.
3. Frederick D. Provenza et al., "Antiemetic Drugs Attenuate Food Aversions in Sheep," *Journal of Animal Science* 72 (1994): 1989–94.
4. John Garcia et al., "A General Theory of Aversion Learning," in *Experimental Assessments and Clinical Applications of Conditioned Food Aversions,* eds. N. S. Braveman and P. Bronstein (1985): 8–21.
5. Juan J. Villalba and Frederick D. Provenza, "Preference for Wheat Straw by Lambs Conditioned with Intraruminal Infusions of Starch," *British Journal of Nutrition* 77 (1997): 287–97; Elizabeth A. Burritt and Frederick D. Provenza, "Lambs Form Preferences for Nonnutritive Flavors Paired with Glucose," *Journal of Animal Science* 70 (1992): 1133–36.
6. Frederick D. Provenza, "Postingestive Feedback as an Elementary Determinant of Food Preference and Intake in Ruminants," *Journal of Range Management* 48 (1995): 2–17; Frederick D. Provenza, "Acquired Aversions as the Basis for Varied Diets of Ruminants Foraging on Rangelands," *Journal of Animal Science* 74 (1996): 2010–20; Frederick D. Provenza et al., "Linking Herbivore Experience, Varied Diets, and Plant Biochemical Diversity," *Small Ruminant Research* 49 (2003): 257–74.
7. W. Freeland and Daniel H. Janzen, "Strategies in Herbivory by Mammals: The Role of Plant Secondary Compounds," *American Naturalist* 108 (1994): 269–89.
8. James A. Pfister et al., "Tall Larkspur Ingestion: Can Cattle Regulate Intake below Toxic Levels?" *Journal of Chemical Ecology* 23 (1997): 759–77.
9. Charles A. Petersen et al., "Influence of Experience on Browsing Sagebrush by Cattle and Its Impacts on Plant Community Structure," *Rangeland Ecology and Management* 67 (2014): 78–87.
10. Luthando E. Dziba and Frederick D. Provenza, "Dietary Monoterpene Concentrations Influence Feeding Patterns of Lambs," *Applied Animal Behaviour Science* 109 (2007).
11. Luthando E. Dziba et al., "Feeding Behavior of Lambs in Relation to Kinetics of 1,8-Cineole Dosed Intravenously or Into the Rumen," *Journal of Chemical Ecology* 32 (2006): 391–408.
12. Garcia et al., "A General Theory of Aversion Learning."
13. T.R. Scott, "The Effect of Physiological Need on Taste," in *Taste, Experience, and Feeding,* eds. E. D. Capaldi and T. L. Powley (Washington, D.C.: American Psychological Association, 1990), 45–61.
14. James A. Young et al., *Halogeton: A History of Mid-20th Century Range Conservation in the Intermountain Area* (Washington, D.C.: United States Department of Agriculture, 1999).
15. J. James et al., "Effects of *Halogeton glomeratus* on pH Values in Rumen of Sheep," *American Journal of Veterinary Research* 29 (1968): 915–18; M. J. Allison et al., "Changes in Ruminal Oxalate Degradation Rates Associated with Adaptation to Oxalate Ingestion," *Journal of Animal Science* 45 (1977): 1173–79.
16. M. J. Allison et al., "Detection of Ruminal Bacteria that Degrade Toxic Dihydroxypyridine Compounds Produced from Mimosine," *Applied and Environmental Microbiology* 56 (1990): 590–94; Dearing et al., "The Influence of Plant Secondary Metabolites."
17. V. Norris et al., "Hypothesis: Bacteria Control Host Appetites," *Journal of Bacteriology* 195 (2013): 411–16; J. Alcock et al., "Is Eating Behavior Manipulated by the Gastrointestinal Microbiota? Evolutionary Pressures and Potential Mechanisms," *Bioessays* 36 (2014): 940–49; E. A. Mayer et al., "Gut microbes and the Brain: Paradigm Shift in Neuroscience," *The Journal of Neuroscience* 34 (2014): 15490–96.
18. M. I. Queipo-Ortuño et al., "Gut Microbiota Composition in Male Rat Models Under Different Nutritional Status and Physical Activity and Its Association with Serum Leptin and Ghrelin Levels," *PLoS ONE* 8 (2013): e65465; Y. Ravussin et al., "Responses of Gut Microbiota to Diet Composition and Weight Loss in Lean and Obese Mice," *Obesity* 20 (2012): 738–47.

19. J. M. Yano et al., "Indigenous Bacteria from the Gut Microbiota Regulate Host Serotonin Biosynthesis," *Cell* 161 (2015): 264–76; G. M. Mawe and J. M. Hoffman, "Serotonin Signalling in the Gut – Functions, Dysfunctions, and Therapeutic Targets," *Nature Reviews Gastroenterology and Hepatology* 10 (2013): 473–86; N. L. Baganz and R. D. Blakely, "A Dialogue Between the Immune System and Brain, Spoken in the Language of Serotonin," *ACS Chemical Neuroscience* 4 (2013): 48–63.
20. N. W. Bellono et al., "Enterochromaffin Cells Are Gut Chemosensors that Couple to Sensory Neural Pathways," *Cell* 170, no. 1 (2017): 185–98.
21. J. A. Bravo et al., "Ingestion of *Lactobacillus* Strain Regulates Emotional Behavior and Central GABA Receptor Expression in a Mouse via the Vagus Nerve," *Proceedings of the National Academy of Sciences* 108 (2011): 16050–55.
22. J. B. Russell et al., "A Net Carbohydrate and Protein System for Evaluating Cattle Diets: I. Ruminal Fermentation," *Journal of Animal Science* 70 (1992): 3551–61; L. A. Sinclair et al., "Effect of Synchronizing the Rate of Dietary Energy and Nitrogen Release on Rumen Fermentation and Microbial Protein Synthesis in Sheep," *Journal of Agricultural Science (Cambridge)* 120 (1993): 251–63; J. Hill et al., "Do Ruminants Alter Their Preference for Pasture Species in Response to the Synchronization of Delivery and Release of Nutrients?," *Rangeland Ecology and Management* 62 (2009): 418–27.
23. I. Kyriazakis and J. D. Oldham, "Diet Selection in Sheep: The Ability of Growing Lambs to Select a Diet that Meets Their Crude Protein (Nitrogen x 6.25) Requirements," *British Journal of Nutrition* 69 (1993): 617–29.
24. S. D. B. Cooper et al., "The Effect of Late Pregnancy on the Diet Selections Made by Ewes," *Livestock Production Science* 40 (1994): 263–75; I. Kyriazakis et al., "The Effect of Subclinical Intestinal Nematode Infection on the Diet Selection of Growing Sheep," *British Journal of Nutrition* 72 (1994): 665–77.
25. Lindsey L. Scott and Frederick D. Provenza, "Lambs Fed Protein or Energy Imbalanced Diets Forage in Locations and on Foods that Rectify Imbalances," *Applied Animal Behaviour Science* 68 (2000): 293–305.
26. Darrell L. Emmick, "Foraging Behavior of Dairy Cattle on Pastures," (PhD diss., Utah State University, 2007).
27. L. C. Pinheiro Machado Filho et al., "How Sustainable Is Grain Supplementation of Grazing Dairy Cows on Family Farms in the South of Brazil?," *Animal* 4 (2014): 463–75.
28. Q. R. Rogers and A. R. Egan, "Amino Acid Imbalance in the Liquid-Fed Lamb," *Australian Journal of Biological Science* 28 (1975): 169–181; A. R. Egan and Q. R. Rogers, "Amino Acid Imbalance in Ruminant Lambs," *Australian Journal of Agricultural Research* 29 (1978): 1263–79.
29. Juan J. Villalba and Frederick D. Provenza, "Nutrient-Specific Preferences by Lambs Conditioned with Intraruminal Infusions of Starch, Casein, and Water," *Journal of Animal Science* 77 (1999): 378–87.
30. I. Kyriazakis and J. D. Oldham, "Food Intake and Diet Selection of Sheep: The Effect of Manipulating the Rates of Digestion of Carbohydrates and Protein of the Foods Offered as a Choice," *British Journal of Nutrition* 77 (1997): 243–54; Juan J. Villalba and Frederick D. Provenza, "Preference for Flavored Foods by Lambs Conditioned With Intraruminal Administrations of Nitrogen," *British Journal of Nutrition* 78 (1997): 545–61; Juan J. Villalba and Frederick D. Provenza, "Preference for Flavored Wheat Straw by Lambs Conditioned with Intraruminal Infusions of Acetate and Propionate," *Journal of Animal Science* 75 (1997): 2905–14.
31. D. E. Amanoel et al., "Sheep Deficient in Vitamin E Preferentially Select for a Feed with a Higher Concentration of Vitamin E," *Animal* 10 (2016): 183–91.
32. Juan J. Villalba et al., "Learned Appetites for Calcium, Phosphorus, and Sodium in Sheep," *Journal of Animal Science* 86 (2008): 738–47.

33. Dawn R. Bazely, "Carnivorous Herbivores: Mineral Nutrition and the Balanced Diet," *Trends in Ecology and Evolution* 41 (1989): 55–156; A. J. Sutcliffe, "Further Notes on Bones and Antlers Chewed by Deer and Other Ungulates," *Journal of the British Deer Society* 4 (1977): 73–82; R. W. Fumess, "Predation on Ground-Nesting Seabirds by Island Populations of Red Deer *Cervus elaphus* and sheep *Ovis*," *Journal of Zoology, London* 216 (1988): 565–73.
34. K. P. Coates et al., "Use of Rodent Middens as Mineral Licks by Bighorn Sheep," in *Seventh Biennial Symposium Northern Wild Sheep and Goat Council*, ed. J.A. Bailey (Alberta: Alberta Fish and Wildlife, 1991), 206–9.
35. D. A. Denton et al., "Physiological Analysis of Bone Appetite (Osteophagia)," *Bioessays* 4 (1986): 40–42; J. R. Blair-West et al., "Behavioral and Tissue Response to Severe Phosphorous Depletion in Cattle," *American Journal of Physiology* 263 (1992): R656–63; M. F. Wallis de Vries "Foraging in a Landscape Mosaic: Diet Selection and Performance of Free-Ranging Cattle in Heathland and Riverine Grassland," (PhD diss., Wageningen, 1994).
36. Juan J. Villalba et al., "Phosphorus Appetite in Sheep: Dissociating Taste from Postingestive Effects," *Journal of Animal Science* 84 (2006): 2213–23.
37. Frederick D. Provenza, "Biological Manipulation of Blackbrush (*Coleogyne ramosissima* Torr.) by Browsing with Goats," (thesis, Utah State University, 1977).
38. Anna G. Thorhallsdottir et al., "Ability of Lambs to Learn about Novel Foods while Observing or Participating with Social Models," *Applied Animal Behaviour Science* 25 (1990): 25–33; Sarwat N. Mirza and Frederick D. Provenza, "Preference of the Mother Affects Selection and Avoidance of Foods by Lambs Differing in Age," *Applied Animal Behaviour Science* 28 (1990): 255–63.
39. Martin R. Yeomans, "Flavour-Nutrient Learning in Humans: An Elusive Phenomenon?" *Physiology and Behavior* 106 (2012): 345–55; Per Møller, "Satisfaction, Satiation and Food Behaviour," *Current Opinion in Food Science* 3 (2015): 59–64; Barbara V. Andersen and Grethe Hyldig, "Consumers' View on Determinants to Food Satisfaction: A Qualitative Approach," *Appetite* 95 (2015): 9–16; Barbara V. Andersen et al., "Integration of the Sensory Experience and Post-Ingestive Measures for Understanding Food Satisfaction. A Case Study on Sucrose Replacement by *Stevia rebaudiana* and Addition of Beta Glucan in Fruit Drinks," *Food Quality and Preference* 58 (2017): 76–84; E. Boelsma et al., "Measures of Postprandial Wellness after Single Intake of Two Protein-Carbohydrate Meals," *Appetite* 54 (2010): 454–65.
40. L. L. Birch and D. W. Marlin, "I Don't Like It; I Never Tried It: Effects of Exposure on Two-Year-Old Children's Food Preferences," *Appetite* 4 (1982): 353–60; C. Sanudo et al., "Regional Variation in the Hedonic Evaluation of Lamb Meat from Diverse Production Systems by Consumers in Six European Countries," *Meat Science* 75 (2007): 610–21.
41. J. M. Brunstrom et al., "In Search of Flavour-Nutrient Learning. A Study of the Samburu Pastoralists of North-Central Kenya," *Appetite* 91 (2015): 415–25.
42. Davis, "Results of the Self-Selection of Diets by Young Children"; Lind, *A Treatise on the Scurvy*; M. G. Tordoff, "The Case for a Calcium Appetite in Humans," in *Calcium in Human Health*, eds. C. M. Weaver and R. P. Heaney (Totowa, NJ: Humana Press, 2006), 247–66; D. A. McCarron et al., "Can Dietary Sodium Intake Be Modified by Public Policy?" *Clinical Journal of the American Society of Nephrology* 4 (2009): 1878–82; E. A. Rose et al., "Pica: Common but Commonly Missed," *Journal of the American Board of Family Practice* 13 (2000): 353–58.
43. S. A. Goff and H. J. Klee, "Plant Volatile Compounds: Sensory Cues for Health and Nutritional Value?" *Science* 311 (2006): 815–19.
44. Jay Schulkin et al., "Curt P. Richter 1894–1988," *Biographical Memoirs, National Academy of Sciences* 65 (1994): 311–20.
45. Richter, "Total Self-Regulatory Functions in Animals and Human Beings"; S. C. Woods and D. S. Ramsay, "Homeostasis: Beyond Curt Richter," *Appetite* 49 (2007): 388–98.

46. C. A. Deans et al., "Nutrition Affects Insect Susceptibility to Bt Toxins," *Scientific Reports* 7 (2017): 39705.
47. Roger E. Banner et al., "Supplemental Barley and Charcoal Increase Intake of Sagebrush by Lambs," *Journal of Range Management* 53 (2000): 415–20; Juan J. Villalba et al., "Influence of Macronutrients and Polyethylene Glycol on Intake of a Quebracho Tannin Diet by Sheep and Goats," *Journal of Animal Science* 80 (2002): 3154–64; J. L. Williams et al., "Snakeweed (*Gutierrezia* spp.) Toxicosis in Beef Heifers," *Proceedings of the Western Section of the American Society of Animal Science* 43 (1992): 67–69; J. R. Strickland et al., "Effects of Nutrient Supplementation in Beef Cows of Poor Body Condition Fed Snakeweed (*Gutierrezia* spp)," *Veterinary and Human Toxicology* 40 (1998): 278–84; Jian Wang and Frederick D. Provenza, "Food Deprivation Affects Preference of Sheep for Foods Varying in Nutrients and a Toxin," *Journal of Chemical Ecology* 22 (1996): 2011–21.
48. Juan J. Villalba et al., "Consequences of Nutrient-Toxin Interactions for Herbivore Selectivity: Benefits or Detriments for Plants?," *Oikos* 97 (2002): 282–92; Juan J. Villalba et al., "Influence of Macronutrients and Activated Charcoal on Intake of Sagebrush by Sheep and Goats," *Journal of Animal Science* 80 (2002): 2099–2109.
49. Juan J. Villalba and Frederick D. Provenza, "Foraging in Chemically Diverse Environments: Energy, Protein, and Alternative Foods Influence Ingestion of Plant Secondary Metabolites by Lambs," *Journal of Chemical Ecology* 31 (2005): 123–38.
50. William J. Foley et al., "Consequences of Biotransformation of Plant Secondary Metabolites on Acid-Base Metabolism in Mammals – A Final Common Pathway?," *Journal of Chemical Ecology* 21 (1995): 721–43; A. W. Illius and N. S. Jessop, "Modeling Metabolic Costs of Allelochemical Ingestion by Foraging Herbivores," *Journal of Chemical Ecology* 21 (1995): 693–719.
51. Tiffany D. Lyman et al., "Cattle Preferences Differ When Endophyte-Infected Tall Fescue, Birdsfoot Trefoil, and Alfalfa Are Grazed in Difference Sequences," *Journal of Animal Science* 89 (2011): 1131–37; Tiffany D. Lyman et al., "Phytochemical Complementarities among Endophyte-Infected Tall Fescue, Reed Canarygrass, Birdsfoot Trefoil and Alfalfa Affect Cattle Foraging," *Animal* 6 (2012): 676–82; Jake Owens et al., "Supplementing Endophyte-Infected Tall Fescue or Reed Canarygrass with Alfalfa or Birdsfoot Trefoil Increases Forage Intake and Digestibility by Sheep," *Journal of the Science of Food and Agriculture* 92 (2012): 987–92.
52. S. S. Seefeldt, "Consequences of Selecting Ramboulliet Ewes for Mountain Big Sagebrush (*Artemisia tridentata* ssp. *vaseyana*) Dietary Preference," *Rangeland Ecology and Management* 58 (2005): 380–84; Travis Mote et al., "Foraging Sequence Influences the Ability of Lambs to Consume Foods Containing Tannins and Terpenes," *Applied Animal Behaviour Science* 113 (2008): 57–68.
53. J. Cliff et al., "Association of High Cyanide and Low Sulphur Intake in Cassava-Induced Spastic Paraparesis," *Lancet* 326 (1985): 1211–13; W. P. Howlett et al., "Konzo, an Epidemic Upper Motor Neuron Disease Studied in Tanzania," *Brain* 113 (1990): 223–35.
54. J. P. Banea et al., "Control of Konzo in DRC Using the Wetting Method on Cassava Flour," *Food and Chemical Toxicology* 50 (2012): 1517–23.
55. T. Johns, *The Origins of Human Diet and Medicine* (Tucson: The University of Arizona Press, 1990); R. Wrangham and N. L. Conklin-Brittain, "Cooking as a Biological Trait," *Comparative Biochemistry and Physiology - Part A* 136 (2003): 35–46; Nabhan, *Why Some Like It Hot*; G. J. Armelagos, "Brain Evolution, the Determinates of Food Choice, and the Omnivore's Dilemma," *Critical Reviews in Food Science and Nutrition* 54 (2014): 1330–41.
56. S. Janssen and I. Depoortere, "Nutrient Sensing in the Gut: New Roads to Therapeutics?," *Trends in Endocrinology and Metabolism* 24 (2013): 92–100; I. Depoortere "Taste Receptors of the Gut: Emerging Roles in Health and Disease," *Gut* 63 (2014): 179–90; J. B. Furness et al., "The Gut as a Sensory Organ," *Nature Reviews Gastroenterology and Hepatology* 10 (2013): 729–40.

## Chapter 5: More Than One Kind of Memory

1. G. C. Green et al., "Long-Term Effects of Early Experience to Supplementary Feeding in Sheep," *Proceedings of the Australian Society of Animal Production* 15 (1984): 373–75; Michael H. Ralphs, "Persistence of Aversion to Larkspur in Naive and Native Cattle," *Journal of Range Management* 50 (1997): 367–70; Randall D. Wiedmeier et al., "Exposure to ammoniated Wheat Straw as Suckling Calves Improves Performance of Mature Beef Cows Wintered on Ammoniated Wheat Straw," *Journal of Animal Science* 80 (2002): 2340–48.
2. Wiedmeier et al., "Exposure to Ammoniated Wheat Straw as Suckling Calves Improves Performance of Mature Beef Cows Wintered on Ammoniated Wheat Straw."
3. Frederick D. Provenza et al., "How Goats Learn to Distinguish between Novel Foods That Differ in Postingestive Consequences," *Journal of Chemical Ecology* 20 (1994): 609–24.
4. Elizabeth A. Burritt and Frederick D. Provenza, "Ability of Lambs to Learn with a Delay Between Food Ingestion and Consequences Given Meals Containing Novel and Familiar Foods," *Applied Animal Behaviour Science* 32 (1991): 179–89.
5. Villalba and Provenza, "Nutrient-Specific Preferences by Lambs Conditioned with Intraruminal Infusions of Starch, Casein, and Water"; "Learned Appetites for Calcium, Phosphorus and Sodium in Sheep."
6. Elizabeth A. Burritt and Frederick D. Provenza, "Effect of an Unfamiliar Location on the Consumption of Novel and Familiar Foods by Sheep," *Applied Animal Behaviour Science* 54 (1997): 317–25.
7. Lyman et al., "Cattle Preferences Differ When Endophyte-Infected Tall Fescue, Birdsfoot Trefoil, and Alfalfa Are Grazed in Difference Sequences."
8. Karen L. Launchbaugh and Frederick D. Provenza, "Can Plants Practice Mimicry to Avoid Grazing by Mammalian Herbivores?," *Oikos* 66 (1993): 501–4.
9. Richard Dafters and George Anderson, "Conditional Tolerance to the Tachycardia Effect of Ethanol in Humans," *Psychopharmacology* 78 (1982): 365–67; B. M. Jones, "Circadian Variation in the Effect of Alcohol on Cognitive Performance," *Quarterly Journal of Studies on Alcohol* 35 (1974): 1212–19; Allan P. Shapiro and Peter E. Nathan, "Human Tolerance to Alcohol: The Role of Pavlovian Conditioning Processes," *Psychopharmacology* 88 (1986): 90–95; S. Siegel and K. Sdao-Jarvie, "Attenuation of Ethanol Tolerance by a Novel Stimulus," *Psychopharmacology* 88 (1986): 258–61.
10. S. Siegel and D. W. Ellsworth, "Pavlovian Conditioning and Death from Apparent Overdose of Medically Prescribed Morphine: A Case Report," *Bulletin of the Psychonomic Society* 24 (1986): 278–80.
11. S. Siegel et al., "Heroine 'Overdose' Death: Contribution of Drug-Associated Environmental Cues," *Science* 216 (1982): 436–37.
12. Ludwig, "Technology, Diet, and the Burden of Chronic Disease."
13. Susan E. Swithers, "Artificial Sweeteners Produce the Counterintuitive Effect of Inducing Metabolic Derangements," *Trends in Endocrinology and Metabolism* 24 (2013): 431–41.
14. Frederick D. Provenza, "Tracking Variable Environments: There Is More Than One Kind of Memory," *Journal of Chemical Ecology* 21 (1995): 911–23.
15. Jill Bolte Taylor, "My Stroke of Insight: A Brain Scientist's Personal Journey," *TED Talk*, March 13, 2008, http://www.ted.com; Taylor, *My Stroke of Insight*.
16. Jared A. Nielsen et al., "An Evaluation of the Left-Brain vs. Right-Brain Hypothesis with Resting State Functional Connectivity Magnetic Resonance Imaging," *PLoS ONE* 8, no. 8 (2013): e71275.
17. Elizabeth A. Burritt and Frederick D. Provenza, "Ability of Lambs to Learn with a Delay Between Food Ingestion and Consequences Given Meals Containing Novel and Familiar Foods," *Applied Animal Behaviour Science* 32 (1991): 179–89.
18. Frederick D. Provenza et al., "Food Aversion Conditioned in Anesthetized Sheep," *Physiology and Behavior* 55 (1994): 429–32.

## Chapter 6: Undermining the Wisdom Body


1. D. S. Ludwig, "Technology, Diet, and the Burden of Chronic Disease," *Journal of the American Medical Association* 305 (2011): 1352–53.
2. Paul Rozin et al., "Attitudes to Food and the Role of Food in Life in the U.S.A., Japan, Flemish Belgium and France: Possible Implications for the Diet–Health Debate," *Appetite* 33 (1999): 163–80.
3. Paul Rozin and James W. Kalat, "Specific Hungers and Poison Avoidance as Adaptive Specializations of Learning," *Psychological Review* 78 (1971): 459–86; Paul Rozin, "Adaptive Food Sampling Patterns in Vitamin Deficient Rats," *Journal of Comparative and Physiological Psychology* 69 (1969): 126–32.
4. John H. Ternouth, "The Kinetics and Requirements of Phosphorus in Ruminants," in *Recent Advances on the Nutrition of Herbivores,* eds. Yin W. Ho, et al. (Kuala Lumpur: Malaysian Society of Animal Production, 1991), 143–51.
5. Frederick D. Provenza et al., "Antiemetic Drugs Attenuate Food Aversions in Sheep," *Journal of Animal Science* 72 (1994): 1989–94.
6. Tim S. Phy and Frederick D. Provenza, "Sheep Fed Grain Prefer Foods and Solutions That Attenuate Acidosis," *Journal of Animal Science* 76 (1998): 954–60; Tim S. Phy and Frederick D. Provenza, "Eating Barley Too Frequently or in Excess Decreases Lambs' Preference for Barley but Sodium Bicarbonate and Lasalocid Attenuate the Response," *Journal of Animal Science* 76 (1998): 1578–83; Juan J. Villalba et al., "Sheep Self-Medicate When Challenged with Illness-Inducing Foods," *Animal Behaviour* 71 (2006): 1131–39.
7. Sheldon B. Atwood et al., "Intake of Lambs Offered Ad Libitum Access to One of Three *Iso*-Caloric and *Iso*-Nitrogenous Mixed Rations or a Choice of All Three Foods," *Livestock Science* 101 (2006): 142–49.
8. Lindsey L. Scott and Frederick D. Provenza, "Variation in Food Selection among Lambs: Effects of Basal Diet and Foods Offered in a Meal," *Journal of Animal Science* 77 (1999): 2391–97.
9. Sheldon B. Atwood et al., "Influence of Free-Choice vs Mixed-Ration Diets on Food Intake and Performance of Fattening Calves," *Journal of Animal Science* 79 (2001): 3034–40.
10. Sheldon B. Atwood et al., "Intake of Lambs Offered Ad Libitum Access to One of Three Iso-Caloric and Iso-Nitrogenous Mixed Rations or Choice of All Three Foods," *Livestock Science* 101 (2006): 142–49.
11. Juan J. Villalba et al., "Links between Ruminants' Food Preference and Their Welfare," *Animal* 4 (2010): 1240–47.
12. Francisco Catanese et al., "The Importance of Diet Choice on Stress-Related Responses by Lambs," *Applied Animal Behaviour Science* 148 (2013): 37–45.
13. Ryan A. Shaw, "Social Organization and Decision Making in North American Bison: Implications for Management," (PhD diss., Utah State University, 2012).
14. B. Burritt et al., "Finishing Bison by Offering a Choice of Feeds and Room to Roam," *Journal of the NACAA* 6 (2013): 1–6.
15. A. J. F. Webster, "Energy Partitioning, Tissue Growth and Appetite Control," *Proceedings of the Nutrition Society* 52 (1993): 69–76.
16. Stephen J. Simpson et al., "Geometric Analysis of Macronutrient Intake in Humans: The Power of Protein?" *Appetite* 41 (2003): 123–40; Stephen J. Simpson and David Raubenheimer, "Obesity: The Protein Leverage Hypothesis," *Obesity Reviews* 6 (2005): 133–42; A. K. Gosby et al., "Protein Leverage and Energy Intake," *Obesity Reviews* 15 (2014): 183–91.
17. David A. Booth, "Macronutrient-Specific Hungers and Satieties and Their Neural Bases, Learnt from Pre- and Postingestional Effects of Eating Particular Foodstuffs," in *Neural and Metabolic Control of Macronutrient Intake,* H. R. Berthoud and R. J. Seeley eds. (Boca Raton: CRC Press, 2000), 61–91; J. M. de Castro, "Macronutrient Selection in Free-Feeding Humans: Evidence for Long-Term Regulation," in *Neural Control of Macronutrient Selection,* H. R. Berthoud and R. J.

Seeley eds. (Boca Raton: CRC Press, 2000), 43–59; D. W. Gietzen, "Neural Mechanisms in the Responses to Amino Acid Deficiency," *Journal of Nutrition* 123 (1993): 610–25.

18. O. P. Garcia et al., "Impact of Micronutrient Deficiencies on Obesity," *Nutrition Reviews* 67 (2009): 559–72.
19. L. H. Epstein et al., "Long-Term Habituation to Food in Obese and Nonobese Women," *American Journal of Clinical Nutrition* 94 (2011): 371–76.
20. S. G. Lemmens et al., "Eating What You Like Induces a Stronger Decrease of 'Wanting' to Eat," *Physiology and Behavior* 98 (2009): 318–25; M. L. Pelchat et al., "Images of Desire: Food-Craving Activation during fMRI," *NeuroImage* 23 (2004): 1486–93; Per Møller, "Taste and Appetite," *Flavour* 4 (2015): 4; Per Møller, "Satisfaction, Satiation and Food Behaviour," *Current Opinion in Food Science* 3 (2015): 59–64.
21. Per Møller, "Gastrophysics in the Brain and Body," *Flavour* 2 (2013): 8.
22. B. V. Andersen et al., "Cayenne Pepper in a Meal: Effect of Oral Heat on Feelings of Appetite, Sensory Specific Desires and Well-Being," *Food Quality and Preference* 60 (2017): 1–8.
23. P. W. Sherman and J. Billing, "Darwinian Gastronomy: Why We Use Spices," *BioScience* 49 (1999): 453–63; Linda C. Tapsell et al., "Health Benefits of Herbs and Spices: The Past, the Present, the Future," *The Medical Journal of Australia* 185 (2006): S4–S24.
24. G. A. Armelagos, "Brain Evolution, the Determinates of Food Choice, and the Omnivore's Dilemma," *Critical Reviews in Food Science and Nutrition* 54 (2014): 1330–41.
25. A. Tremblay and H. Arguin, "Functional Foods, Satiation and Satiety," in *Satiation, Satiety and the Control of Food Intake*, J. E. Blundell and F. Bellisle eds. (Oxford: Woodhead Publishing, 2013), 202–18.
26. Darrell L. Emmick, "Foraging Behavior of Dairy Cattle on Pastures," (PhD diss., Utah State University, 2007).
27. Juan J. Villalba et al., "Learned Appetites for Calcium, Phosphorus, and Sodium in Sheep," *Journal of Animal Science* 86 (2008): 738–47; J. Hills et al., "Conditioned Feeding Responses in Sheep to Flavoured Foods Associated with Sulphur Doses," *Animal Science* 69 (1999): 313–25.
28. Juan J. Villalba and Frederick D. Provenza, "Preference for Flavored Wheat Straw by Lambs Conditioned with Intraruminal Administrations of Sodium Propionate," *Journal of Animal Science* 74 (1996): 2362–68.
29. Bill J. Freeland and Daniel H. Janzen, "Strategies in Herbivory By Mammals: The Role of Plant Secondary Compounds," *American Naturalist* 108 (1994): 269–86.
30. Juan J. Villalba et al., "Preference for Diverse Pastures by Sheep in Response to Intraruminal Administrations of Tannins, Saponins, and Alkaloids," *Grass and Forage Science* 66 (2011): 224–36.
31. Seung Hee Lee-Kwan et al., "Disparities in State-Specific Adult Fruit and Vegetable Consumption – United States, 2015," *Morbidity and Mortality Weekly Report* 66 (2017): 1241–47.
32. Julie A. Mennella et al., "The Development of Sweet Taste: From Biology to Hedonics," *Reviews in Endocrine and Metabolic Disorders* 17 (2016): 171–78.
33. Paul M. Wise et al., "Reduced Dietary Intake of Simple Sugars Alters Perceived Sweet Taste Intensity but Not Perceived Pleasantness," *American Journal of Clinical Nutrition* 103 (2016): 50–60; M. Cabanac et al., "Influence of Internal Factors on the Pleasantness of a Gustative Sweet Sensation," *Communications in Behavioral Biology, Part A* 1 (1968): 77–82.
34. B. A. Lorson et al., "Correlates of Fruit and Vegetable Intakes in US Children," *Journal of the American Dietetic Association* 109 (2009): 474–78; Kevin C. Mathias et al., "Does Serving Children Larger Portions of Fruit Affect Vegetable Intake?" *Obesity* 17 (2009): S90; T. V. Kral et al., "Effects of Doubling the Portion Size of Fruit and Vegetable Side Dishes on Children's Intake at a Meal," *Obesity* 18 (2010): 521–27; J. S. Savage et al., "Serving Smaller Age-Appropriate Entree Portions to Children Aged 3–5 Y Increases Fruit and Vegetable Intake and Reduces Energy Density and Energy Intake at Lunch," *American Journal of Clinical Nutrition* 95 (2012): 335–41.

35. M. Bartholomew, "James Lind's *Treatise of the Scurvy* (1753)," *Postgraduate Medical Journal* 78 (2002): 695–96.
36. Carmen Piernas et al., "Does Diet-Beverage Intake Affect Dietary Consumption Patterns? Results from the Choose Healthy Options Consciously Everyday (CHOICE) Randomized Clinical Trial," *American Journal of Clinical Nutrition* 97 (2013): 604–11; Valisa E. Hedrick et al., "Dietary Quality Changes in Response to a Sugar-Sweetened Beverage–Reduction Intervention: Results from the Talking Health Randomized Controlled Clinical Trial," *American Journal of Clinical Nutrition* 105 (2017): 824–33.
37. Shi-Sheng Zhou and Yiming Zhou, "Excess Vitamin Intake: An Unrecognized Risk Factor for Obesity," *World Journal of Diabetes* 5 (2014): 1–13.
38. C. J. Greenbaum et al., "Nicotinamide's Effects on Glucose Metabolism in Subjects at Risk for IDDM," *Diabetes* 45 (1996): 1631–34; J. J. Kelly et al., "Effects of Nicotinic Acid on Insulin Sensitivity and Blood Pressure in Healthy Subjects," *Journal of Human Hypertension* 14 (2000): 567–72.
39. Robert E. Hodges et al., "Experimental Scurvy in Man," *American Journal of Clinical Nutrition* 22 (1969): 535–48.
40. J. J. B. Anderson et al., "Calcium Intake from Diet and Supplements and the Risk of Coronary Artery Calcification and Its Progression among Older Adults: 10-Year Follow-Up of the Multi-Ethnic Study of Atherosclerosis (MESA)," *The Journal of the American Heart Association* 5 (2016): e003815.
41. D. R. Jacobs and L. C. Tapsell, "Food, Not Nutrients, is the Fundamental Unit in Nutrition," *Nutrition Review* 65 (2007): 439–50; H. Macpherson et al., "Multivitamin-Multimineral Supplementation and Mortality: A Meta-Analysis of Randomized Controlled Trials," *American Journal of Clinical Nutrition* 97 (2013): 437–44.
42. L. Allen et al., eds., *Guidelines on Food Fortification with Micronutrients* (World Health Organization and Food and Agriculture Organization of the United Nations, 2006).
43. R. Sangani and A. Ghio, "Iron, Human Growth, and the Global Epidemic of Obesity," *Nutrients* 5 (2013): 4231–49.
44. Balch, *Prescription for Herbal Healing*, 107.
45. Harri Hemilä and Elizabeth Chalker, "Vitamin C for Preventing and Treating the Common Cold," *The Cochrane Database of Systematic Reviews* 1 (2013): CD000980.
46. S. F. Choumenkovitch et al., "Folic Acid Intake from Fortification in United States Exceeds Predictions," *Journal of Nutrition* 132 (2002): 2792–98; A. D. Smith et al., "Is Folic Acid Good for Everyone?" *American Journal of Clinical Nutrition* 87 (2008): 517–33.
47. T. M. Brasky et al., "Long-Term, Supplemental, One-Carbon Metabolism–Related Vitamin B Use in Relation to Lung Cancer Risk in the Vitamins and Lifestyle (VITAL) Cohort," *Journal of Clinical Oncology* 35, no. 30 (2017): 3440–48.
48. Neil B. Metcalfe and Carlos Alonso-Alvarez, "Oxidative Stress as a Life-History Constraint: The Role of Reactive Oxygen Species in Shaping Phenotypes from Conception to Death," *Functional Ecology* 24 (2012): 984–96; Carlo Catoni et al., "Life History Trade-Offs Are Influenced by the Diversity, Availability and Interactions of Dietary Antioxidants," *Animal Behaviour* 76 (2008): 1107–19.
49. Goran Bjelakovic et al., "Meta-Regression Analyses, Meta-Analyses, and Trial Sequential Analyses of the Effects of Supplementation with Beta-Carotene, Vitamin A, and Vitamin E Singly or in Different Combinations on All-Cause Mortality: Do We Have Evidence for Lack of Harm?" *PLoS ONE* 8, no. 9 (2013): e74558; J. N. Hathcock, "Vitamins and Minerals: Efficacy and Safety," *American Journal of Clinical Nutrition* 66 (1997): 427–37; E. R. Miller et al., "Meta-Analysis: High-Dosage Vitamin E Supplementation May Increase All-Cause Mortality," *Annals of Internal Medicine* 142 (2005): 37–46.
50. C. A. Mulholland and D. J. Benford, "What Is Known about the Safety of Multivitamin-Multimineral Supplements for the Generally Healthy Population? Theoretical Basis for Harm,"

*American Journal of Clinical Nutrition* 85 (2007): 318S–22; U. Ramakrishnan, "Prevalence of Micronutrient Malnutrition Worldwide," *Nutrition Reviews* 60 (2002): S46–52; C. L. Rock, "Multivitamin-Multimineral Supplements: Who Uses Them?" *American Journal of Clinical Nutrition* 85 (2007): 277S–79.

51. Gøran Paulsen et al., "Vitamin C and E Supplementation Hampers Cellular Adaptation to Endurance Training in Humans: A Double-Blind, Randomised, Controlled Trial," *Journal of Physiology* 592, no. 8 (2014): 1887–901.
52. W. G. Christen et al. for The Steering Committee of Physicians' Health Study II, "Design of Physicians' Health Study II – A Randomized Trial of Beta-Carotene, Vitamins E and C, and Multivitamins, in Prevention of Cancer, Cardiovascular Disease, and Eye Disease, and Review of Results of Completed Trials," *Annals of Epidemiology* 10 (2000): 125–34.
53. Howard D. Sesso et al., "Vitamins E and C in the Prevention of Cardiovascular Disease in Men: The Physicians' Health Study II Randomized Controlled Trial," *Journal of the American Medical Association* 300 (2008): 2123–33; J. Michael Gaziano et al., "Vitamins E and C in the Prevention of Prostate and Total Cancer in Men: The Physicians' Health Study II Randomized Controlled Trial," *Journal of the American Medical Association* 301 (2009): 52–62.
54. David J. A. Jenkins et al., "Supplemental Vitamins and Minerals for CVD prevention and Treatment," *Journal of the American College of Cardiology* 71 (2018): 2570–84.
55. S. T. Mayne et al., "Diet, Nutrition, and Cancer: Past, Present and Future, *Nature Reviews Clinical Oncology* 13 (2016): 504–15.
56. S. T. Mayne et al., "Lessons Learned from Randomized Clinical Trials of Micronutrient Supplementation for Cancer Prevention," *Annual Review of Nutrition* 32 (2012): 369–90; The Alpha-Tocopherol, Beta Carotene Cancer Prevention Study Group, "The Effect of Vitamin E and Beta Carotene on the Incidence of Lung Cancer and Other Cancers in Male Smokers," *New England Journal of Medicine* 330 (1994): 1029–35.
57. C. Yfanti et al., "Antioxidant Supplementation Does Not Alter Endurance Training Adaptation," *Medicine and Science in Sports and Exercise* 42 (2010): 1388–95; C. Yfanti et al., "Role of Vitamin C and E Supplementation on IL-6 in Response to Training," *Journal of Applied Physiology* 112 (2012): 990–1000; C. Kang et al., "Exercise Activation of Muscle Peroxisome Proliferator-Activated Receptor-Gamma Coactivator-1alpha Signaling Is Redox Sensitive," *Free Radical Biology and Medicine* 47 (2009): 1394–400; N. A. Strobel et al., "Antioxidant Supplementation Reduces Skeletal Muscle Mitochondrial Biogenesis," *Medicine and Science in Sports and Exercise* 43 (2011): 1017–24; Cleva Villanueva and Robert D. Kross, "Antioxidant-Induced Stress," *International Journal of Molecular Sciences* 13 (2012): 2091–109; H. Feng et al., "Training-Induced Mitochondrial Adaptation: Role of Peroxisome Proliferator-Activated Receptor Gamma Coactivator-1alpha, Nuclear Factor-KappaB and Beta-Blockade," *Experimental Physiology* 98 (2013): 784–95.
58. M. Ristow et al., "Antioxidants Prevent Health-Promoting Effects of Physical Exercise in Humans," *Proceedings of the National Academy of Sciences* 106 (2009): 8665–70; G. Paulsen et al., "Vitamin C and E Supplementation Hampers Cellular Adaptation to Endurance Training in Humans: A Double-Blind Randomized Controlled Trial," *The Journal of Physiology* 592 (2014): 1887–901; M. Mattson and E. Calabrese, "Best in Small Doses," *New Scientist* 2672 (2008): 36–39; K. T. Howitz and D. A. Sinclair, "Xenohormesis: Sensing the Chemical Cues of Other Species," *Cell* 133 (2008): 387–91.
59. M. P. Mattson and A. Cheng, "Neurohormetic Phytochemicals: Low-Dose Toxins That Induce Adaptive Neuronal Stress Responses," *Trends in Neurosciences* 29 (2006): 632–39.
60. D. E. Amanoel et al., "Sheep Deficient in Vitamin E Preferentially Select for a Feed with a Higher Concentration of Vitamin E," *Animal* 10 (2016): 183–91; C. Catoni et al., "Fruit for Health: The Effect of Flavonoids on Humoral Immune Response and Food Selection in a Frugivorous Bird," *Functional Ecology* 22 (2008): 649–54.

## Chapter 7: Medicating in Nature's Pharmacy


1. M. Cabanac et al., "Influence of Internal Factors on the Pleasantness of a Gustative Sweet Sensation," *Communications in Behavioral Biology, Part A* 1 (1968): 77–82.
2. Paul M. Wise et al., "Reduced Dietary Intake of Simple Sugars Alters Perceived Sweet Taste Intensity but Not Perceived Pleasantness," *American Journal of Clinical Nutrition* 103 (2016): 50–60.
3. Howard, *The Owner's Manual for the Brain*, 29.
4. K. F. Green and J. Garcia, "Recuperation from Illness: Flavor Enhancement for Rats," *Science* 173 (1971): 749–51.
5. Elizabeth A. Bernays and Michael S. Singer, "Taste Alteration and Endoparasites," *Nature* 436 (2005): 476.
6. M. S. Singer et al., "Self-Medication as Adaptive Plasticity: Increased Ingestion of Plant Toxins by Parasitized Caterpillars," *PLoS ONE* 4 (2009): e4796; Richard Karban and Gregory English-Loeb, "Tachinid Parasitoids Affect Host Plant Choice by Caterpillars to Increase Caterpillar Survival," *Ecology* 78 (1997): 603–11.
7. T. Lefèvre et al., "Evidence for Trans-Generational Medication in Nature," *Ecology Letters* 13 (2010): 1485–93.
8. Michael Simone-Finstrom and Marla Spivak, "Propolis and Bee Health: The Natural History and Significance of Resin Use by Honey Bees," *Apidologie (Celle)* 41 (2010): 295–311.
9. Harmen P. Hendriksma and Sharoni Shafir, "Honey Bee Foragers Balance Colony Nutritional Deficiencies," *Behavioral Ecology and Sociobiology* 70 (2016): 509–17.
10. W. Mao et al., "Honey Constituents Up-Regulate Detoxification and Immunity Genes in the Western Honey Bee, *Apis mellifera*," *Proceedings of the National Academy of Sciences* 110, (2013): 8842–46.
11. V. O. Ezenwa, "Selective Defecation and Selective Foraging: Antiparasite Behavior in Wild Ungulates?" *Ethology* 110 (2004): 851–62; Benjamin L. Hart, "Behavioural Defence against Parasites: Interaction with Parasite Invasiveness," *Parasitology* 109 (1994): S139–S151.
12. M. R. Hutchings et al., "Can Animals Use Foraging Behaviour to Combat Parasites?" *Proceedings of the Nutrition Society* 62 (2003): 361–70; M. R. Hutchings et al., "Grazing in Heterogeneous Environments: Infra- and Supra-Parasite Distributions Determine Herbivore Grazing Decisions," *Oecologia* 132 (2002): 453–60.
13. Gregoire Castella et al., "Prophylaxis with Resin in Wood Ants," *Animal Behaviour* 75 (2008): 1591–96.
14. Monserrat Suárez-Rodríguez and Constantino M. Garcia, "An Experimental Demonstration That House Finches Add Cigarette Butts in Response to Ectoparasites," *Journal of Avian Biology* 48 (2017): 1316–21.
15. Serge Landau et al., "Anthelmintic Activity of *Pistacia lentiscus* Foliage in Two Middle Eastern Breeds of Goats Differing in Their Propensity to Consume Tannin-Rich Browse," *Veterinary Parasitology* 173 (2010): 280–86; M. Amit et al., "Self-Medication with Tannin-Rich Browse in Goats Infected with Gastro-Intestinal Nematodes," *Veterinary Parasitology* 198 (2013): 305–11.
16. K. E. Glander, "Nonhuman Primate Self-Medication with Wild Plant Foods," in *Eating on the Wild Side: The Pharmacologic, Ecologic, and Social Implication of Using Noncultigens*. Nina L. Etkin ed. (Tucson: University of Arizona Press, 1994), 227–39.
17. Michael A. Huffman, "Primate Self-Medication," in *Primates in Perspective (2nd Edition)* Christina Campbell, et al. eds. (Oxford: Oxford University Press, 2011) 563–73; Jacobus C. de Roode et al., "Self-Medication in Animals," *Science* 340 (2013): 150–51; Jessica Juhnke et al., "Preference for Condensed Tannins by Sheep in Response to Challenge Infection with *Haemonchus contortus*," *Veterinary Parasitology* 188, (2012): 104–14; Juan J. Villalba et al., "Preference for Tanniferous (*Onobrychis viciifolia*) and Non-Tanniferous (*Astragalus cicer*) Forage Plants by

Sheep in Response to Challenge Infection with *Haemonchus contortus*," *Small Ruminant Research* 112 (2013): 199–207.

18. George A. Lozano, "Parasitic Stress and Self-Medication in Wild Animals," *Advances in the Study of Behavior*, 291–317.
19. Michael A. Huffman and M. A. Seifu, "Observations on the Illness and Consumption of a Possibly Medicinal Plant *Vernonia amygdalina* (Del.) by a Wild Chimpanzee in the Mahale Mountains National Park, Tanzania," *Primates* 30 (1989): 51–63; Michael A. Huffman et al., "Further Observations on the Use of *Vernonia amygdalina* (Del.) by a Wild Chimpanzee, Its Possible Effect on Parasite Load, and Its Phytochemistry," *African Study Monograph* 14 (2012): 227–40; Koichi Koshimizu et al., "Use of *Vernonia amygdalina* by Wild Chimpanzee: Possible Roles of Its Bitter and Related Constituents," *Physiology and Behavior* 56 (1994): 1209–16.
20. Sabrina Krief et al., "Novel Antimalarial Compounds Isolated in a Survey of Self-Medicative Behavior of Wild Chimpanzees in Uganda," *Antimicrobiological Agents and Chemotherapy* 48 (2004): 3196–99; S. Krief et al., "Ethnomedicinal and Bioactive Properties of Plants Ingested by Wild Chimpanzees in Uganda," *Journal of Ethnopharmacology* 101 (2005): 1–15.
21. J. E. Phillips-Conroy, "Baboons, Diet, and Disease: Food Plant Selection and Schistosomiasis," in *Current Perspectives in Primate Social Dynamics* D. Taub and F. King eds. (New York: Van Nostrand-Reinhold, 1986): 287–304.
22. Andrew J. J. MacIntosh and Michael A. Huffman, "Toward Understanding the Role of Diet in Host-Parasite Interactions: The Case for Japanese Macaques," in *The Japanese Macaques* N. Nakagawa et al., eds. (Tokyo: Springer, 2010): 323–44.
23. D. Janzen, "Complications in Interpreting the Chemical Defenses of Trees against Tropical Arboreal Plant-Eating Vertebrates," in *The Ecology of Arboreal Folivores* G. Montgomery ed. (Washington D.C.: Smithsonian Institution Press, 1978): 73–84.
24. Frederick D. Provenza and Juan J. Villalba, "The Role of Natural Plant Products in Modulating the Immune System: An Adaptable Approach for Combating Disease in Grazing Animals," *Small Ruminant Research* 89 (2010): 131–39; S. Athanasiadou and I. Kyriazakis, "Plant Secondary Metabolites: Antiparasitic Effects and Their Role in Ruminant Production Systems," *Proceedings of the Nutrition Society* 63 (2004): 631–39; B.R. Min et al., "The Effect of Condensed Tannins in *Lotus corniculatus* upon Reproductive Efficiency and Wool Production in Ewes during Autumn," *Animal Feed Science and Technology* 92 (2001): 185–202; 279–83.
25. V. Carrai et al., "Increase in Tannin Consumption by Sifaka (*Propithecus verreauxi verreauxi*) Females during the Birth Season: A Case for Self-Medication in Prosimians?" *Primates* 44 (2003): 61–66.
26. Thomas P. Clausen et al., "Ecological Implications of Condensed Tannin Structure: A Case Study," *Journal of Chemical Ecology* 16 (1990): 2381–92.
27. Hideyuki Ito et al., "Anti-Tumor Promoting Activity of Polyphenols from *Cowania mexicana* and *Coleogyne ramosissima*," *Cancer Letters* 143 (1999): 5–13.
28. Michael A. Huffman, "Current Evidence for Self-Medication in Primates: A Multidisciplinary Perspective," *American Journal of Physical Anthropology* 40 (1997): 171–200.
29. D. H. Al-Rajhy et al., "Acaricidal Effects of Cardiac Glycosides, Azadirachtin and Neem Oil against the Camel Tick, *Hyalomma dromedarii* (Acari: Ixodidae)," *Pest Management Science* 59 (2003): 1250–54.
30. Serge Y. Landau et al., "Neem-Tree (*Azadirachta indica* Juss.) Extract as a Feed-Additive against the American Dog Tick (*Dermacentor variabilis*) in Sheep (*Ovis aries*)," *Veterinary Parasitology* 165 (2009): 311–17.
31. R. D. Madden et al., "Dietary Modification of Host Blood Lipids Affects Reproduction in the Lone Star Tick, *Amblyomma americanum* (L.)," *Journal of Parasitology* 82 (1996): 203–9.

32. B. R. Min et al., "Wheat Pasture Bloat Dynamics, In Vitro Ruminal Gas Production, and Potential Bloat Mitigation with Condensed Tannins," *Journal of Animal Science* 83 (2005): 1322–31; B. R. Min et al., "Effects of Condensed Tannins Supplementation Level on Weight Gain and In Vitro and In Vivo Bloat Precursors in Steers Grazing Winter Wheat," *Journal of Animal Science* 84 (2006): 2546–54.
33. Juan J. Villalba et al., "Rumen Distension and Contraction Influence Feed Preference by Sheep," *Journal of Animal Science* 87 (2009): 340–50.
34. Tim S. Phy and Frederick D. Provenza, "Sheep Fed Grain Prefer Foods and Solutions that Attenuate Acidosis," *Journal of Animal Science* 76 (1998): 954–60.
35. Juan J. Villalba and Frederick D. Provenza, "Preference for Polyethylene Glycol by Sheep Fed a Quebracho Tannin Diet," *Journal of Animal Science* 79 (2001): 2066–74.
36. Frederick D. Provenza et al., "Self-Regulation of Intake of Polyethylene Glycol by Sheep Fed Diets Varying in Tannin Concentrations," *Journal of Animal Science* 78 (2000): 1206–12.
37. Juan J. Villalba et al., "Sheep Self-Medicate When Challenged with Illness-Inducing Foods," *Animal Behaviour* 71 (2006): 1131–39.
38. J. T. Gradé et al., "Four-Footed Pharmacists: Indications of Self-Medicating Livestock in Karamoja, Uganda," *Economic Botany* 63 (2009): 29–42.
39. Michael A. Huffman, "An Ape's Perspective on the Origins of Medicinal Plant Use in Humans," in *Wild Harvest,* 55–70; Michael A. Huffman, "Culture, Religion, and Belief Systems: Animals as Sources of Medicinal Wisdom in Traditional Societies," in *Encyclopedia of Human-Animal Relationships: A Global Exploration of Our Connections with Animals,* Marc Bekoff ed. (Westport, CT: Greenwood Press, 2007): 434–41; Michael A. Huffman, "Animal Self-Medication and Ethno-Medicine: Exploration and Exploitation of the Medicinal Properties of Plants," *Proceedings of the Nutrition Society* 62 (2003): 371–81; Michael A. Huffman, "Self-Medicative Behavior in the African Great Apes: An Evolutionary Perspective into the Origins of Human Traditional Medicine," *Bioscience* 51 (2001): 651–61; Michael A. Huffman et al., "African Great Ape Self-Medication: A New Paradigm for Treating Parasite Disease with Natural Medicines?" in *Towards Natural Medicine Research in the 21st Century* H. Ageta et al., eds. (Amsterdam: Elsevier Science, 1998): 113–23.
40. J. C. de Roode et al., "Self-Medication in Animals," *Science* 340 (2013): 150–51; J. R. Mukherjee et al., "Do Animals Eat What We Do? Observations on Medicinal Plants Used By Humans and Animals of Mundanthurai Range, Tamil Nadu," in *Medicinal Plants and Sustainable Development* C.P. Kala ed. (New York: Nova Science Publishers, 2013): 3–27.
41. J. Qiu, "Traditional Medicine: A Culture in the Balance," *Nature* 448 (2007): 126–28; D. X. Kong et al., "Where Is the Hope for Drug Discovery? Let History Tell the Future," *Drug Discovery Today* 14 (2009): 115–19.
42. D. X. Kong et al., "Where Is the Hope for Drug Discovery? Let History Tell the Future," *Drug Discovery Today* 14 (2009): 115–19.
43. A. Touwaide et al., "Medicinal Plants for the Treatment of Urogenital Tract Pathologies According to Dioscorides' De Materia Medica," *American Journal of Nephrology* 17 (1997): 241–47; D. E. Moerman, "Symbols and Selectivity: A Statistical Analysis of Native American Medical Ethnobotany," *Journal of Ethnopharmacology* 1 (1979): 111–19; D. E. Moerman, "An Analysis of the Food Plants and Drug Plants of Native North America," *Journal of Ethnopharmacology* 52 (1996): 1–22; E. Thomas et al., "The Relationship between Plant Use and Plant Diversity in the Bolivian Andes, with Special Reference to Medicinal Plant Use," *Human Ecology* 36 (2008): 861–79; M. Leonti et al., "A Comparison of Medicinal Plant Use in Sardinia and Sicily – De Materia Medica Revisited?" *Journal of Ethnopharmacology* 121 (2009): 255–67.
44. A. T. Borchers et al., "Inflammation and Native American Medicine: The Role of Botanicals," *American Journal of Clinical Nutrition* 72 (2000): 339–47; L. Gray, "Reading the Mind of Nature:

Ecophysiology and Indigenous Wisdom," in *Ecological Medicine: Healing the Earth, Healing Ourselves,* K. Ausubel and J. P. Harpignies eds. (San Francisco: Sierra Club Books, 2004) 223–27.
45. Frederick D. Provenza, "What Does It Mean to Be Locally Adapted and Who Cares Anyway?" *Journal of Animal Science* 86 (2008): E271–84.
46. K. Osoro et al., "Anthelmintic and Nutritional Effects of Heather Supplementation on Cashmere Goats Grazing Perennial Ryegrass-White Clover Pastures," *Journal of Animal Science* 85 (2007): 861–70.
47. Larry D. Lisonbee et al., "Tannins and Self-Medication: Implications for Sustainable Parasite Control in Herbivores," *Behavioral Processes* 82 (2009): 184–89; Larry D. Lisonbee et al., "Effects of Tannins on Selection by Sheep of Forages Containing Alkaloids, Tannins and Saponins," *Journal of the Science of Food and Agriculture* 89 (2009): 2668–77; Juan J. Villalba et al., "Selection of Tannins by Sheep in Response to Gastrointestinal Nematode Infection," *Journal of Animal Science* 88 (2010): 2189–98; Juan J. Villalba et al., "Preference for Tanniferous (*Onobrychis viciifolia*) and Non-Tanniferous (*Astragalus cicer*) Forage Plants by Sheep in Response to Challenge Infection with *Haemonchus contortus*," *Small Ruminant Research* 112 (2013): 199–207.
48. C. Martinez Ortiz de Montellano et al., "Effect of a Tropical Tannin-Rich Plant *Lysiloma latisiliquum* on Adult Populations of *Haemonchus contortus* in Sheep," *Veterinary Parasitology* 172 (2010): 283–90.
49. Kenny Ausubel, "Hoxsey: When Healing Becomes a Crime," in *Ecological Medicine: Healing the Earth, Healing Ourselves* K. Ausubel and J.P. Harpignies eds. (San Francisco: Sierra Club Books, 2004): 92–105.
50. James Duke, "The Herbal Shotgun Shell," *American Botanical Council HerbalGram,* no. 18–19, (Fall 1988/Winter 1989): 12–13; Keith I. Block et al. "A Broad-Spectrum Integrative Design for Cancer Prevention and Therapy," *Seminars in Cancer Biology*, 35, Suppl (2015): S276–304; S. S. Mayne et al., Diet, "Nutrition, and Cancer: Past, Present and Future," *Nature Reviews* Clinical Oncology 13 (2016): 504–15.
51. S. Zhang et al., "Anti-Cancer Potential of Sesquiterpene Lactones: Bioactivity and Molecular Mechanisms," *Current Medicinal Chemistry Anticancer Agents* 5 (2005): 239–49; C. Y. Looi et al., "Induction of Apoptosis in Human Breast Cancer Cells via Caspase Pathway by Vernodalin Isolated from *Centratherum anthelminticum* (L.) Seeds," *PLoS ONE* 8(2) (2013): e56643; S. K. A. Sadagopan et al., "Forkhead Box Transcription Factor (FOXO3a) Mediates the Cytotoxic Effect of Vernodalin In Vitro and Inhibits the Breast Tumor Growth In Vivo," *Journal of Experimental and Clinical Cancer Research* 34 (2015): 147.
52. Kenny Ausubel, "Hoxsey: When Healing Becomes a Crime," in *Ecological Medicine: Healing the Earth, Healing Ourselves,* K. Ausubel and J.P. Harpignies eds. (San Francisco: Sierra Club Books, 2004): 92–105.
53. Editorial, "Hoxsey – Cancer Charlatan," *Journal of the American Medical Association*, 133 (1947): 774–75.
54. Mary A. Richardson et al., "Assessment of Outcomes at Alternative Medicine Cancer Clinics: A Feasibility Study," *The Journal of Alternative and Complementary Medicine* 7 (2004); Skyler B. Johnson et al., "Use of Alternative Medicine for Cancer and Its Impact on Survival," *Journal of the National Cancer Institute* 110, no. 1 (2018): djx145.
55. R. Doll and R. Petro, "The Causes of Cancer: Quantitative Estimates of Avoidable Risks of Cancer in the United States Today," *Journal of the National Cancer Institute* 66 (1981): 1191–308.
56. Jason Lazarou et al., "Incidence of Adverse Drug Reactions in Hospitalized Patients: A Meta-Analysis of Prospective Studies," *Journal of the American Medical Association* 279 (1998): 1200–5.
57. Andrew Weil, "Healing, Nature, and Modern Medicine," in *Ecological Medicine: Healing the Earth, Healing Ourselves,* K. Ausubel and J.P. Harpignies eds. (San Francisco: Sierra Club Books, 2004) 109–24.

58. Chris Hobbs, "Community Herbalism in Modern Health Care," in *Ecological Medicine: Healing the Earth, Healing Ourselves,* K. Ausubel and J.P. Harpignies eds. (San Francisco: Sierra Club Books, 2004): 132–37.
59. C. Gratus et al., "The Use of Herbal Medicines by People with Cancer: A Qualitative Study," *BMC Complementary and Alternative Medicine* (2009).
60. K. H. Neldner, "Complementary and Alternative Medicine," *Dermatologic Clinics* 18 (2000): 189–93.
61. R. Graziose et al., "Merging Traditional Chinese Medicine with Modern Drug Discovery Technologies to Find Novel Drugs and Functional Foods," *Current Drug Discovery Technology* 7 (2010): 2–12.
62. C. Milesi et al., "A Strategy for Mapping and Modeling the Ecological Effects of US Lawns," in 3rd International Symposium Remote Sensing and Data Fusion over Urban Areas (URBAN 2005), 5th International Symposium Remote Sensing of Urban Areas (URS 2005), Tempe, AZ, USA; March 14–16, 2005.
63. Andrew Weil, "Healing, Nature, and Modern Medicine," in *Ecological Medicine: Healing the Earth, Healing Ourselves,* K. Ausubel and J.P. Harpignies eds. (San Francisco: Sierra Club Books, 2004): 109–24.
64. Juan J. Villalba and Frederick D. Provenza, "Self-Medication and Homeostatic Endeavor in Herbivores: Learning about the Benefits of Nature's Pharmacy," *Animal* 1 (2007): 1360–70.
65. R. Hocquemiller et al., "Isolation and Synthesis of Espintanol, a New Antiparasitic Monoterpene," *Journal of Natural Products* 54 (1991): 445–52.
66. O. Kayser et al., "Natural Products as Antiparasitic Drugs," *Parasitology Research* 90 (2003): S55–S62.
67. Kevin D. Welch et al., "The Effect of 7, 8-Methylenedioxylycoctonine-Type Diterpenoid Alkaloids on the Toxicity of Tall Larkspur (*Delphinium* spp.) in Cattle," *Journal of Animal Science* 90 (2012): 2394–401.
68. L. A. Dyer et al., "Synergistic Effects of Three Piper Amides on Generalist and Specialist Herbivores," *Journal of Chemical Ecology* 29 (2003): 2499–514.
69. R. C. Kreck and P. J. Waller, "Towards the Implementation of the 'Basket of Options' Approach to Helminth Parasite Control of Livestock: Emphasis on the Tropics/Subtropics," *Veterinary Parasitology* 139 (2006): 270–82; R. M. Kaplan, "Drug Resistance in Nematodes of Veterinary Importance: A Status Report," *Trends in Parasitology* 20 (2004): 477–81; P. J. Waller, "Sustainable Nematode Parasite Control Strategies for Ruminant Livestock by Grazing Management and Biological Control," *Animal Feed Science and Technology* 126 (2006): 277–89; F. Jackson and J. Miller, "Alternative Approaches to Control – Quo Vadit?" *Veterinary Parasitology* 139 (2006): 371–84.
70. B. Patwardhan et al., "Reverse Pharmacology and Systems Approaches for Drug Discovery and Development," *Current Bioactive Compounds* 4 (2008): 201–12.
71. Jürg Gertsch, "Botanical Drugs, Synergy, and Network Pharmacology: Forth and Back to Intelligent Mixtures," *Planta Medica* 77 (2011): 1086–98.
72. Andrew L. Hopkins, "Network Pharmacology," *Nature Biotechnology* 25 (2007): 1110–11; Andrew L. Hopkins, "Network Pharmacology: The Next Paradigm in Drug Discovery," *Nature Chemical Biology* 4 (2008): 682–90.
73. C. Eng, "Are Herbal Medicines Ripe for the Cancer Clinic?" *Science Translational Medicine* 2 (2010): 45ps41.
74. H. Wagner, "Synergy Research: Approaching a New Generation of Phytopharmaceuticals," *Fitoterapia* 82 (2011): 34–37.
75. Z. Pancer and M. D. Cooper, "The Evolution of Adaptive Immunity," *Annual Review of Immunology* 24 (2006): 497–518.
76. F. Saji et al., "Dynamics of Immunoglobulins at the Feto-Maternal Interface," *Reviews of Reproduction* 4 (1999): 81–89.
77. P. Van de Perre, "Transfer of Antibody via Mother's Milk," *Vaccine* 21 (2003): 3374–76.

## Chapter 8: Delighting in the Colors


1. Temple Grandin, "The Effect of Stress on Livestock and Meat Quality Prior to and During Slaughter," *International Journal for the Study of Animal Problems* 1 (1980): 313–37.
2. V. A. Kuttappan et al., "Influence of Growth Rate on the Occurrence of White Striping in Broiler Breast Fillets," *Poultry Science* 91 (2012): 2677–85; S. Barbut et al., "Progress in Reducing the Pale, Soft and Exudative (PSE) Problem in Pork and Poultry Meat," *Meat Science* 79 (2008): 46–63; M. Petracci et al., "Effect of White Striping on Chemical Composition and Nutritional Value of Chicken Breast Meat," *Italian Journal of Animal Science* 13 (2014): 179–83; V. V. Tijare et al., "Meat Quality of Broiler Breast Fillets with White Striping and Woody Breast Muscle Myopathies," *Poultry Science* 95 (2016): 2167–73.
3. Frederick D. Provenza et al., "Linking Herbivore Experience, Varied Diets, and Plant Biochemical Diversity," *Small Ruminant Research* 49 (2003): 257–74.
4. Frederick D. Provenza, "What Does It Mean to Be Locally Adapted and Who Cares Anyway?" *Journal of Animal Science* 86 (2008): E271–E284.
5. G. Urbach, "Effect of Feed on Flavor in Dairy Foods," *Journal of Dairy Science* 73 (1990): 3639–50.
6. J. O. Bosset et al., "Effect of Botanical Composition of Grazing Areas on Some Components of L'Etivaz or Gruyere-Type Cheeses," *Revue Suisse Agriculture* 30, (1998): 167–71; B. Jeangros et al., "Comparison of the Botanical and Chemical Characteristics of Grazed Pastures, in Lowlands and in Mountains," *Fourrages* 159 (1999): 277–92; J. E. O'Connell and P. F. Fox, "Significance and Applications of Phenolic Compounds in the Production and Quality of Milk and Dairy Products: A Review," *International Dairy Journal* 11 (2001): 103–20.
7. S. Carpino et al., "Composition and Aroma Compounds of Ragusano Cheese: Native Pasture and Total Mixed Rations," *Journal of Dairy Science* 87 (2004): 816–30.
8. S. Carpino et al., "Contribution of Native Pasture to the Sensory Properties of Ragusano Cheese," *Journal of Dairy Science* 87 (2004): 308–15.
9. S. Bowen and T. Mutersbaugh, "Local or Localized? Exploring the Contributions of Franco-Mediterranean Agrifood Theory to Alternative Food Research," *Agriculture and Human Values* 31 (2013): 201–13.
10. K. N. Kilcawley et al., "Factors Influencing the Flavour of Bovine Milk and Cheese from Grass Based Versus Non-Grass Based Milk Production Systems," *Foods* 7, 37 (2018).
11. C. Sanudo et al., "Fatty Acid Composition and Sensory Characteristics of Lamb Carcasses from Britain and Spain," *Meat Science* 54 (2000): 339–46; C. Sanudo et al., "Regional Variation in the Hedonic Evaluation of Lamb Meat from Diverse Production Systems by Consumers in Six European Countries," *Meat Science* 75 (2007): 610–21.
12. B. M. Sitz et al., "Consumer Sensory Acceptance and Value of Domestic, Canadian, and Australian Grass-Fed Beef Steaks," *Journal of Animal Science* 83 (2005): 2863–68; J. D. Wood et al., "Effects of Fatty Acids on Meat Quality: Review," *Meat Science* 66 (2003): 21–32.
13. K. M. Killinger et al., "A Comparison of Consumer Sensory Acceptance and Value of Domestic Beef Steaks and Steaks from a Branded, Argentine Beef Program," *Journal Animal Science* 82 (2004): 3302–7; C. R. Calkins and J. M. Hodgen, "A Fresh Look at Meat Flavor," *Meat Science* 77 (2007): 63–80.
14. Schatzker, *Steak*; C. Maughan et al., "Development of a Beef Flavor Lexicon and Its Application to Compare the Flavor Profile and Consumer Acceptance of Rib Steaks from Grass- or Grain-Fed Cattle," *Meat Science* 90 (2011): 116–21; B. Maughan et al., "Importance of Grass-Legume Choices on Cattle Grazing Behavior, Performance and Meat Characteristics," *Journal of Animal Science* 92 (2014): 2309–24; M. E. Van Elswyk and S. H. McNeill, "Impact of Grass/Forage Feeding versus Grain Finishing on Beef Nutrients and Sensory Quality: The U.S. Experience," *Meat Science* 96 (2014): 535–40.

15. S. Prache et al., "Traceability of Animal Feeding Diet in the Meat and Milk of Small Ruminants," *Small Ruminant Research* 59 (2005): 157–68; F. Camin et al., "Multi-Element (H, C, N, S) Stable Isotope Characteristics of Lamb Meat from Different European Regions," *Analytical and Bio-Analytical Chemistry* 389 (2007): 309–20.
16. V. Vasta et al., "Alternative Feed Resources and Their Effects on the Quality of Meat and Milk from Small Ruminants," *Animal Feed Science and Technology* 147 (2008): 223–46; A. Priolo et al., "Meat Odour and Flavour and Indoles Concentration in Ruminal Fluid and Adipose Tissue of Lambs Fed Green Herbage or Concentrates with or without Tannins," *Animal* 3 (2009): 454–60; V. Vasta and G. Luciano, "The Effects of Dietary Consumption of Plants Secondary Compounds on Small Ruminants' Products Quality," *Small Ruminant Research* 101 (2011): 150–59; E. A. Bjorklund et al., "Fatty Acid Profiles, Meat Quality, and Sensory Attributes of Organic versus Conventional Dairy Beef Steers," *Journal of Dairy Science* 97 (2014): 1828–34; J. O. Monteschio et al., "Clove and Rosemary Essential Oils and Encapsuled Active Principles (Eugenol, Thymol and Vanillin Blend) on Meat Quality of Feedlot-Finished Heifers," *Meat Science* 130 (2017): 50–57.
17. J. Neethling et al., "Factors Influencing the Flavour of Game Meat: A Review," *Meat Science* 113 (2016): 139–53.
18. D. Aune et al., "Meat Consumption and the Risk of Type 2 Diabetes: A Systematic Review and Meta-Analysis of Cohort Studies," *Diabetologia* 52 (2009): 2277–87; A. Pan et al., "Red Meat Consumption and Mortality: Results from 2 Prospective Cohort Studies," *Archives of Internal Medicine* 172 (2012): 555–63.
19. A. Etemadi et al., "Mortality from Different Causes Associated with Meat, Heme Iron, Nitrates, and Nitrites in the NIH-AARP Diet and Health Study: Population Based Cohort Study," *BMJ* (2017): 357.
20. B. Halliwell and J. M. Gutteridge, "Role of Free Radicals and Catalytic Metal Ions in Human Disease: An Overview," *Methods in Enzymology* 186 (1990): 1–85; S. A. Bingham et al., "Effect of White Versus Red Meat on Endogenous N-Nitrosation in the Human Colon and Further Evidence of a Dose Response," *Journal of Nutrition* 132, Suppl (2002): 3522S–25; A. J. Cross et al., "A Prospective Study of Red and Processed Meat Intake in Relation to Cancer Risk," *PLoS Medicine* (2007): 4:e325; S. Rajpathak et al., "Iron Intake and the Risk of Type 2 Diabetes in Women: A Prospective Cohort Study," *Diabetes Care* 29 (2006):1370–76; W. Yang et al., "Is Heme Iron Intake Associated with Risk of Coronary Heart Disease? A Meta-Analysis of Prospective Studies," *European Journal of Nutrition* 53 (2014): 395–400; Z. Zhao et al., "Body Iron Stores and Heme-Iron Intake in Relation to Risk of Type 2 Diabetes: A Systematic Review and Meta-Analysis," *PLoS ONE* 7, no. 7 (2012): e41641.
21. E. Hopps et al., "A Novel Component of the Metabolic Syndrome: The Oxidative Stress," *Nutrition, Metabolism and Cardiovascular Disease* 20 (2010): 72–77; K. Esposito and D. Giugliano, "Diet and Inflammation: A Link to Metabolic and Cardiovascular Diseases," *European Heart Journal* 27 (2006): 15–20; A. N. Margioris, "Fatty Acids and Postprandial Inflammation," *Current Opinion in Clinical Nutrition and Metabolic Care* 12: (2009): 129–37; J. H. O'Keefe and D. S. Bell, "Postprandial Hyperglycemia/Hyperlipidemia (Postprandial Dysmetabolism) Is a Cardiovascular Risk Factor," *American Journal Cardiology* 100 (2007): 899–904; G. S. Hotamisligil, "Inflammation and Metabolic Disorders," Nature 444 (2006): 860–67; J. H. O'Keefe and D. S. Bell, "Postprandial Hyperglycemia/Hyperlipidemia (Postprandial Dysmetabolism) Is a Cardiovascular Risk Factor," *American Journal Cardiology* 100 (2007): 899–904.
22. L. Cordain et al., "The Paradoxical Nature of Hunter-Gatherer Diets: Meat-Based, Yet Non-Atherogenic," *European Journal of Clinical Nutrition* 56 (2002): S42–S52
23. D. A. Raichlen et al., "Physical Activity Patterns and Biomarkers of Cardiovascular Disease Risk in Hunter-Gatherers," *American Journal of Human Biology* 29, no. 2 (2016):e22919; Edward Archer, "In Defense of Sugar: A Critique of Diet-Centrism," *Progress in Cardiovascular Diseases* (2018).

24. A. B. Falowo et al., "Natural Antioxidants against Lipid-Protein Oxidative Deterioration in Meat and Meat Products: A Review," *Food Research International* 64 (2014): 171–81.
25. D. E. Corpet, "Red Meat and Colon Cancer: Should We Become Vegetarians, or Can We Make Meat Safer?" *Meat Science* 89 (2011): 310–16; K. Esposito and D. Giugliano, "Diet and Inflammation: A Link to Metabolic and Cardiovascular Diseases," *European Heart Journal* 27 (2006): 15–20; A. N. Margioris, "Fatty Acids and Postprandial Inflammation," *Current Opinion in Clinical Nutrition and Metabolic Care* 12 (2009): 129–37; I. Zanotti et al., "Atheroprotective Effects of (Poly)phenols: A Focus on Cell Cholesterol Metabolism," *Food and Function* 6 (2015): 13–31.
26. M. Herieka and C. Erridge, "High-Fat Meal Induced Postprandial Inflammation," *Molecular Nutrition and Food Research* 58 (2014): 136–46.
27. M. Martinez-Tome et al., "Antioxidant Properties of Mediterranean Spice Compared with Common Food Additives," *Journal of Food Protection* 64 (2001): 1412–19.
28. Z. Li et al., "Antioxidant-Rich Spice Added to Hamburger Meat during Cooking Results in Reduced Meat, Plasma, and Urine Malondialdehyde Concentrations," *American Journal of Clinical Nutrition* 91 (2010): 1180–84.
29. S. Gorelik et al., "A Novel Function of Red Wine Polyphenols in Humans: Prevention of Absorption of Cytotoxic Lipid Peroxidation Products," *FASEB Journal* 22 (2008): 41–46; A. M. Cuevas et al., "A High-Fat Diet Induces and Red Wine Counteracts Endothelial Dysfunction in Human Volunteers," *Lipids* 35 (2000): 143–48.
30. J. Kanner et al., "Protection by Polyphenols of Postprandial Human Plasma and Low-Density Lipoprotein Modification: The Stomach as a Bioreactor," *Journal of Agricultural and Food Chemistry* 60 (2012): 8790–96.
31. A. M. Descalzo and A. M. Sancho, "A Review of Natural Antioxidants and Their Effects on Oxidative Status, Odor and Quality of Fresh Beef Produced in Argentina," *Meat Science* 79 (2008): 423–36; D. Del Rio et al., "Dietary (Poly) Phenolics in Human Health: Structures, Bioavailability, and Evidence of Protective Effects against Chronic Diseases. Antioxidants & Redox Signaling," 18 (2013): 1818–92; A. B. Falowo et al., "Natural Antioxidants against Lipid-Protein Oxidative Deterioration in Meat and Meat Products: A Review," *Food Research International* 64 (2014): 171–81; C. Papuc et al., "Mechanisms of Oxidative Processes in Meat and Toxicity Induced by Postprandial Degradation Products: A Review," *Comprehensive Reviews in Food Science and Food Safety* 16 (2017): 96–123.
32. F. Arya et al., "Differences in Postprandial Inflammatory Responses to a 'Modern' v. Traditional Meat Meal: A Preliminary Study," *British Journal Nutrition* 104 (2010): 724–28.
33. W. J. Craig, "Health-Promoting Properties of Common Herbs," *American Journal of Clinical Nutrition* 70 (1999): 491S–99; S. S. Rathore et al., "Potential Health Benefits of Major Seed Spices," *International Journal of Seed Spices* 3 (2013): 1–12; H. M. Woo et al., "Active Spice-Derived Components Can Inhibit Inflammatory Responses of Adipose Tissue in Obesity by Suppressing Inflammatory Actions of Macrophages and Release of Monocyte Chemoattractant Protein-1 from Adipocytes," *Life Science* 80 (2007): 926–31; P. W. Sherman and J. Billing, "Darwinian Gastronomy: Why We Use Spices," *BioScience* 49 (1999): 453–63.
34. C. M. Alfaia et al., "Effect of Cooking Methods on Fatty Acids, Conjugated Isomers of Linoleic Acid and Nutritional Quality of Beef Intramuscular Fat," *Meat Science* 84 (2010): 769–77.
35. R. Cheung and P. McMahon, "Back to Grass: The Market Potential for US Grassfed Beef," Stone Barns Center for Food and Agriculture, Pocantico Hills, New York, 2017.
36. Frederick D. Provenza et al., "Our Landscape, Our Livestock, Ourselves: Restoring Broken Linkages among Plants, Herbivores, and Humans with Diets That Nourish and Satiate," *Appetite* 95 (2015): 500–19.
37. Charles Massy, *Call of the Reed Warbler: A New Agriculture, A New Earth* (St. Lucia: University of Queensland Press, 2017).

38. Baskin, *Underground*; Meuret and Provenza, *The Art and Science of Shepherding*; Michel Meuret and Frederick D. Provenza, "When Art and Science Meet: Integrating Knowledge of French Herders with Science of Foraging Behavior," *Rangeland Ecology and Management* 68 (2015): 1–17; W. R. Teague et al., "Multi-Paddock Grazing on Rangelands: Why the Perceptual Dichotomy between Research Results and Rancher Experience?" *Journal of Environmental Management* 128 (2013): 699–717.
39. D. Tilman and M. Clark, "Global Diets Link Environmental Sustainability and Human Health," *Nature* 515 (2014): 518–522; John D. Potter, "Red and Processed Meat, and Human and Planetary Health," *BMJ* (2017): 357; W. J. Ripple et al., "Ruminants, Climate Change and Climate Policy," *Nature Climate Change* 4 (2014): 2–5.
40. W. R. Teague et al., "The Role of Ruminants in Reducing Agriculture's Carbon Footprint in North America," *Journal of Soil and Water Conservation* 71 (2016): 156–64.
41. Eric Toensmeier, *The Carbon Farming Solution: A Global Toolkit of Perennial Crops and Regenerative Agriculture Practices for Climate Change Mitigation and Food Security* (White River Junction, VT: Chelsea Green, 2016); Steve Gabriel, *Silvopasture: A Guide to Managing Grazing Animals, Forage Crops, and Trees in a Temperate Farm Ecosystem* (White River Junction, VT: Chelsea Green, 2018).
42. Paul Hawken, *Drawdown: The Most Comprehensive Plan Ever Proposed to Reverse Global Warming* (New York: Penguin Books, 2017).

## Chapter 9: Creating Nourishing Bouquets

1. K. L. Hoehn et al., "Insulin Resistance Is a Cellular Antioxidant Defense Mechanism," *Proceedings of the National Academy of Sciences* 106 (2009): 17787–92.
2. N. F. Butte, "Carbohydrate and Lipid Metabolism in Pregnancy: Normal Compared with Gestational Diabetes Mellitus," *American Journal of Clinical Nutrition* 71 (2000): 1256S–61.
3. S. B. Haugaard et al., "Dietary Intervention Increases N-3 Long-Chain Polyunsaturated Fatty Acids in Skeletal Muscle Membrane Phospholipids of Obese Subjects. Implications for Insulin Sensitivity," *Clinical Endocrinology* 64 (2006): 169–78.
4. Gian Luigi Russo, "Dietary n-6 and n-3 Polyunsaturated Fatty Acids: From Biochemistry to Clinical Implications in Cardiovascular Prevention," *Biochemical Pharmacology* 77 (2009): 937–46.
5. Barbara J. Rolls, "Sensory-Specific Satiety," *Nutrition Reviews* 44 (1986): 93–101; Frederick D. Provenza, "Acquired Aversions as the Basis for Varied Diets of Ruminants Foraging on Rangelands," *Journal of Animal Science* 74 (1996): 2010–20.
6. Sheldon B. Atwood et al., "Changes in Preferences of Gestating Heifers Fed Untreated or Ammoniated Straw in Different Flavors," *Journal of Animal Science* 79 (2001): 3027–33; David Early and Frederick D. Provenza, "Food Flavor and Nutritional Characteristics Alter Dynamics of Food Preference in Lambs," *Journal of Animal Science* 76 (1998): 728–34.
7. Mark Westoby, "What Are the Biological Bases of Varied Diets?" *American Naturalist* 112 (1978): 627–31.
8. Derek W. Bailey et al., "Effect of Resource and Terrain Heterogeneity on the Feeding Site Selection and Livestock Movement Patterns," *Animal Production Science* 55 (2015): 298–308.
9. William J. Freeland and Dan H. Janzen, "Strategies in Herbivory by Mammals: The Role of Plant Secondary Compounds," *American Naturalist* 108 (1994): 269–86.
10. Frederick D. Provenza et al., "The Value to Herbivores of Plant Physical and Chemical Diversity in Time and Space," *Crop Science* 47 (2007): 382–98; Derek W. Bailey and Frederick D. Provenza, "Mechanisms Determining Large-Herbivore Distribution," in *Resource Ecology: Spatial and Temporal Dynamics of Foraging* H. H. T. Prins and F. van Langevelde eds. (The Netherlands: Springer, 2008), 7–28.

11. Frederick D. Provenza, "Acquired Aversions as the Basis for Varied Diets of Ruminants Foraging on Rangelands," *Journal of Animal Science* 74 (1996): 2010–20.
12. Arnold E. Gade and Frederick D. Provenza, "Nutrition of Sheep Grazing Crested Wheatgrass versus Crested Wheatgrass-Shrub Pastures during Winter," *Journal of Range Management* 39 (1986): 527–30.
13. A. J. Parsons et al., "Diet Preference of Sheep: Effect of Recent Diet, Physiological State and Species Abundance," *Journal of Animal Ecology* 63 (1994): 465–78; S. D. B. Cooper et al., "Diet Selection in Sheep: The Role of the Rumen Environment in the Selection of a Diet from Two Feeds that Differ in Their Energy Density," *British Journal of Nutrition* 74 (1995): 39–54; S. A. Francis, "Investigating the Role of Carbohydrates in the Dietary Choices of Ruminants with an Emphasis on Dairy Cows," (PhD diss., University of Melbourne, 2003); G. E. Lobley and G. D. Milano, "Regulation of Hepatic Nitrogen Metabolism in Ruminants," *Proceedings of the Nutrition Society* 57 (1997): 547–63.
14. S. S. Seefeldt, "Consequences of Selecting Ramboulliet Ewes for Mountain Big Sagebrush (*Artemisia tridentata* ssp. *vaseyana*) Dietary Preference," *Rangeland Ecology and Management* 58 (2005): 380–84.
15. Travis Mote et al., "Foraging Sequence Influences the Ability of Lambs to Consume Foods Containing Tannins and Terpenes," *Applied Animal Behaviour Science* 113 (2008): 57–68.
16. Jake Owens et al., "Supplementing Endophyte-Infected Tall Fescue or Reed Canarygrass with Alfalfa or Birdsfoot Trefoil Increases Forage Intake and Digestibility by Sheep," *Journal of Science of Food and Agriculture* 92 (2012): 987–92.
17. Tiffany D. Lyman et al., "Cattle Preferences Differ When Endophyte-Infected Tall Fescue, Birdsfoot Trefoil, and Alfalfa Are Grazed in Difference Sequences," *Journal of Animal Science* 89 (2011): 1131–37; Tiffany D. Lyman et al., "Phytochemical Complementarities among Endophyte-Infected Tall Fescue, Reed Canarygrass, Birdsfoot Trefoil, and Alfalfa Affect Cattle Foraging," *Animal* 6 (2012): 676–82.
18. Allen Savory and Jodi Butterfield, *Holistic Management: A New Framework for Decision-Making* (Washington, D.C.: Island Press, 1999); Jim Gerrish, *Management-Intensive Grazing, the Grassroots of Grass Farming* (Ridgeland: Green Park Press, 2004).
19. Michel Meuret and Frederick D. Provenza, "When Art and Science Meet: Integrating Knowledge of French Herders with Science of Foraging Behavior," *Rangeland Ecology and Management* 68 (2015): 1–17.
20. E. A. Gluesing and D. F. Balph, "An Aspect of Feeding Behavior and Its Importance to Grazing Systems," *Journal of Range Management* 33 (1980): 426–27.
21. L. Brondel et al., "Variety Enhances Food Intake in Humans: Role of Sensory-Specific Satiety," *Physiology and Behavior* 97 (2009): 44–51; L. B. Sørensen et al., "Effect of Sensory Perception of Foods on Appetite and Food Intake: A Review of Studies on Humans," *International Journal of Obesity* 27 (2003): 1152–66; B. J. Rolls et al., "Variety in a Meal Enhances Food Intake in Man," *Physiology and Behavior* 26 (1981): 215–21.
22. B. J. Rolls, "How Variety and Palatability Can Stimulate Appetite," *Nutrition Bulletin* 5 (1979): 78–86; B. J. Rolls et al., "The Influence of Variety on Human Food Selection and Intake," in *The Psychobiology of Human Food Selection*, L. M. Baker ed. (Westport: AVI, 1982) 101–22; A. K. Remick et al., "Internal and External Moderators of the Effect of Variety on Food Intake," *Psychological Bulletin* 135 (2009): 434–51.
23. Blundell and Bellisle eds. *Satiation, Satiety and the Control of Food Intake*; D. Chapelot, "Quantifying Satiation and Satiety," in *Satiation, Satiety and the Control of Food Intake,* J. E. Blundell and F. Bellisle eds. (Oxford: Woodhead Publishing 2013), 12–39.
24. David A. Booth, "Learnt Reduction in the Size of a Meal. Measurement of the Sensory-Gastric Inhibition from Conditioned Satiety," *Appetite* 52 (2009): 745–49; Frederick D. Provenza

et al., "Our Landscape, Our Livestock, Ourselves: Restoring Broken Linkages among Plants, Herbivores, and Humans with Diets That Nourish and Satiate," *Appetite* 95 (2015): 500–519.

25. S. H. A. Holt et al., "A Satiety Index of Common Foods," *European Journal of Clinical Nutrition* 49 (1995): 675–90; S. H. A. Holt et al., "Interrelationships among Postprandial Satiety, Glucose and Insulin Responses and Changes in Subsequent Food Intake," *European Journal of Clinical Nutrition* 50 (1996): 788–97; S. H. A. Holt et al., "The Effects of Equal-Energy Portions of Different Breads on Blood Glucose Levels, Feelings of Fullness and Subsequent Food Intake," *Journal of the American Dietetic Association* 101 (2001): 767–73.
26. A. Johnstone, "Protein and Satiety," in *Satiation, Satiety and the Control of Food Intake,* J. E. Blundell and F. Bellisle eds. (Oxford: Woodhead Publishing, 2013), 128–42; S. D. Poppitt, "Carbohydrates and Satiety," in *Satiation, Satiety and the Control of Food Intake,* J. E. Blundell and F. Bellisle eds. (Oxford: Woodhead Publishing, 2013), 166–81; S. D. Hennink and P. W .J. Maljaars, "Fats and Satiety," in *Satiation, Satiety and the Control of Food Intake,* J. E. Blundell and F. Bellisle eds. (Oxford: Woodhead Publishing, 2013), 143–65.
27. Barbara J. Rolls, "The Relationship between Dietary Energy Density and Energy Intake," *Physiology and Behavior* 97 (2009): 609–15.
28. M. K. Spill et al., "Energy Density: An Effective Strategy to Increase Children's Vegetable Intake and Reduce Energy Intake," *American Journal of Clinical Nutrition* 94 (2011): 735–41.
29. A. D. Blatt et al., "Hidden Vegetables: An Effective Strategy to Reduce Energy Intake and Increase Vegetable Intake in Adults," *American Journal of Clinical Nutrition* 93 (2011): 756–63.
30. J. A. Ello-Martin et al., "Dietary Energy Density in the Treatment of Obesity: A Year-Long Trial Comparing Two Weight-Loss Diets," *American Journal of Clinical Nutrition* 85 (2007): 1465–77.
31. M. K. Spill et al., "Serving Large Portions of Vegetable Soup at the Start of a Meal Affected Children's Energy and Vegetable Intake," *Appetite* 57 (2011): 213–19; M. K. Spill et al., "Eating Vegetables First: The Use of Portion Size to Increase Vegetable Intake in Preschool Children," *American Journal of Clinical Nutrition* 91 (2010): 1237–43; Barbara J. Rolls et al., "Salad and Satiety: Energy Density and Portion Size of a First-Course Salad Affect Energy Intake at Lunch," *Journal of the American Dietetic Association* 104 (2004): 1570–76.
32. S. Imai and S. Kajiyama, "Eating Order Diet Reduced the Postprandial Glucose and Glycated Haemoglobin Levels in Japanese Patients with Type 2 Diabetes," *Journal of Rehabilitation and Health Science* 8 (2010): 1–7; A. P. Shukla et al., "Food Order Has a Significant Impact on Postprandial Glucose and Insulin Levels," *Diabetes Care* 38 (2015): e98–e99.
33. D. Jakubowicz et al., "High-Energy Breakfast with Low-Energy Dinner Decreases Overall Daily Hyperglycaemia in Type 2 Diabetic Patients: A Randomised Clinical Trial," *Diabetologia* 58 (2015): 912–19; Barbara J. Rolls et al., "Salad and Satiety: Energy Density and Portion Size of a First-Course Salad Affect Energy Intake at Lunch," *Journal of the American Dietetic Association* 104 (2004): 1570–76; F. Q. Nuttall and M. C. Gannon, "Metabolic Response of People with Type 2 Diabetes to a High Protein Diet," *Nutrition and Metabolism* 1 (2004):6; A. H. Frid et al., "Effect of Whey on Blood Glucose and Insulin Responses to Composite Breakfast and Lunch Meals in Type 2 Diabetic Subjects," *American Journal of Clinical Nutrition* 82 (2005): 69–75; D. Jakubowicz et al., "Incretin, Insulinotropic and Glucose-Lowering Effects of Whey Protein Pre-Load in Type 2 Diabetes: A Randomised Clinical Trial," *Diabetologia* 57 (2014): 1807–11.
34. A. J. Stull et al., "Bioactives in Blueberries Improve Insulin Sensitivity in Obese, Insulin-Resistant Men and Women," *Journal of Nutrition* 140 (2010): 1764–68; R. Deng, "A Review of the Hypoglycemic Effects of Five Commonly Used Herbal Food Supplements," *Recent Patents on Food Nutrition and Agriculture* 4 (2012): 50–60; R. W. Allen et al., "Cinnamon Use in Type 2 Diabetes: An Updated Systematic Review and Meta-Analysis," *Annals of Family Medicine* 11 (2013): 452–59; V. Vuksan et al., "American Ginseng (*Panax quinquefolius* L.) Reduces

Postprandial Glycemia in Nondiabetic Subjects and Subjects with Type 2 Diabetes Mellitus," *Archives of Internal Medicine* 160 (2000): 1009–13; V. Vuksan et al., "American Ginseng Improves Glycemia in Individuals with Normal Glucose Tolerance: Effect of Dose and Time Escalation," *Journal of the American College of Nutrition* 19 (2000): 738–44.

35. E. I. Petsiou et al., "Effect and Mechanisms of Action of Vinegar on Glucose Metabolism, Lipid Profile, and Body Weight," *Nutrition Reviews* 72 (2014): 651–61.
36. D. J. A. Jenkins et al., "Glycemic Index of Foods: A Physiological Basis for Carbohydrate Exchange," *American Journal of Clinical Nutrition* 34 (1981): 362–66; D. J. A. Jenkins et al., "Glycemic Index: Overview of Implications in Health and Disease," *American Journal of Clinical Nutrition* 76 (2002): 266S–73; D. A. J. Jenkins et al., "Type 2 Diabetes and the Vegetarian Diet," *American Journal of Clinical Nutrition* 78 (2003): 610S–16; D. S. Ludwig, "The Glycemic Index: Physiological Mechanisms Relating to Obesity, Diabetes, and Cardiovascular Disease," *Journal of the American Medical Association* 287 (2002): 2414–23; C.-J. Chiu et al., "Informing Food Choices and Health Outcomes by Use of the Dietary Glycemic Index," *Nutrition Reviews* 69 (2011): 231–42.
37. F. R. J. Bornet et al., "Insulinemic and Glycemic Indexes of Six Starch-Rich Foods Taken Alone and in a Mixed Meal by Type 2 Diabetics," *American Journal of Clinical Nutrition* 45 (1987): 588–95; D. Estrich et al., "Effects of Co-Ingestion of Fat and Protein Upon Carbohydrate-Induced Hyperglycemia," *Diabetes* 16 (1967): 232–37; I. M. Welch et al., "Duodenal and Ileal Lipid Suppresses Postprandial Blood Glucose and Insulin Responses in Man: Possible Implications for the Dietary Management of Diabetes Mellitus," *Clinical Science* 72 (1987): 209–16.
38. T. M. Larsen et al., "Diets with High or Low Protein Content and Glycemic Index for Weight-Loss Maintenance," *New England Journal of Medicine* 363 (2010): 2102–13; C. B. Ebbeling et al., "Effects of Dietary Composition on Energy Expenditure during Weight-Loss Maintenance," *Journal of the American Medical Association* 307 (2012): 2627–34.
39. B. S. Lennerz et al., "Effects of Dietary Glycemic Index on Brain Regions Related to Reward and Craving in Men," *American Journal of Clinical Nutrition* 98 (2013): 641–47.
40. D. Mozaffarian et al., "Changes in Diet and Lifestyle and Long-Term Weight Gain in Women and Men," *New England Journal of Medicine* 364 (2011): 2392–404; S. Liu et al., "Relation between Changes in Intakes of Dietary Fiber and Grain Products and Changes in Weight and Development of Obesity among Middle-Aged Women," *American Journal of Clinical Nutrition* 78 (2003): 920–27; P. Koh-Banerjee and E. B. Rimm, "Whole Grain Consumption and Weight Gain: A Review of the Epidemiological Evidence, Potential Mechanisms and Opportunities for Future Research," *Proceedings of the Nutrition Society* 62 (2003): 25–29; P. Koh-Banerjee et al., "Changes in Whole-Grain, Bran, and Cereal Fiber Consumption in Relation to 8-Y Weight Gain among Men," *American Journal of Clinical Nutrition* 80 (2004): 1237–45; M. Fogelholm et al., "Dietary Macronutrients and Food Consumption as Determinants of Long-Term Weight Change in Adult Populations: A Systematic Literature Review," *Food and Nutrition Research* 56 (2012); M. S. Westerterp-Plantenga et al., "Dietary Protein – Its Role in Satiety, Energetics, Weight Loss and Health," *British Journal of Nutrition* 108 (2012): S105–12.
41. J. D. Smith et al., "Changes in Intake of Protein Foods, Carbohydrate Amount and Quality, and Long-Term Weight Change: Results from 3 Prospective Cohorts," *American Journal of Clinical Nutrition* 101 (2015): 1–9.
42. R. J. Scharf et al., "Longitudinal Evaluation of Milk Type Consumed and Weight Status in Preschoolers," *Archives of Diseases in Childhood* 98 (2013): 335–40; D. S. Ludwig and W. C. Willett, "Three Daily Servings of Reduced-Fat Milk: An Evidence-Based Recommendation?" *Journal of the American Medical Association Pediatrics* 167 (2013): 788–89.
43. Frank B. Hu et al., "Prospective Study of Major Dietary Patterns and Risk of Coronary Heart Disease in Men," *American Journal of Clinical Nutrition* 72 (2000): 912–21.

44. D. Anue et al., "Nut Consumption and Risk of Cardiovascular Disease, Total Cancer, All-Cause and Cause-Specific Mortality: A Systematic Review and Dose-Response Meta-Analysis of Prospective Studies," *BMC Medicine* 14 (2016): 207.
45. R. Doll and R. Petro, "The Causes of Cancer: Quantitative Estimates of Avoidable Risks of Cancer in the United States Today," *Journal of the National Cancer Institute* 66 (1981): 1191–308.
46. D. B. Boyd, "Insulin and Cancer," *Integrative Cancer Therapies* 2 (2003): 315–329.
47. S. Djiogue et al., "Insulin Resistance and Cancer: The Role of Insulin and IGFs," *Endocrine-Related Cancer* 20 (2013): R1–17.
48. C. C. Deocaris et al., "Merger of Ayurveda and Tissue Culture-Based Functional Genomics: Inspirations from Systems Biology," *Journal of Translational Medicine* 6 (2008): 14 doi:10.1186/1479-5876-6-14; J. W. H. Liand and J. C. Vederas, "Drug Discovery and Natural Products: End of an Era or an Endless Frontier?" *Science* 325 (2009): 161–65; K. I. Block et al. "A Broad-Spectrum Integrative Design for Cancer Prevention and Therapy," *Seminars in Cancer Biology*, 35, (Suppl) (2015): S276–304; S. S. Mayne et al., "Diet, Nutrition, and Cancer: Past, Present and Future," *Nature Reviews Clinical Oncology* 13 (2016): 504–15.
49. W. Vanden Berghe, "Epigenetic Impact of Dietary Polyphenols in Cancer Chemoprevention: Lifelong Remodeling of Our Epigenomes," *Pharmacological Research* 65 (2012): 565–76; A. Link et al., "Cancer Chemoprevention by Dietary Polyphenols: Promising Role for Epigenetics," *Biochemical Pharmacology* 80 (2010): 1771–92; D. Hanahanand and R. A. Weinberg, "Hallmarks of Cancer: The Next Generation," *Cell* 144 (2011): 646–74.
50. M. B. Sporn, "Perspective: The Big C – For Chemoprevention," *Nature* 471 (2011): S10–11; T. M. Kok et al., "Mechanisms of Combined Action of Different Chemopreventive Dietary Compounds," *European Journal of Nutrition* 47 (2008): 51–59; A. Harvey, "Natural Products in Drug Discovery," *Drug Discovery Today* 13 (2008): 894–901; S. M. Meeran et al., "Epigenetic Targets of Bioactive Dietary Components for Cancer Prevention and Therapy," *Clinical Epigenetics* 1 (2010): 101–16; M. A. Parasramka et al., "MicroRNAs, Diet, and Cancer: New Mechanistic Insights on the Epigenetic Actions of Phytochemicals," *Molecular Carcinogenesis* 51 (2012): 213–30; K. Szarc vel Szic et al., Nature or Nurture: Let Food Be Your Epigenetic Medicine in Chronic Inflammatory Disorders," *Biochemical Pharmacology* 80 (2010): 1816–32; N. P. Seeram et al., "In Vitro Antiproliferative, Apoptotic and Antioxidant Activities of Punicalagin, Ellagic Acid and a Total Pomegranate Tannin Extract Are Enhanced in Combination with Other Polyphenols as Found in Pomegranate Juice," *Journal of Nutritional Biochemistry* 16 (2005): 360–67; Y. J. Surh, "Cancer Chemoprevention with Dietary Phytochemicals," *Nature Reviews Cancer* 3 (2003): 768–80.
51. C. Faulk and D. C. Dolinoy, "Timing is Everything: The When and How of Environmentally Induced Changes in the Epigenome of Animals," *Epigenetics* 6 (2011): 791–97.
52. F. Visioli et al., "Dietary Intake of Fish vs. Formulations Leads to Higher Plasma Concentrations of N-3 Fatty Acids," *Lipids* 38 (2003): 415–18; D. R. Jacobs and L. C. Tapsell, "Food, Not Nutrients, Is the Fundamental Unit in Nutrition," *Nutrition Review* 65 (2007): 439–50.
53. G. C. Burdge et al., "Nutrition in Early Life, and Risk of Cancer and Metabolic Disease: Alternative Endings in an Epigenetic Tale?" *British Journal of Nutrition* 101 (2009): 619–30.
54. K. Wolfe et al., "Antioxidant Activity of Apple Peels," *Journal of Agriculture and Food Chemistry* 51 (2003): 609–14; X. He and R. H. Liu, "Triterpenoids Isolated from Apple Peels Have Potent Antiproliferative Activity and May Be Partially Responsible for Apple's Anticancer Activity," *Journal of Agriculture and Food Chemistry* 55 (2007): 4366–70; M. V. Eberhardt et al., "Antioxidant Activity of Fresh Apples," *Nature* (2000): 405, 903–4; R. H. Liu, J. Liu, B. Chen, "Apples Prevent Mammary Tumors in Rats," *Journal of Agriculture and Food Chemistry* 53 (2005): 2341–43.
55. E. Giovannucci and S. K. Clinton, "Tomatoes, Lycopene, and Prostate Cancer," *Proceedings of the Society for Experimental Biology and Medicine* 218 (1998): 129–39; K. Canene-Adams et al.,

"Combinations of Tomato and Broccoli Enhance Antitumor Activity in Dunning R3327-H Prostate Adenocarcinomas," *Cancer Research* 67 (2007): 836–43.

56. B. Fuhrman et al., "Lycopene Synergistically Inhibits LDL Oxidation in Combination with Vitamin E, Glabridin, Rosmarinic Acid, Carnosic Acid, or Garlic," *Antioxidants and Redox Signaling* 2 (2000): 491–506; P. Ninfali et al., "Antioxidant Capacity of Vegetables, Spices and Dressings Relevant to Nutrition," *British Journal of Nutrition* 93 (2005): 257–66.
57. A. S. Keck et al., "Food Matrix Effects on Bioactivity of Broccoli-Derived Sulforaphane in Liver and Colon of F344 Rats," *Journal of Agriculture and Food Chemistry* 51 (2003): 3320–27; M. A. Wallig et al., "Synergy among Phytochemicals within Crucifers: Does It Translate into Chemoprotection?" *Journal of Nutrition* 135 (2005): S2972–977.
58. Linda C. Tapsell et al., "Health Benefits of Herbs and Spices: The Past, the Present, the Future," *The Medical Journal of Australia* 185 (2006): S1–24; E. I. Opara and M. Chohan, "Culinary Herbs and Spices: Their Bioactive Properties, the Contribution of Polyphenols and the Challenges in Deducing Their True Health Benefits," *International Journal of Molecular Sciences* 15 (2014): 19183–202; S. Singh, "From Exotic Spice to Modern Drug?" *Cell* 130 (2007): 765–68; C. M. Kaefer and J. A. Milner, "The Role of Herbs and Spices in Cancer Prevention," *Journal of Nutritional Biochemistry* 19 (2008): 347–61; I. Zanotti et al., "Atheroprotective Effects of (Poly) phenols: A Focus on Cell Cholesterol Metabolism," *Food and Function* 6 (2015): 13–31.
59. P. Ninfali et al., "Antioxidant Capacity of Vegetables, Spices and Dressings Relevant to Nutrition," *British Journal of Nutrition* 93 (2005): 257–66.
60. C. Manach et al., "Bioavailability and Bioefficacy of Polyphenols in Humans. I. Review of 97 Bioavailability Studies," *American Journal of Clinical Nutrition* 81 (2005): 230S–42; G. Williamson and C. Manach, "Bioavailability and Bioefficacy of Polyphenols in Humans. II. Review of 93 Intervention Studies," *American Journal of Clinical Nutrition* 81 (2005): 243S–55.
61. M. S. Farvid et al., "Dietary Fiber Intake in Young Adults and Breast Cancer Risk," *Pediatrics* 137, no. 3 (2016): e20151226.
62. D. R. Jacobs et al., "Fiber from Whole Grains, but Not Refined Grains, Is Inversely Associated with All-Cause Mortality in Older Women: The Iowa Women's Health Study," *Journal of the American College of Nutrition* 19 (2000): S326–30.
63. L. Harnack et al., "An Evaluation of the Dietary Guidelines for Americans in Relation to Cancer Occurrence," *American Journal of Clinical Nutrition* 76 (2002): 889–96.

## Chapter 10: Painting Your Canvas

1. C. K. Miller et al., "Comparison of a Mindful Eating Intervention to a Diabetes Self-Management Intervention among Adults with Type 2 Diabetes: A Randomized Controlled Trial," *Health Education and Behavior* 41 (2014): 145–54.
2. James E. Painter et al., "How Visibility and Convenience Influence Candy Consumption," *Appetite* 38 (2002): 237–38.
3. Brian Wansink et al., "Proximity's Influence on Estimated and Actual Candy Consumption," *International Journal of Obesity* 20 (2006): 871–75.
4. A. J. Parsons et al., "Diet Preference of Sheep: Effect of Recent Diet, Physiological State and Species Abundance," *Journal of Animal Ecology* 63 (1994): 465–78.
5. Michael H. Ralphs, "Persistence of Aversion to Larkspur in Naive and Native Cattle," *Journal of Range Management* 50 (1997): 367–70; Michael H. Ralphs and Frederick D. Provenza, Conditioned Food Aversions: Principles and Practices, with Special Reverence to Social Facilitation," *Proceedings of the Nutrition Society* 58 (1999): 813–20.
6. Brian Wansink, *Mindless Eating: Why We Eat More Than We Think* (New York: Hay House, 2009).
7. Brian Wansink and Collin R. Payne, "Counting Bones: Environmental Cues That Decrease Food Intake," *Perceptual and Motor Skills* 104 (2007): 273–76.

8. Wansink, *Mindless Eating*, 47–52.
9. Leann L. Birch and Jennifer O. Fisher, "Mother's Child-Feeding Practices Influence Daughters' Eating and Weight," *American Journal of Clinical Nutrition*, 71 (2000): 1054–61; Leann L. Birch et al., "Clean Up Your Plate: Effects of Child Feeding Practices on the Conditioning of Meal Size," *Learning and Motivation* 18 (1987): 301–17.
10. Brian Wansink and SeaBum Park, "At the Movies: How External Cues and Perceived Taste Impact Consumption Volume," *Food Quality and Preference* 12 (2001): 69–74.
11. C. A. Anastasiou et al., "Weight Regaining: From Statistics and Behaviors to Physiology and Metabolism," *Metabolism* 64 (2015): 1395–407.
12. D. M. Thomas et al., "Why Do Individuals Not Lose More Weight from an Exercise Intervention at a Defined Dose? An Energy Balance Analysis," *Obesity Reviews* 13 (2012): 835–47.
13. C. D. Gardner et al., "Comparison of the Atkins, Zone, Ornish, and LEARN Diets for Change in Weight and Related Risk Factors among Overweight Premenopausal Women the A TO Z Weight Loss Study: A Randomized Trial," *Journal of the American Medical Association* 297 (2007): 969–77.
14. M. V. Stanton et al., "DIETFITS Study (Diet Intervention Examining the Factors Interacting with Treatment Success) – Study Design and Methods," *Contemporary Clinical Trials* 53: (2017): 151–61; C. D. Gardner et al., "Effect of Low-Fat vs Low-Carbohydrate Diet on 12-Month Weight Loss in Overweight Adults and the Association With Genotype Pattern or Insulin Secretion: The DIETFITS Randomized Clinical Trial," *Journal of the American Medical Association* 319 (2018): 667-79.
15. Michael D. Jensen et al., "AHA/ACC/TOS Guideline for the Management of Overweight and Obesity in Adults: A Report of the American College of Cardiology/American Heart Association Task Force on Practice Guidelines and the Obesity Society," *Circulation* 29 (2014): S102–S138.
16. N. A. King et al., "Individual Variability Following 12 Weeks of Supervised Exercise: Identification and Characterization of Compensation for Exercise-Induced Weight Loss," *International Journal of Obesity* 32 (2008): 177–84.
17. A. McTiernan et al., "Exercise Effect on Weight and Body Fat in Men and Women," *Obesity* 15 (2007): 1496–1512.
18. T. S. Church et al., "Trends over 5 Decades in U.S. Occupation-Related Physical Activity and Their Associations with Obesity," *PLoS ONE* 6(5) (2011): e19657.
19. Edward Archer, "In Defense of Sugar: A Critique of Diet-Centrism," *Progress in Cardiovascular Diseases* (2018).
20. D. M. Thomas et al., "Why Do Individuals Not Lose More Weight from an Exercise Intervention at a Defined Dose? An Energy Balance Analysis," *Obesity Reviews* 13 (2012): 835–47.
21. K. M. Beavers et al., "Effect of an 18-Month Physical Activity and Weight Loss Intervention on Body Composition in Overweight and Obese Older Adults," *Obesity* 22 (2014): 325–31; R. A. Washburn et al., "Does the Method of Weight Loss Affect Long-Term Changes in Weight, Body Composition or Chronic Disease Risk Factors in Overweight or Obese Adults? A Systematic Review," *PLoS ONE* 9(10) (2014): e109849.
22. Y. Gepner et al., "Effect of Distinct Lifestyle Interventions on Mobilization of Fat Storage Pools: The CENTRAL MRI Randomized Controlled Trial," *Circulation* (2017).
23. T. Moro et al., "Effects of Eight Weeks of Time-Restricted Feeding (16/8) on Basal Metabolism, Maximal Strength, Body Composition, Inflammation, and Cardiovascular Risk Factors in Resistance-Trained Males," *Journal of Translational Medicine* 14 (2016).
24. J. F. Trepanowski et al., "Effect of Alternate-Day Fasting on Weight Loss, Weight Maintenance, and Cardioprotection among Metabolically Healthy Obese Adults: A Randomized Clinical Trial," *Journal of the American Medical Association Internal Medicine*, 177 (2017): 930–38.

25. M. P. Mattson et al., "Impact of Intermittent Fasting on Health and Disease Processes," *Ageing Research Reviews* 39 (2017): 46–58; M. Wei et al., "Fasting-Mimicking Diet and Markers/Risk Factors for Aging, Diabetes, Cancer, and Cardiovascular Disease," *Science Translational Medicine* 9 (2017).
26. B. Martin et al., "'Control' Laboratory Rodents Are Metabolically Morbid: Why It Matters," *Proceedings of the National Academy of Science* 107 (2010): 6127–33.
27. L. Belkacemi et al., "Intermittent Fasting Modulation of the Diabetic Syndrome in Streptozotocin-Injected Rats," *International Journal of Endocrinology* 2012 (2012); M. Hatori et al., "Time-Restricted Feeding without Reducing Caloric Intake Prevents Metabolic Diseases in Mice Fed a High-Fat Diet," *Cell Metabolism* 15 (2012): 848–60.
28. G. I. Fond et al., "Fasting in Mood Disorders: Neurobiology and Effectiveness: A Review of the Literature," *Psychiatry Research* 209 (2013): 253–58.
29. T. W. McDade, "Early Environments and the Ecology of Inflammation," *Proceedings of the National Academy of Science USA* 109, Supplement 2, (2012): 17281–88; Cynthia M. Kroeger et al., "Improvement in Coronary Heart Disease Risk Factors during an Intermittent Fasting/Calorie Restriction Regimen: Relationship to Adipokine Modulations," *Nutrition and Metabolism (London)* 9 (2012): 98.
30. C. Patel et al., "Prolonged Reactive Oxygen Species Generation and Nuclear Factor-Kβ Activation after a High-Fat, High-Carbohydrate Meal in the Obese," *Journal of Clinical Endocrinology and Metabolism* 92 (2007): 4476–79.
31. David S. Ludwig and Mark I. Friedman, "Increasing Adiposity: Consequence or Cause of Overeating?" *Journal of the American Medical Association* 311 (2014): 2167–68; C. B. Ebbeling et al., "Effects of Dietary Composition on Energy Expenditure during Weight-Loss Maintenance," *Journal of the American Medical Association* 307 (2013): 2627–34.
32. C. E. Forsythe et al., "Comparison of Low Fat and Low Carbohydrate Diets on Circulating Fatty Acid Composition and Markers of Inflammation," *Lipids* 43 (2008): 65–77.
33. M. P. Mattson et al., "Meal Frequency and Timing in Health and Disease," *Proceedings of the National Academy of Science* 111 (2014): 16647–53; M. Uno et al., "A Fasting-Responsive Signaling Pathway that Extends Life Span in *C. elegans*," *Cell Reports* 3 (2013): 79–91; C. L. Goodrick et al., "Effects of Intermittent Feeding upon Growth and Life Span in Rats," *Gerontology* 28 (1982): 233–41; C. L. Goodrick et al., "Effects of Intermittent Feeding upon Growth, Activity, and Lifespan in Rats Allowed Voluntary Exercise," *Experimental Aging Research* 9 (1983): 203–9.
34. Leanne M. Redman et al., "Metabolic Slowing and Reduced Oxidative Damage with Sustained Caloric Restriction Support the Rate of Living and Oxidative Damage Theories of Aging," *Cell Metabolism* 27 (2018): 805–15.
35. N. Vogelzangs et al., "Urinary Cortisol and Six-Year Risk of All-Cause and Cardiovascular Mortality," *Journal of Clinical Endocrinology and Metabolism* 95 (2010): 4959–64.
36. T. C. Adam et al., "Cortisol Is Negatively Associated with Insulin Sensitivity in Overweight Latino Youth," *Journal of Clinical Endocrinology and Metabolism* 95 (2010): 4729–35; H. B. Holt et al., "Cortisol Clearance and Associations with Insulin Sensitivity, Body Fat and Fatty Liver in Middle-Aged Men," *Diabetologia* 50 (2007): 1024–32; J. Q. Purnell et al., "Enhanced Cortisol Production Rates, Free Cortisol, and 11beta-HSD-1 Expression Correlate with Visceral Fat and Insulin Resistance in Men: Effect of Weight Loss," *American Journal of Physiology-Endocrinology and Metabolism* 296 (2009): E351–E357.
37. V. Hainer and A. Aldhoon-Hainerov, "Obesity Paradox Does Exist," *Diabetes Care* 36 (2013): S276–81; J.-M. Kvamme et al., "Body Mass Index and Mortality in Elderly Men and Women: The Tromsø and HUNT Studies," *Journal of Epidemiology and Community Health* 66 (2012): 611–17; J. E. Winter et al., "BMI and All-Cause Mortality in Older Adults: A Meta-Analysis,"

*American Journal of Clinical Nutrition* 99 (2014): 875–90; C. J. Lavie et al., "Implications Regarding Fitness, Fatness, and Severity in the Obesity Paradox," *Journal of the American College of Cardiology* 63 (2014): 1345–54.
38. William Falk, "Editor's Letter," *The Week* (March 30, 2018): 3.
39. A. Kouris-Blazos, "Morbidity Mortality Paradox of 1st Generation Greek Australians," *Asia Pacific Journal of Clinical Nutrition* 11 Supplement 3 (2002): S569–75.
40. Q. Li et al., "Visiting a Forest, But Not a City, Increases Human Natural Killer Activity and Expression of Anti-Cancer Proteins," *International Journal of Immunopathology and Pharmacology* 21 (2008): 117–27.
41. N. Takayama et al., "Emotional, Restorative and Vitalizing Effects of Forest and Urban Environments at Four Sites in Japan," *International Journal of Environmental Research and Public Health* 11 (2014): 7207–30.
42. M. G. Berman et al., "Interacting with Nature Improves Cognition and Affect for Individuals with Depression," *Journal of Affective Disorders* 140 (2012): 300–5.
43. O. R. W. Pergams and P. A. Zaradic, "Evidence for a Fundamental and Pervasive Shift Away from Nature-Based Recreation," *Proceedings of the National Academy of Sciences USA* 105 (2008): 2295–300.
44. P. Kareiva, "Ominous Trends in Nature Recreation," *Proceedings of the National Academy of Sciences* 105 (2008): 2757–58.

## Chapter 11: Linking Palates with Landscapes

1. J. M. Davis and J. A. Stamps, "The Effect of Natal Experience on Habitat Preferences," *Trends in Ecology and Evolution* 19 (2004): 411–16.
2. V. Altbäcker et al., "Rabbit Mothers' Diet Influence the Pups' Later Food Choice," *Ethology* 95 (1995): 107–16.
3. K. Mátrai et al., "Seasonal Diet of Rabbits and Their Browsing Effect on Juniper in Bugac Juniper Forest (Hungary)," *Acta Theriologica* 43 (1998): 107–12.
4. R. Hudson et al., "Transmission of Olfactory Information from Mother to Young in European Rabbit," in *Mammalian Social Learning: Comparative and Ecological Perspectives,* Hilary O. Box and Kathleen R. Gibson eds. (Cambridge: Cambridge University Press, 1999), 141–57.
5. R. Hudson and V. Altbäcker, "Development of Feeding and Food Preference in the European Rabbit: Environmental and Maturational Determinants," in *Behavioral Aspects of Feeding: Basic and Applied Research in Mammals* B. G. Galef et al., eds. (Abingdon: Harwood Academic Publishers, 1992), 125–45.
6. Kathy Voth, *Cows Eat Weeds: How to Turn Your Cows into Weed Managers* kvoth@livestockforlandscapes.com 2010.
7. A. Ducs et al., "Milkweed Control by Food Imprinted Rabbits," *Behavioral Processes* 130 (2016): 75–80.
8. A. Bilkó et al., "Transmission of Food Preference in the Rabbit: The Means of Information Transfer," *Physiology and Behavior* 56 (1994): 907–12.
9. R. Andersen, "Habitat Deterioration and the Migratory Behaviour of Moose (*Alces alces* L.) in Norway," *Journal of Applied Ecology* 28 (1991): 102–8.
10. Dax L. Mangus, "Reducing Reliance on Supplemental Winter Feeding in Elk (*Cervus canadensis*): An Applied Management Experiment at Deseret Land and Livestock Ranch" (MS thesis, Utah State University, 2011).
11. Jordan F. Smith, *Engineering Eden: The True Story of a Violent Death, a Trial, and the Fight over Controlling Nature* (New York: Crown, 2016).
12. Roy L. Roath and William C. Krueger, "Cattle Grazing and Behavior on a Forested Range," *Journal of Range Management* 35 (1982): 332–38.

13. Larry D. Howery et al., "Differences in Home Range and Habitat Use among Individuals in a Cattle Herd," *Applied Animal Behaviour Science* 49 (1996): 305–20; Larry D. Howery et al., "Social and Environmental Factors Influence Cattle Distribution on Rangeland," *Applied Animal Behaviour Science* 55 (1998): 231–44.
14. R. F. Hunter and C. Milner, "The Behavior of Individual, Related and Groups of South Country Cheviot Hill Sheep," *Animal Behaviour* 11 (1963): 507–13; C. Key and R. M. Maclver, "The Effects of Maternal Influences on Sheep: Breed Differences in Grazing, Resting and Courtship Behavior," *Applied Animal Ethology* 6 (1980): 33–48.
15. P. E. Simitzis et al., "Feeding Preferences in Lambs Influenced by Prenatal Flavour Exposure," *Physiology and Behavior* 93 (2008): 529–36; Dale L. Nolte et al., "Garlic in the Ovine Fetal Environment," *Physiology and Behavior* 52 (1992): 1091–93; Dale L. Nolte and Frederick D. Provenza, "Food Preferences in Lambs after Exposure to Flavors in Milk," *Applied Animal Behaviour Science* 32 (1992): 381–89.
16. G. C. Green et al., "Long-Term Effects of Early Experience to Supplementary Feeding in Sheep," *Proceedings of the Australian Society of Animal Production* 15 (1984): 373–75.
17. Anna G. Thorhallsdottir et al., "Ability of Lambs to Learn about Novel Foods while Observing or Participating with Social Models," *Applied Animal Behaviour Science* 25 (1990): 25–33; Sarwat N. Mirza and Frederick D. Provenza, "Preference of the Mother Affects Selection and Avoidance of Foods by Lambs Differing in Age," *Applied Animal Behaviour Science* 28 (1990): 255–63; Sarwat N. Mirza and Frederick D. Provenza, "Effects of Age and Conditions of Exposure on Maternally Mediated Food Selection in Lambs," *Applied Animal Behaviour Science* 33 (1992): 35–42.
18. Frederick D. Provenza et al., "The Relative Importance of Mother and Toxicosis in the Selection of Foods by Lambs," *Journal of Chemical Ecology* 19 (1993): 313–23.
19. Randall D. Wiedmeier et al., "Eating a High-Fiber Diet during Pregnancy Increases Intake and Digestibility of a High-Fiber Diet by Offspring in Cattle," *Animal Feed Science and Technology* 177 (2012): 144–51.
20. Randall D. Wiedmeier et al., "Exposure to Ammoniated Wheat Straw as Suckling Calves Improves Performance of Mature Beef Cows Wintered on Ammoniated Wheat Straw," *Journal of Animal Science* 80 (2002): 2340–48.
21. Randall D. Wiedmeier et al., "Heritability of Low-Quality Forage Utilization in Beef Cattle," *Proceedings of the Western Section of the American Society of Animal Science* 46 (1995): 404–6.
22. Charles A. Petersen et al., "Influence of Experience on Cattle Browsing Sagebrush and Its Impacts on Plant Community Structure," *Rangeland Ecology and Management* 67 (2014): 78–87.
23. S. Biquand and V. Biquand-Guyot, "The Influence of Peers, Lineage and Environment on Food Selection of the Criollo Goat (*Capra hircus*)," *Applied Animal Behaviour Science* 34 (1992): 231–45.
24. Frederick D. Provenza and David F. Balph, "Diet Learning by Domestic Ruminants: Theory, Evidence and Practical Implications," *Applied Animal Behaviour Science* 18 (1987): 211–32.
25. S. Landau and G. Molle, "Grazing Livestock, Our Connection to Grass: A Mediterranean Insight. Why They Eat What They Eat, and How It Affects Us," in *All Flesh is Grass. Plant-Animal Interrelationships* J. Seckbach and A. Dubinsky eds. (New York: Springer, 2009), 217–36.
26. T.A. Glasser et al., "Breed and Maternal Effects on the Intake of Tannin-Rich Browse by Juvenile Goats (*Capra hircus*)," *Applied Animal Behaviour Science* 119 (2009): 71–77.
27. Roberto A. Distel and Frederick D. Provenza, "Experience Early in Life Affects Voluntary Intake of Blackbrush by Goats," *Journal of Chemical Ecology* 17 (1991): 431–50.
28. Megan A. Chadwick et al., "Programming Sheep Production on Saltbrush: Adaptations of Offspring from Ewes That Consumed High Amounts of Salt during Pregnancy and Early Lactation," *Animal Production Science* 49 (2009): 311–17.
29. Roberto A. Distel et al., "Effects of Early Experience on Voluntary Intake of Low-Quality Roughage by Sheep," *Journal of Animal Science* 72 (1994): 1191–95.

30. M. Du et al., "Fetal Programming of Skeletal Muscle Development in Ruminant Animals," *Journal of Animal Science* 88, E. Suppl. (2010): E51–60.
31. P. D. Gluckman et al., "Predictive Adaptive Responses and Human Evolution," *Trends in Ecology and Evolution* 20 (2005): 527–33.
32. Frederick D. Provenza et al., "Mechanisms of Learning in Diet Selection with Reference to Phytotoxicosis in Herbivores," *Journal of Range Management* 45 (1992): 36–45.
33. A. Tallian et al., "Predator Foraging Response to a Resurgent Dangerous Prey," *Functional Ecology* (2017); C. Riginos, "Climate and the Landscape of Fear in an African Savanna," *Journal of Animal Ecology* 84 (2015): 124–33.
34. Henry F. Dobyns, *Their Number Become Thinned: Native American Population Dynamics in Eastern North America (Native American Historic Demography Series)* (Knoxville: University of Tennessee Press, 1983); Frank H. Mayer and Charles B. Roth, *The Buffalo Harvest* (Tennessee: Pioneer Press, 1995).
35. W. C. H. Green et al., "Post-Weaning Associations among Bison Mothers and Daughters," *Animal Behaviour* 38 (1989): 847–58; J. S. Brookshier and W. S. Fairbanks, "The Nature and Consequences of Mother-Daughter Associations in Naturally and Forcibly Weaned Bison," *Canadian Journal of Zoology* 81 (2003): 414–23; T. H. Clutton-Brock et al., *Red Deer: Behavior and Ecology of Two Sexes* (Chicago: University of Chicago Press, 1982); I. Douglas-Hamilton, "On the Ecology and Behaviour of the Lake Manyara Elephants," *East African Wildlife Journal* 11 (1973): 401–3; C. J. Moss and J. H. Poole, "Relationships and Social Structure of African Elephants," in *Primate Social Relationships: An Integrated Approach*, R.A. Hinde ed. (Sinauer: Sutherland, 1983), 315–25; J. L. Aycrigg and W. F. Porter, "Sociospatial Dynamics of White-Tailed Deer in the Central Adirondack Mountains, New York," *Journal of Mammalogy* 7 (1997): 468–82; M. Festa-Bianchet, "Seasonal Dispersion of Overlapping Mountain Sheep Ewe Groups," *Journal of Wildlife Management* 50 (1986): 325–30; R. F. Hunter and C. Milner, "The Behavior of Individual, Related and Groups of South Country Cheviot Hill Sheep," *Animal Behaviour* 11 (1963): 507–13; A. Lazo, "Social Segregation and the Maintenance of Social Stability in a Feral Cattle Population," *Animal Behaviour* 48 (1994): 1133–41; V. Reinhardt and A. Reinhardt, "Cohesive Relationships in a Cattle Herd (*Bos indicus*)," *Behaviour* 77 (1981): 121–51; D. G. Tulloch, "The Water Buffalo, *Bubalus bubalis*, in Australia: Home Range," *Australian Wildlife Research* 5 (1978): 327–54.
36. C. J. Moss and P. C. Lee, "The Social Context for Learning and Behavioral Development among Wild African Elephants," in *Mammalian Social Learning: Comparative and Ecological Perspectives*, H. O. Box and K. R. Gibson eds. (Cambridge University Press, Cambridge, 1999), 102–25; K. McComb et al., "Matriarchs as Repositories of Social Knowledge in African Elephants," *Science* 292 (2001): 491–94.
37. T. N. C. Vidya and R. Sukumar, "Social and Reproductive Behaviour in Elephants," *Animal Behaviour* 89 (2005): 1200–07; E. A. Archie et al., "The Ties That Bind: Genetic Relatedness Predicts the Fission and Fusion of Social Groups in Wild African Elephants," *Proceedings of the Royal Society* 273 (2006): 513–22.
38. Poole et al., *The Status of Kenya's Elephants*; G. Wittemeyer et al., "Where Sociality and Relatedness Diverge: The Genetic Basis for Hierarchical Social Organization in African Elephants," *Proceedings of the Royal Society Bulletin* 276 (2009): 3513–21.
39. C. J. Moss, "The Demography of an African Elephant (*Loxodonta africana*) Population in Amboseli, Kenya," *Journal of Zoology* 255 (2001): 145–56.
40. L. J. N. Brent et al., "Ecological Knowledge, Leadership, and the Evolution of Menopause in Killer Whales," *Current Biology* 25 (2015): 1–5; G. DeLuca et al., "Fishing out Collective Memory of Migratory Schools," *Journal of the Royal Society Interface* 11, no. 95 (2014).
41. E. A. Zimmerman, "Desert Ranching in Central Nevada," *Rangelands* 2 (1980): 184–86.

42. N. G. Blurton Jones et al., "Antiquity of Postreproductive Life: Are There Modern Impacts on Hunter-Gatherer Postreproductive Life Spans?" *American Journal of Human Biology* 14 (2002): 184–205; M. Lahdenpera et al., "Fitness Benefits of Prolonged Post-Reproductive Lifespan in Women," *Nature* 428 (2004): 178–81; K. Hawkes et al., "Hadza Women's Time Allocation, Offspring Provisioning, and the Evolution of Long Postmenopausal Life Spans," *Current Anthropology* 38 (1997): 551–77; Jared Diamond, "Unwritten Knowledge," *Nature* 410 (2001): 521; L. L. Carstensen and C. E. Lockenhoff, "Aging, Emotion, and Evolution: The Bigger Picture," *Annals of the New York Academy of Sciences* 1000 (2003): 152–79.
43. Provenza, *Foraging Behavior*; Frederick D. Provenza, "What Does It Mean to Be Locally Adapted and Who Cares Anyway?" *Journal of Animal Science* 86 (2008): E271–E284.
44. Frederick D. Provenza, "Twenty-Five Years of Paradox in Plant-Herbivore Interactions and 'Sustainable' Grazing Management," *Rangelands* 25 (2003): 4–15.
45. Ryan A. Shaw, "Social Organization and Decision Making in North American Bison: Implications for Management," (PhD diss. Utah State University, 2012).
46. Elizabeth Burritt et al., "Finishing Bison by Offering a Choice of Feeds and Room to Roam," *Journal of the NACAA* 6 (2013): 1–6.
47. Paul Rozin, "Social Learning about Food by Humans," in *Social Learning: Psychological and Biological Perspectives* T. R. Zentall and B. G. Galef, Jr. eds. (Hillsdale: Lawrence Erlbaum Associates, 1988), 165–87; Paul Rozin, "Sociocultural Influences on Human Food Selection," in *Why We Eat What We Eat: The Psychology of Eating* E. D. Capaldi ed. (Washington, DC: American Psychological Association, 1996), 233–63.
48. Campbell and Campbell II, *The China Study*; Taubes, *Good Calories Bad Calories*; Teicholz, *The Big Fat Surprise*.
49. J. C. Brand Miller and S. Colagiuri, "Evolutionary Aspects of Diet and Insulin Resistance," *World Review of Nutrition and Dietetics* 84 (1999): 74–105; L. Cordain et al., "Macronutrient Estimations in Hunter-Gatherer Diets," *American Journal of Clinical Nutrition* 72 (2000): 1589–90; L. Cordain et al., "Plant-Animal Subsistence Ratios and Macronutrient Energy Estimations in Worldwide Hunter-Gatherer Diets," *American Journal of Clinical Nutrition* 71 (2000): 682–92; K. Milton, "The Critical Role Played by Animal Source Foods in Human (*Homo*) Evolution," *Journal of Nutrition* 133 (2003): 3886S–92; A. Ströhle and A. Hahn, "Diets of Modern Hunter-Gatherers Vary Substantially in Their Carbohydrate Content Depending on Ecoenvironments: Results from an Ethnographic Analysis," *Nutrition Research* 31 (2011): 429–35.
50. Julie Mennella, "Ontogeny of Taste Preferences: Basic Biology and Implications for Health," *American Journal of Clinical Nutrition* 99 (2014): 704S–11; A. K. Ventura and J. Worobey, "Early Influences on the Development of Food Preferences," *Current Biology* 23 (2013): R401–8; Frederick D. Provenza et al., "Our Landscapes, Our Livestock, Ourselves: Restoring Broken Linkages among Plants, Herbivores, and Humans with Diets That Nourish and Satiate," *Appetite* 95 (2015): 500–519.
51. J. A. Mennella and J. C. Trabulsi, "Complementary Foods and Flavor Experiences: Setting the Foundation," *Annals of Nutrition and Metabolism* 60 (2012): 40–50.
52. R. Wrangham and N. L. Conklin-Brittain, "'Cooking as a Biological Trait'," *Comparative Biochemistry and Physiology Part A* 136 (2003): 35–46; G. A. Armelagos, "Brain Evolution, the Determinates of Food Choice, and the Omnivore's Dilemma," *Critical Reviews in Food Science and Nutrition* 54 (2014): 1330–41.
53. Clara M. Davis, "Self-Selection of a Diet by Newly Weaned Infants," *American Journal Disabled Children* 36 (1928): 651–79.
54. Clara M. Davis, "Results of the Self-Selection of Diets by Young Children," *Canadian Medical Association Journal* 41 (1939): 257–61.

55. Stephan Strauss, "Clara M. Davis and the Wisdom of Letting Children Choose Their Own Diets," *Canadian Medical Association Journal* 175 (2006): 1199–201.

56. A. Olsen et al., "Early Origins of Overeating: Early Habit Formation and Implications for Obesity in Later Life," *Current Obesity Reports* 2 (2013): 157–64; Edward Archer, "The Childhood Obesity Epidemic as a Result of Nongenetic Evolution: The Maternal Resources Hypothesis," *Mayo Clinic Proceedings* 90 (2015): 77–92; A. Gonzalez-Bulnes et al., "Transgenerational Inheritance in the Offspring of Pregnant Women with Metabolic Syndrome," *Current Pharmaceutical Biotechnology* 15 (2014): 13–23.

57. B. E. Levin and E. Govek, "Gestational Obesity Accentuates Obesity in Obesity Prone Progeny," *American Journal of Physiology* 275 (1998): R1374–79; B. E. Levin, "The Obesity Epidemic: Metabolic Imprinting on Genetically Susceptible Neural Circuits," *Obesity Research* 8 (2000): 342–47; M. S. Martin-Gronert and S. E. Ozanne, "Programming of Appetite and Type 2 Diabetes," *Early Human Development* 81 (2005): 981–88; I. C. McMillen and J. S. Robinson, "Developmental Origins of the Metabolic Syndrome: Prediction, Plasticity, and Programming," *Physiological Reviews* 85 (2005): 571–633; A. Plagemann, "Perinatal Nutrition and Hormone-Dependent Programming of Food Intake," *Hormone Research* 65 (2006): 83–89; P. D. Taylor and L. Poston, "Developmental Programming of Obesity in Mammals," *Experimental Physiology* 92 (2007): 287–98; P. D. Gluckman et al., "Effect of in Utero and Early-Life Conditions on Adult Health and Disease," *New England Journal of Medicine* 359 (2008): 61–73; P. D. Gluckman et al., "Predictive Adaptive Responses and Human Evolution," *Trends in Ecology and Evolution* 20 (2005): 527–33; P. Iozzo et al., "Developmental ORIgins of Healthy and Unhealthy AgeiNg: The Role of Maternal Obesity – Introduction to DORIAN," *Obesity Facts* 7 (2014): 130–51.

58. J. G. Kral et al., "Large Maternal Weight Loss from Obesity Surgery Prevents Transmission of Obesity to Children Who Were Followed for 2 to 18 Years," *Pediatrics* 118 (2006): 1644–49; J. Smith et al., "Effects of Maternal Surgical Weight Loss in Mothers on Intergenerational Transmission of Obesity," *Journal of Clinical Endocrinology and Metabolism*, 94 (2009): 4275–83.

59. L. E. Grivetti, "Edible Wild Plants as Food and as Medicine: Reflections on Thirty Years of Fieldwork," in *Eating and Healing: Traditional Food as Medicine*, A. Pieroni and L. L. Price eds. (New York: The Haworth Press, 2006), 11–38.

60. A. Drewnowski and S. E. Specter, "Poverty and Obesity: The Role of Energy Density and Energy Costs," *American Journal of Clinical Nutrition* 79 (2004): 6–16; A. Drewnowski and N. Darmon, "The Economics of Obesity: Dietary Energy Density and Energy Cost," *American Journal of Clinical Nutrition* 82 (2005): 265S–73; A. Drewnowski and N. Darmon, "Food Choices and Diet Costs: An Economic Analysis," *Journal of Nutrition* 135 (2005): 900–4; B. M. Popkin et al., "NOW AND THEN: The Global Nutrition Transition: The Pandemic of Obesity in Developing Countries," *Nutrition Reviews* 70 (2012): 3–21.

## Chapter 12: How to Poison a Rat, Cow, or Human

1. Peter R. Cheeke, *Natural Toxicants in Feeds, Forages, and Poisonous Plants* (Danville: Interstate Publishing, 1998).

2. Frederick D. Provenza et al., "Antiemetic Drugs Attenuate Food Aversions in Sheep," *Journal of Animal Science* 72 (1994): 1989–94.

3. Elizabeth A. Burritt and Frederick D. Provenza, "Ability of Lambs to Learn With a Delay between Food Ingestion and Consequences Given Meals Containing Novel and Familiar Foods," *Applied Animal Behaviour Science* 32 (1991): 179–89.

4. Jon Krakauer, "How Chris McCandless Died: An Update," *The New Yorker*, February 11, 2015.

5. John Garcia et al., "A General Theory of Aversion Learning," in *Experimental Assessments and Clinical Applications of Conditioned Food Aversions*, N.S. Braveman and P. Bronstein, eds. (New York: New York Academy of Science, 1985), 8–21.

6. A. Etemadi et al., "Mortality from Different Causes Associated with Meat, Heme Iron, Nitrates, and Nitrites in the NIH-AARP Diet and Health Study: Population Based Cohort Study," *BMJ* (2017): 357.
7. E. M. Schulte et al., "Which Foods May Be Addictive? The Roles of Processing, Fat Content, and Glycemic Load," *PLoS ONE* 10 (2) (2015): e0117959.
8. J. A. Welsh et al., "Consumption of Added Sugars Is Decreasing in the United States," *American Journal of Clinical Nutrition* 94 (2011): 726–34.
9. B. J. Tepper and R. B. Kanarek, "Dietary Self-Selection Patterns of Rats with Mild Diabetes," *The Journal of Nutrition* 115 (1985): 699–709.
10. J. M. Hanna and C. A. Hornick, "Use of Coca Leaf in Southern Peru: Adaptation or Addiction," *Bulletin on Narcotics* 29 (1977): 63–74.
11. K. Verebey and M. S. Gold, "From Coca Leaves to Crack: The Effects of Dose and Routes of Administration in Abuse Liability," *Psychiatric Annals* 18 (1988): 513–20.
12. Nicole M. Avena et al., "Further Developments in the Neurobiology of Food and Addiction: Update on the State of the Science," *Nutrition* 28 (2012): 341–43.
13. Margaret L. Westwater et al., "Sugar Addiction: The State of the Science," *European Journal of Nutrition* 55, Suppl 2 (2016): S55–69.
14. Nicole M. Avena et al., "Evidence for Sugar Addiction: Behavioral and Neurochemical Effects of Intermittent, Excessive Sugar Intake," *Neuroscience and Biobehavioral Reviews* 32 (2008): 20–39.
15. Nicole M. Avena et al., "Sugar and Fat Bingeing Have Notable Differences in Addictive-Like Behavior," *Journal of Nutrition* 139 (2009): 623–28.
16. Z. Vucetic et al., "Maternal High-Fat Diet Alters Methylation and Gene Expression of Dopamine and Opioid-Related Genes," *Endocrinology* 151 (2010): 4756–64; B. M. Geiger et al., "Deficits of Mesolimbic Dopamine Neurotransmission in Rat Dietary Obesity," *Neuroscience* 159 (2009): 1193–99.
17. S. U. Maier et al., "Acute Stress Impairs Self-Control in Goal-Directed Choice by Altering Multiple Functional Connections within the Brain's Decision Circuits," *Neuron* 87 (2015): 621–31.
18. T. C. Adam and E. S. Epel, "Stress, Eating and the Reward System," *Physiology and Behavior* 91 (2007): 449–58.
19. K. L. Teff et al., "Dietary Fructose Reduces Circulating Insulin and Leptin, Attenuates Postprandial Suppression of Ghrelin, and Increases Triglycerides in Women," *Journal of Clinical Endocrinology and Metabolism* 89 (2004): 2963–72; K. L. Teff et al., "Endocrine and Metabolic Effects of Consuming Fructose- and Glucose-Sweetened Beverages with Meals in Obese Men and Women: Influence of Insulin Resistance on Plasma Triglyceride Responses," *Journal of Clinical Endocrinology and Metabolism* 94 (2009): 1562–69.
20. P. J. Havel et al., "High-Fat Meals Reduce 24-H Circulating Leptin Concentrations in Women," *Diabetes* 48 (1999): 334–41.
21. A. Drewnowski and N. Darmon, "The Economics of Obesity: Dietary Energy Density and Energy Cost," *American Journal of Clinical Nutrition* 82 (2005): 265S–73.
22. David S. Ludwig and Mark I. Friedman, "Increasing Adiposity: Consequence or Cause of Overeating?" *Journal of the American Medical Association* 311 (2014): 2167–68.
23. B. S. Lennerz et al., "Effects of Dietary Glycemic Index on Brain Regions Related to Reward and Craving in Men," *American Journal of Clinical Nutrition* 98 (2013): 641–47.
24. B. A. Swinburn et al., "The Global Obesity Pandemic: Shaped by Global Drivers," *Lancet* 378 (2011): 804–14.
25. Jeffrey Sachs, "Mick Mulvaney Delivers the Chilling Truth," CNN, April 25, 2018.
26. Brian Wansink, *Mindless Eating: Why We Eat More Than We Think* (New York: Bantam Books, 2007); Brian Wansink, *Slim by Design: Mindless Eating Solutions for Everyday Life* (New York:

HarperCollins, 2014); William D. Pierce and Carl D. Cheney, *Behavior Analysis and Learning: A Biobehavioral Approach* (New York: Taylor & Francis, 2017).

## Chapter 13: When Authority Trumps Wisdom

1. Michael F. Holick, "Vitamin D Deficiency," *New England Journal of Medicine* 357 (2007): 266–81.
2. Michael F. Holick et al., "Evaluation, Treatment, and Prevention of Vitamin D Deficiency: An Endocrine Society Clinical Practice Guideline," *Journal of Clinical Endocrinology and Metabolism* 96 (2011): 1911–30.
3. R. Scragg et al., "Effect of Monthly High-Dose Vitamin D Supplementation on Cardiovascular Disease in the Vitamin D Assessment Study a Randomized Clinical Trial," *Journal of the American Medical Association Cardiology* (2017).
4. J. Lappe et al., "Effect of Vitamin D and Calcium Supplementation on Cancer Incidence in Older Women: A Randomized Clinical Trial," *Journal of the American Medical Association* 317 (2017): 1234–43.
5. S. L. McDonnell et al., "Serum 25-Hydroxyvitamin D Concentrations 40 Ng/ Ml Are Associated with >65% Lower Cancer Risk: Pooled Analysis of Randomized Trial and Prospective Cohort Study," *PLoS ONE* 11(4) (2016): e0152441.
6. M. Leshem et al., "Calcium Taste Preference and Sensitivity in Humans: II. Hemodialysis Patients," *Physiology and Behavior* 78 (2003): 409–14.
7. F. R. Greer et al., "Elevated Serum Parathyroid Hormone, Calcitonin, and 1,25-Dihydroxyvitamin D in Lactating Women Nursing Twins," *American Journal of Clinical Nutrition* 40 (1984): 562–68.
8. M. G. Tordoff, "The Case for a Calcium Appetite in Humans," in *Calcium in Human Health*, C. M. Weaver and R. P. Heaney eds. (Totowa, NJ: Humana Press, 2006), 247-2–66.
9. R. Brommage and H. F. DeLuca, "Self-Selection of a High Calcium Diet by Vitamin D-Deficient Lactating Rats Increases Food Consumption and Milk Production," *Journal of Nutrition* 114 (1984): 1377–85.
10. Michael G. Tordoff et al., "Calcium Depravation Increases Salt Intake," *American Journal of Physiology* 259 (1990): R411–19; Juan J. Villalba et al., "Learned Appetites for Calcium, Phosphorus, and Sodium in Sheep," *Journal of Animal Science* 86 (2008): 738–47.
11. F. M. Sacks et al., on behalf of the American Heart Association, "Dietary Fats and Cardiovascular Disease a Presidential Advisory from the American Heart Association," *Circulation* 136 (2017): e1-e23; A. Ascherio and W. C. Willett, "Health Effects of Trans Fatty Acids," *American Journal of Clinical Nutrition* 66 (1997): 1006S–10; F. A. Kummerow, "The Negative Effects of Hydrogenated Trans Fats and What to Do about Them," *Atherosclerosis* 205 (2009): 458–65.
12. Frank B. Hu et al., "Types of Dietary Fat and Risk of Coronary Heart Disease: A Critical Review," *Journal of the American College of Nutrition* 20 (2001): 5–19.
13. M. U. Jakobsen et al., "Major Types of Dietary Fat and Risk of Coronary Heart Disease: A Pooled Analysis of 11 Cohort Studies," *American Journal of Clinical Nutrition* 89 (2009): 1425–32.
14. P. W. Siri-Tarino et al., "Meta-Analysis of Prospective Cohort Studies Evaluating the Association of Saturated Fat with Cardiovascular Disease," *American Journal of Clinical Nutrition* 91 (2010): 535–46; R. Chowdhury et al., "Association of Dietary, Circulating, and Supplement Fatty Acids with Coronary Risk: A Systematic Review and Meta-Analysis," *Annals of Internal Medicine* 160 (2014): 398–406.
15. Frank B. Hu et al., "Types of Dietary Fat and Risk of Coronary Heart Disease: A Critical Review," *Journal of the American College of Nutrition* 20 (2001): 5–19; M. U. Jakobsen et al., "Major Types of Dietary Fat and Risk of Coronary Heart Disease: A Pooled Analysis of 11 Cohort Studies," *American Journal of Clinical Nutrition* 89 (2009): 1425–32; Y. Li et al., "Saturated Fats Compared with Unsaturated Fats and Sources of Carbohydrates in Relation to Risk of Coronary Heart Disease: A Prospective Cohort Study," *Journal of the American College of Cardiology* 66 (2015): 1538–48.

16. D. Mozaffarian et al., "Effects on Coronary Heart Disease of Increasing Polyunsaturated Fat in Place of Saturated Fat: A Systematic Review and Meta-Analysis of Randomized Controlled Trials," *PLoS Medicine* 7 (2010):e1000252.
17. L. Hooper et al., "Reduction in Saturated Fat Intake for Cardiovascular Disease," *Cochrane Database Systematic Review* CD011737 (2015).
18. Frank B. Hu, "Are Refined Carbohydrates Worse Than Saturated Fat?" *American Journal of Clinical Nutrition* 91 (2010): 1541–42; David S. Ludwig and Walter C. Willett, "Three Daily Servings of Reduced-Fat Milk: An Evidence-Based Recommendation?" *Journal of the American Medical Association Pediatrics* 167 (2013): 788–89.
19. M. U. Jakobsen et al., "Intake of Carbohydrates Compared with Intake of Saturated Fatty Acids and Risk of Myocardial Infarction: Importance of the Glycemic Index," *American Journal of Clinical Nutrition* 91 (2010): 1764–68.
20. M. Dehghan et al., on behalf of the Prospective Urban Rural Epidemiology (PURE) study investigators, "Associations of Fats and Carbohydrate Intake with Cardiovascular Disease and Mortality in 18 Countries from Five Continents (PURE): A Prospective Cohort Study," *The Lancet* (2017).
21. R. H. Eckel et al., "2013 AHA/ACC Guideline on Lifestyle Management to Reduce Cardiovascular Risk: A Report of the American College of Cardiology/American Heart Association Task Force on Practice Guidelines Circulation," 129 (2014): S76–S99. doi:10.1161/01.cir.0000437740.48606.d1. [Published Corrections Appear in *Circulation* 129 (2014): S100–01 and *Circulation* (2015): 131 e326.]
22. C. D. Rehm et al., "Dietary Intake among US Adults, 1999–2012," *Journal of the American Medical Association* 315 (2016): 2542–53.
23. Ansel Keys, *Seven Countries: A Multivariate Analysis of Death and Coronary Heart Disease* (Cambridge: Harvard University Press, 1980).
24. Jane B. Morgan et al., "Healthy Eating for Infants – Mothers' Attitudes," *Acta Paediatrica* 84 (1995): 512–15.
25. Paul Rozin et al., "Lay American Conceptions of Nutrition: Dose Insensitivity, Categorical Thinking, Contagion, and the Monotonic Mind," *Health Psychology* 15 (1996): 438–47.
26. K. Hardy et al., "The Importance of Dietary Carbohydrate in Human Evolution," *The Quarterly Review of Biology* 90 (2015): 251–68.
27. R. C. Muhlbauer et al., "Onion and a Mixture of Vegetables, Salads, and Herbs Affect Bone Resorption in the Rat by a Mechanism Independent of Their Base Excess," *Journal of Bone and Mineral Research* 17 (2002): 1230–36; R. C. Muhlbauer et al., "Common Herbs, Essential Oils, and Monoterpenes Potently Modulate Bone Metabolism," *Bone* 32 (2003): 372–80; S. E. Putnam et al., "Natural Products as Alternative Treatments for Metabolic Bone Disorders and for Maintenance of Bone Health," *Phytotherapy Research* 21 (2007): 99–112.
28. Jeff Leach, "Please Pass the Microbes," Human Food Project, October 2, 2013.
29. D. A. McCarron et al., "Can Dietary Sodium Intake Be Modified by Public Policy?" *Clinical Journal of the American Society of Nephrology* 4 (2009): 1878–82.
30. US Department of Health and Human Services and USDA, "Dietary Guidelines for Americans, 2015–2020," 8th Edition (2015). Available at https://health.gov/dietaryguidelines/2015/resources/2015-2020_Dietary_Guidelines.pdf.
31. Institute of Medicine, Food and Nutrition Board, *Dietary Reference Intakes for Water, Potassium, Sodium, Chloride, and Sulfate*, (2004): Available at https://www.nap.edu/read/10925/chapter/8#386.
32. Curt P. Richter, "Increased Salt Appetite in Adrenalectomized Rats," *American Journal of Physiology* 115 (1936): 155–61.
33. J. C. Geerling and A. D. Loewy, "Central Regulation of Sodium Appetite," *Experimental Physiology* 93 (2008): 177–209.

34. Intersalt Cooperative Research Group, "Intersalt: An International Study of Electrolyte Excretion and Blood Pressure. Results for 24 Hour Sodium and Potassium Excretion," *British Medical Journal* 297 (1988): 319–28.
35. B. F. Zhou et al., "INTERMAP Research Group: Nutrient Intakes of Middle-Aged Men and Women in China, Japan, United Kingdom, and United States in the Late 1990s: The INTERMAP Study," *Journal of Human Hypertension* 17 (2003): 623–30; K. T. Khaw et al., "Blood Pressure and Urinary Sodium in Men and Women: The Norfolk Cohort of the European Prospective Investigation into Cancer (EPIC-Norfolk)," *American Journal of Clinical Nutrition* 80 (2004): 1397–403.
36. Trials of Hypertension Prevention Collaborative Research Group, "Effects of Weight Loss and Sodium Reduction Intervention on Blood Pressure and Hypertension Incidence in Overweight People with High Normal Blood Pressure. The Trials of Hypertension Prevention, Phase II," *Archives of Internal Medicine* 157 (1997): 657–67.
37. D. A. McCarron et al., "Blood Pressure and Metabolic Responses to Moderate Sodium Restriction in Isradipine-Treated Hypertensive Patients," *American Journal of Hypertension* 10 (1997): 68–76; D.A. McCarron, "Dietary Sodium and Cardiovascular and Renal Disease Risk Factors: Dark Horse or Phantom Entry?" *Nephrology Dialysis Transplantation* 23 (2008): 2133–37.
38. David A. McCarron, "What Determines Human Sodium Intake: Policy or Physiology?" *Advances in Nutrition* 5 (2014): 578–84.

## Chapter 14: When Beliefs Trump Authority

1. W. J. Lee et al., "You Taste What You See: Do Organic Labels Bias Taste Perceptions?" *Food Quality Preference* 29 (2013): 33–39; P. Sörqvist et al., "Who Needs Cream and Sugar When There Is Eco-Labeling? Taste and Willingness to Pay for 'Eco-Friendly' Coffee," *PLoS ONE* 8, no. 12 (2013): e80719.; B. Bratanova et al., "Savouring Morality. Moral Satisfaction Renders Food of Ethical Origin Subjectively Tastier," *Appetite* 91 (2015): 137–49; F. Grabenhorst et al., "How Cognition Modulates Affective Responses to Taste and Flavor: Top-Down Influences on the Orbitofrontal and Pregenual Cingulate Cortices," *Cerebral Cortex* 18 (2008): 1549–59; H. Plassmann et al., "Marketing Actions Can Modulate Neural Representations of Experienced Pleasantness," *Proceedings National Academy of Science* 105 (2008): 1050–54.
2. E. C. Anderson and L. F. Barrett, "Affective Beliefs Influence the Experience of Eating Meat," *PLoS ONE* 11, no. 8 (2016): e0160424.
3. J. Ng et al., "An fMRI Study of Obesity, Food Reward, and Perceived Caloric Density. Does a Low-Fat Label Make Food Less Appealing?" *Appetite* 57 (2011): 65–72; A. J. Crum et al., "Mind over Milkshakes: Mind-Sets, Not Just Nutrients, Determine Ghrelin Response," *Health Psychology* 30 (2011): 424–29.
4. J. Molina-Infante et al., "Systematic Review: Noncoeliac Gluten Sensitivity," *Alimentary Pharmacology and Therapeutics* 41 (2015): 807–20.
5. Y. Junker et al., "Wheat Amylase Trypsin Inhibitors Drive Intestinal Inflammation via Activation of Toll-Like Receptor 4," *Journal of Experimental Medicine* 209 (2012): 2395–408; J. R. Biesiekierski et al., "No Effects of Gluten in Patients with Self-Reported Non-Celiac Gluten Sensitivity after Dietary Reduction of Fermentable, Poorly Absorbed, Short-Chain Carbohydrates," *Gastroenterology* 145 (2013): 320–28.
6. H. C. van den Broeck et al., "Presence of Celiac Disease Epitopes in Modern and Old Hexaploid Wheat Varieties: Wheat Breeding May Have Contributed to Increased Prevalence of Celiac Disease," *Theoretical and Applied Genetics* 121 (2010): 1527–39.
7. F. Sofi et al., "Effect of *Triticum turgidum* subsp. *turanicum* Wheat on Irritable Bowel Syndrome: A Double-Blinded Randomised Dietary Intervention Trial," *British Journal of Nutrition* 111 (2014): 1992–99.

8. Anna Sapone et al., "Spectrum of Gluten-Related Disorders: Consensus on New Nomenclature and Classification," *BMC Medicine* 10 (2012): 13; F. J. P. H. Brouns et al., "Does Wheat Make Us Fat and Sick?" *Journal of Cereal Science* 58 (2013): 209–15.
9. J. R. Biesiekierski et al., "No Effects of Gluten in Patients with Self-Reported Non-Celiac Gluten Sensitivity after Dietary Reduction of Fermentable, Poorly Absorbed, Short-Chain Carbohydrates," *Gastroenterology* 145 (2013): 320–28.
10. J. Kong et al., "Brain Activity Associated with Expectancy-Enhanced Placebo Analgesia as Measured by Functional Magnetic Resonance Imaging," *Journal of Neuroscience* 26 (2006): 381–88; F. Benedetti, "Mechanisms of Placebo and Placebo-Related Effects across Diseases and Treatments," *Annual Review of Pharmacology and Toxicology* 48 (2008): 33–60; J. Kong et al., "Placebo Analgesia: Findings from Brain Imaging Studies and Emerging Hypotheses," *Review of Neuroscience* 18 (2007): 173–90; J. Kong et al., "A Functional Magnetic Resonance Imaging Study on the Neural Mechanisms of Hyperalgesic Nocebo Effect," *Journal of Neuroscience* 28 (2008): 13354–62.
11. Y. Nestoriuc et al., "Is It Best to Expect the Worst? Influence of Patients' Side-Effect Expectations on Endocrine Treatment Outcome in a 2-Year Prospective Clinical Cohort Study," *Annals of Oncology* (2016): mdw266.
12. A. Tinnermann et al., "Interactions between Brain and Spinal Cord Mediate Value Effects in Nocebo Hyperalgesia," *Science* 358 (2017): 105–8.
13. Andrew Weil, "Healing, Nature, and Modern Medicine," in *Ecological Medicine: Healing the Earth, Healing Ourselves* K. Ausubel and J. P. Harpignies eds. (San Francisco: Sierra Club Books, 2004), 109–24.
14. K. Jensen et al., "Classical Conditioning of Analgesic and Hyperalgesic Pain Responses without Conscious Awareness," *Proceedings of the National Academy of Sciences* 112 (2015): 7863–67.
15. Paul Rozin, "Disorders of Food Selection: The Compromise of Pleasure," *Annals of the New York Academy of Sciences* 575 (1989): 376–85; Paul Rozin, "Towards a Psychology of Food and Eating: From Motivation to Model to Meaning, Morality and Metaphor," *Current Directions in Psychological Science* 5 (1996): 1–7.
16. A. Drewnowski et al., "Diet Quality and Dietary Diversity in France: Implications for the French Paradox," *Journal of the American Dietetic Association* 96 (1996): 663–69.
17. R. J. Samuelson, *The Economist Book of Vital World Statistics* (New York: Random House, 1990).
18. S. Renaud and M. de Lorgeril, "Wine, Alcohol, Platelets, and the French Paradox for Coronary Heart Disease," *Lancet* 339 (1992): 1523–26; M. H. Criqui and B. L. Ringel, "Does Diet or Alcohol Explain the French Paradox?" *Lancet* 344 (1994): 1719–23.
19. Paul Rozin et al., "Attitudes to Food and the Role of Food in Life in the U.S.A., Japan, Flemish Belgium and France: Possible Implications for the Diet–Health Debate," *Appetite* 33 (1999): 163–80;
20. P. Netter, "Health and Pleasure," in *Pleasure and Quality of Life*, D. M. Warburton and N. Sherwood eds. (Chichester: John Wiley, 1996), 81–89.
21. T. J. Kaptchuk et al., "Components of the Placebo Effect: A Randomized Controlled Trial in Irritable Bowel Syndrome," *British Medical Journal* 336 (2008): 999–1003.
22. Michelle L. Dossett et al., "Patient-Provider Interactions Affect Symptoms in Gastroesophageal Reflux Disease: A Pilot Randomized, Double-Blind, Placebo-Controlled Trial," *PLoS ONE* 10, no. 9 (2015).
23. M. E. Wechsler et al., "Active Albuterol or Placebo, Sham Acupuncture, or No Intervention in Asthma," *New England Journal of Medicine* 365 (2011): 119–26.
24. T. J. Kaptchuk et al., "Placebos without Deception: A Randomized Controlled Trial in Irritable Bowel Syndrome," *PLoS ONE* 5, no. 12 (2010): e15591.
25. S. Kam-Hansen et al., "Altered Placebo and Drug Labeling Changes the Outcome of Episodic Migraine Attacks," *Science Translational Medicine* 6, no. 218 (2014): 218ra5.

26. D. G. Finniss et al., "Placebo Effects: Biological, Clinical and Ethical Advances," *Lancet* 375 (2010): 686–95; F. G. Miller et al., "The Placebo Effect: Illness and Interpersonal Healing," *Perspectives in Biology and Medicine* 52 (2009): 518–39; K. T. Hall et al., "Genetics and the Placebo Effect: The Placebome," *Trends in Molecular Medicine* 21 (2015): 285–94; T. J. Kaptchuk and F. G. Miller, "Placebo Effects in Medicine," *New England Journal of Medicine* 373 (2015): 8–9; V. Napadow et al., "The Imagined Itch: Brain Circuitry Supporting Nocebo-Induced Itch in Atopic Dermatitis Patients," *Allergy* (2015).
27. JoBeth McDaniel, "Beating the Odds: Cancer Outliers," AARP Bulletin, March 2016.
28. C. P. Le et al., "Chronic Stress in Mice Remodels Lymph Vasculature to Promote Tumour Cell Dissemination," *Nature Communications* 7 (2016): 10634.
29. E. C. Prophet and E. L. Prophet, *Reincarnation: The Missing Link in Christianity* (Gardiner: Summit University Press, 1997).
30. Capra, *The Tao of Physics*, 143.

## Chapter 15: When Understanding Trumps Beliefs

1. D. Mozaffarian and D. S. Ludwig, "The 2015 US Dietary Guidelines: Lifting the Ban on Total Dietary Fat," *Journal of the American Medical Association* 313 (2015): 2421–22.
2. Gary Taubes, "Why Nutrition Is So Confusing," *New York Times*, February 8, 2014.
3. David S. Ludwig and Mark I. Friedman, "Increasing Adiposity: Consequence or Cause of Overeating?" *Journal of the American Medical Association* 311 (2014): 2167–68.
4. Kevin D. Hall et al., "Calorie for Calorie, Dietary Fat Restriction Results in More Body Fat Loss Than Carbohydrate Restriction in People with Obesity," *Cell Metabolism* 22 (2015): 427–36; Kevin D. Hall et al., "Energy Expenditure and Body Composition Changes after an Isocaloric Ketogenic Diet in Overweight and Obese Men," *American Journal of Clinical Nutrition* 104 (2016): 324–33; Kevin D. Hall and J. Guo, "Obesity Energetics: Body Weight Regulation and the Effects of Diet Composition," *Gastroenterology* 152 (2017): 1718–27.
5. M. J. Leach and S. Kumar, "Cinnamon for Diabetes Mellitus," *Cochrane Database of Systematic Reviews* 9 (2012): CD007170.
6. R. W. Allen et al., "Cinnamon Use in Type 2 Diabetes: An Updated Systematic Review and Meta-Analysis," *Annals of Family Medicine* 11 (2013): 452–59.
7. C. P. Wild, "The Exposome: From Concept to Utility," *International Journal of Epidemiology* 41 (2012): 24–32.
8. J. P. A. Ioannidis, "Why Most Published Research Findings Are False," *PLoS Medicine* 2, no. 8 (2005): e124.
9. J. J. Lara et al., "Intentional Mis-Reporting of Food Consumption and Its Relationship with Body Mass Index and Psychological Scores in Women," *Journal of Human Nutrition and Dietetics* 17 (2004): 209–18.
10. Edward Archer et al., "A Discussion of the Refutation of Memory-Based Dietary Assessment Methods (M-BMs): The Rhetorical Defense of Pseudoscientific and Inadmissible Evidence," *Mayo Clinic Proceedings* 90 (2015): 1736–38.
11. Edward Archer et al., "Validity of U.S. Nutritional Surveillance: National Health and Nutrition Examination Survey Caloric Energy Intake Data, 1971–2010," *PLoS ONE* 8 (2013): e76632; Edward Archer et al., "The Inadmissibility of What We Eat in America and NHANES Dietary Data in Nutrition and Obesity Research and the Scientific Formulation of National Dietary Guidelines," *Mayo Clinic Proceedings* 90 (2015): 911–26.
12. H. O. Bang et al., "Plasma Lipid and Lipoprotein Pattern in Greenlandic West-Coast Eskimos," *Lancet* 297 (1971): 1143–46; H. O. Bang and J. Dyerberg, "Plasma Lipids and Lipoproteins in Greenlandic West Coast Eskimos," *Acta Med Scandinavica* 192 (1972): 85–94; H. O. Bang et al.,

"The Composition of Food Consumed by Greenland Eskimos," *Acta Medica Scandinavica* 200 (1976): 69–73.

13. J. George Fodor et al., "'Fishing' for the Origins of the 'Eskimos and Heart Disease' Story: Facts or Wishful Thinking?" *Canadian Journal of Cardiology* 30 (2014): 864–68.
14. Cynthia A. Daley et al., "A Review of Fatty Acid Profiles and Antioxidant Content in Grass-Fed and Grain-Fed Beef," *Nutrition Journal* 9 (2010); C. M. Benbrook et al., "Organic Production Enhances Milk Nutritional Quality by Shifting Fatty Acid Composition: A United States–Wide, 18-Month Study," *PLoS ONE*, 8, no. 12, (2013): e82429.; A. Elgersma, "Grazing Increases the Unsaturated Fatty Acid Concentration of Milk from Grass-Fed Cows: A Review of the Contributing Factors, Challenges and Future Perspectives," *European Journal of Lipid Science and Technology* 117 (2015): 1345–69.
15. T. L. Blasbalg et al., "Changes in Consumption of Omega-3 and Omega-6 Fatty Acids in the United States during the 20th Century," *American Journal of Clinical Nutrition* 93 (2011): 950–62.
16. A. Simopoulos, "Omega-3 Fatty Acids in Health and Disease and in Growth and Development," *American Journal of Clinical Nutrition* 54 (1991): 438–63.
17. G. H. Johnson and K. Fritsche, "Effect of Dietary Linoleic Acid on Markers of Inflammation in Healthy Persons: A Systematic Review of Randomized Controlled Trials," *Journal of the Academy of Nutrition and Dietetics* 112 (2012): 1029–41; C. E. Forsythe et al., "Comparison of Low Fat and Low Carbohydrate Diets on Circulating Fatty Acid Composition and Markers of Inflammation," *Lipids* 43 (2008): 65–77.
18. J. H. Y. Wu et al. and Cohorts for Heart and Aging Research in Genomic Epidemiology (CHARGE) Fatty Acids and Outcomes Research Consortium (FORCE), "Omega-6 Fatty Acid Biomarkers and Incident Diabetes: Pooled Analysis of Individual-Level Data for 39740 Adults from 20 Prospective Cohort Studies," *The Lancet Diabetes and Endocrinology* 5 (2017): 965–74.
19. L. E. Robinson and V. C. Mazurak, "N-3 Polyunsaturated Fatty Acids: Relationship to Inflammation in Healthy Adults and Adults Exhibiting Features of Metabolic Syndrome," *Lipids* 48 (2013): 319–32.
20. B. B. Albert et al., "Marine Oils: Complex, Confusing, Confounded?" *Journal of Nutrition and Intermediary Metabolism* 5 (2016): 3–10.
21. Ancel Keys, *Seven Countries: A Multivariate Analysis of Death and Coronary Heart Disease* (Cambridge: Harvard University Press, 1980).
22. C. E. Ramsden et al., "N-6 Fatty Acid-Specific and Mixed Polyunsaturated Dietary Interventions Have Different Effects on CHD Risk: A Meta-Analysis of Randomised Controlled Trials," *British Journal of Nutrition* 104 (2010): 1586–1600; C. E. Ramsden et al., "Use of Dietary Linoleic Acid for Secondary Prevention of Coronary Heart Disease and Death: Evaluation of Recovered Data from the Sydney Diet Heart Study and Updated Meta-Analysis," *BMJ* (2013): 346:e8707; C. E. Ramsden et al., "Re-Evaluation of the Traditional Diet-Heart Hypothesis: Analysis of Recovered Data from Minnesota Coronary Experiment (1968–73)," *BMJ* (2016): 353.
23. C. E. Ramsden et al., "Re-Evaluation of the Traditional Diet-Heart Hypothesis: Analysis of Recovered Data from Minnesota Coronary Experiment (1968–73)," *BMJ* (2016): 353.
24. J. E. Bekelman et al., "Scope and Impact of Financial Conflicts of Interest in Biomedical Research: A Systematic Review," *Journal of the American Medical Association* 289 (2003): 454–65; L. I. Lesser et al., "Relationship between Funding Source and Conclusion Among Nutrition-Related Scientific Articles," *PLoS Medicine* 4, no. 1 (2007): e5. D. Fanelli et al., "Meta-Assessment of Bias in Science," *Proceedings National Academy of Sciences* 114 (2017): 3714–19; J. Washburn, *University Inc.: The Corporate Corruption of Higher Education* (New York: Basic Books, 2006).

25. K. L. Stanhope, "Sugar Consumption, Metabolic Disease, and Obesity: The State of the Controversy," *Critical Reviews in Clinical Laboratory Sciences* (2015).
26. C. E. Kearns et al., "Sugar Industry and Coronary Heart Disease Research: A Historical Analysis of Internal Industry Documents," *Journal of the American Medical Association Intern Medicine* (2016): doi:10.1001/jamainternmed.2016.5394; C. E. Kearns et al., "Sugar Industry Sponsorship of Germ-Free Rodent Studies Linking Sucrose to Hyperlipidemia and Cancer: An Historical Analysis of Internal Documents," *PLoS Biology* 15, no. 11 (2017): e2003460.
27. Marie Bes-Rastrollo et al., "Financial Conflicts of Interest and Reporting Bias Regarding the Association between Sugar-Sweetened Beverages and Weight Gain: A Systematic Review of Systematic Reviews," *PLoS Medicine* 10, no. 12 (2013): e1001578; J. Massougbodji et al., "Reviews Examining Sugar-Sweetened Beverages and Body Weight: Correlates of Their Quality and Conclusions," *American Journal of Clinical Nutrition* 99 (2014): 1096–104.
28. Z. Yu et al., "High-Fructose Corn Syrup and Sucrose Have Equivalent Effects on Energy-Regulating Hormones at Normal Human Consumption Levels," *Nutrition Research* 33 (2013): 1043–52; S. Bravo et al., "Consumption of Sucrose and High-Fructose Corn Syrup Does Not Increase Liver Fat or Ectopic Fat Deposition in Muscles," *Applied Physiology, Nutrition and Metabolism* 38 (2013): 681–88.
29. Daniel Sarewitz, "Saving Science," *The New Atlantis* 49 (2016): 4–40.
30. C. G. Begley and J. P. A. Ioannidis, "Reproducibility in Science Improving the Standard for Basic and Preclinical Research," *Circulation Research* 116 (2015): 116–26
31. C. G. Begley and L. M. Ellis, "Drug Development: Raise Standards for Preclinical Cancer Research," *Nature* 483 (2012): 531–33.
32. S. Perrin, "Make Mouse Studies Work," *Nature* 507 (2014): 423–25; J. Seoka et al., and the Inflammation and Host Response to Injury, Large Scale Collaborative Research Program, "Genomic Responses in Mouse Models Poorly Mimic Human Inflammatory Diseases," *Proceedings of the National Academy of Sciences* 110 (2012): 3507–12
33. M. Scudellari, "A Case of Mistaken Identity," *The Scientist* (2016): September 16.
34. A. B. Miller et al., "Twenty Five Year Follow-Up for Breast Cancer Incidence and Mortality of the Canadian National Breast Screening Study: Randomised Screening Trial," *BMJ* (2014): 348.
35. J. Seoka et al., and the Inflammation and Host Response to Injury, Large Scale Collaborative Research Program, "Genomic Responses in Mouse Models Poorly Mimic Human Inflammatory Diseases," *Proceedings of the National Academy of Sciences* 110 (2012): 3507–12.
36. A. Eklunda et al., "Cluster Failure: Why fMRI Inferences for Spatial Extent Have Inflated False-Positive Rates," *Proceedings of the National Academy of Sciences* 113 (2016): 7900–5.
37. Open Science Collaboration, "Estimating the Reproducibility of Psychological Science," *Science* 349 (2015): 943–51.
38. Alvin M. Weinberg, "Science and Trans-Science," *Minerva*, 10, (1972). 209–22.
39. Jerome S. Bruner and Leo Postman, "On the Perception of Incongruity: A Paradigm," *Journal of Personality* 18 (1949): 206–23.
40. John Kenneth Galbraith, *Economics, Peace and Laughter* (Boston: Houghton Mifflin), 50.
41. Frederick D. Provenza et al., "Complex Creative Systems: Principles, Processes, and Practices of Transformation," *Rangelands* 35 (2013): 6–13.

## Chapter 16: The Harmony of Nature

1. Gary Paul Nabhan, "Cultural Parallax in Viewing North American Indians," in *Reinventing Nature? Responses to Postmodern Deconstruction*, M. Soulé and G. Lease, eds. (Washington, DC: Island Press, 1995), 87–101.
2. G. H. Miller et al., "Ecosystem Collapse in Pleistocene Australia and a Human Role in Megafaunal Extinction," *Science* 309 (2005): 287–90.

3. Emma Marris, *Rambunctious Garden: Saving Nature in a Post-Wild World* (New York: Bloomsbury, 2011).
4. Chesterton, *Orthodoxy*, 8.
5. D. Smith and D. Tyers, "The History and Current Status and Distribution of Beavers in Yellowstone National Park," *Northwest Science* 86 (2012): 276–88.
6. Kristin N. Marshall et al., "Interactions among Herbivory, Climate, Topography and Plant Age Shape Riparian Willow Dynamics in National Park, USA," *Journal of Ecology* 102 (2014): 667–77; Ben Goldfarb, *Eager: The Surprising, Secret Life of Beavers and Why They Matter* (White River Junction, VT: Chelsea Green, 2018).
7. Arthur Middleton, "Is the Wolf a Real American Hero?" *New York Times,* March 9, 2014.
8. B. L. Allen et al., "Can We Save Large Carnivores without Losing Large Carnivore Science?" *Food Webs* (2017).
9. Sustainable Human, "How Wolves Change Rivers," narrated by George Monbiot, available at: https://www.youtube.com/user/TheSustainableMan, accessed 15 August 2017.
10. D. Pimentel et al., "Update on the Environmental and Economic Costs Associated with Alien-Invasive Species in the United States," *Ecological Economics* 52 (2005): 273–68.
11. P. Bagla, "Hardy Cotton-Munching Pests Are Latest Blow to GM Crops," *Science* 327 (2010): 1439.
12. Y. Lu et al., "Mirid Bug Outbreaks in Multiple Crops Correlated with Wide-Scale Adoption of Bt Cotton in China," *Science* 328 (2010): 1151–54; S. Wang et al., "Bt Cotton and Secondary Pests," *International Journal of Biotechnology* 10 (2008): 113–21; J. H. Zhao et al., "Benefits of Bt Cotton Counterbalanced by Secondary Pests? Perceptions of Ecological Change in China," *Environmental Monitoring and Assessment* 173 (2010): 985–94.
13. Lucile Muneret et al., "Evidence that Organic Farming Promotes Pest Control," *Nature Sustainability* 1 (2018): 361–368.
14. G. S. Johal and D. M. Huber, "Glyphosate Effects on Diseases of Plants," *European Journal of Agronomy* 31 (2009): 144–52.
15. G. Brookes and P. Barfoot, "Key Environmental Impacts of Global Genetically Modified (GM) Crop Use 1996–2011," *GM Crops and Food – Biotechnology in Agriculture and the Food Chain* 4 (2013): 109–19.
16. Charles Benbrook, "Impacts of Genetically Engineered Crops on Pesticide Use: The First Thirteen Years," *Environmental Sciences Europe* 24 (2012): 24.
17. S. O. Duke and S. B. Powles, "Glyphosate Resistant Crops and Weeds: Now and in the Future," *AgBioForum* 12 (2009): 346–57.
18. I. Heap, "The International Survey of Herbicide Resistant Weeds," Online. Internet. Thursday, January 4, 2018 Available at http://www.weedscience.org.
19. M. R. Behrens et al., "Dicamba Resistance: Enlarging and Preserving Biotechnology-Based Weed Management Strategies," *Science* 316 (2007): 1185–88; T. R. Wright et al., "Robust Crop Resistance to Broadleaf and Grass Herbicides Provided by Aryloxyalkanoate Dioxygenase Transgenes," *Proceedings of the National Academy of Sciences* 107 (2010): 20240–45.
20. S. B. Powles and Q. Yu, "Evolution in Action: Plants Resistant to Herbicides," *Annual Review of Plant Biology* 61 (2010): 317–47.
21. D. A. Mortensen et al., "Navigating a Critical Juncture for Sustainable Weed Management," *BioScience* 62 (2012): 75–84.
22. J. Diels et al., "Association of Financial or Professional Conflict of Interest to Research Outcomes on Health Risks or Nutritional Assessment Studies of Genetically Modified Products," *Food Policy* 36 (2011): 197–203.
23. Gilles-Eric Séralini et al., "Genetically Modified Crops Safety Assessments: Present Limits and Possible Improvements," *Environmental Sciences Europe* 23 (2011): 10.

24. Gilles-Eric Séralini et al., "Long-Term Toxicity of a Roundup Herbicide and a Roundup-Tolerant Genetically Modified Maize" *Food and Chemical Toxicology* 50 (2012): 4221–31. Retracted in *Food and Chemical Toxicology*, 4263, 4244.
25. Gilles-Eric Séralini et al., "Republished Study: Long-Term Toxicity of a Roundup Herbicide and a Roundup-Tolerant Genetically Modified Maize," *Environmental Sciences Europe* 26 (2014): 14.
26. Gilles-Eric Séralini et al., "Conflicts of Interests, Confidentiality and Censorship in Health Risk Assessment: The Example of an Herbicide and a GMO," *Environmental Sciences Europe* 26 (2014): 13.
27. "Retracting Séralini Study Violates Science and Ethics," Science in Society Archive, http://www.i-sis.org.uk/Retracting_Serallini_study_violates_science_and_ethics.php accessed January 4, 2018.
28. David Pimentel et al., "Environmental, Energetic and Economic Comparisons of Organic and Conventional Farming Systems," *BioScience* 55 (2005): 573–82; David Pimentel, "World Food Crisis: Energy and Pests," *Bulletin of the Entomological Society of America* 22 (1976): 20–26.
29. J. Mikola et al., "Biodiversity, Ecosystem Functioning, and Soil Decomposer Food Webs," in *Biodiversity Functioning and Ecosystem Functioning: Synthesis and Perspectives*, M. Loreau et al eds. (New York: Oxford, 2002), 169–80.
30. C. K. Khoury et al., "Increasing Homogeneity in Global Food Supplies and the Implications for Food Security," *Proceedings of the National Academy of Sciences* 111 (2014): 4001–6.
31. C. G. Begley and L. M. Ellis, "Drug Development: Raise Standards for Preclinical Cancer Research," *Nature* 483 (2012): 531–33.
32. D. M. Parkin et al., "The Fraction of Cancer Attributable to Lifestyle and Environmental Factors in the UK in 2010," *British Journal of Cancer* 105 (2011): S1–81.
33. C. Tomasetti and B. Vogelstein, "Cancer Etiology. Variation in Cancer Risk among Tissues Can Be Explained by the Number of Stem Cell Divisions," *Science* 347 (2015): 78–81; C. Tomasetti et al., "Stem Cell Divisions, Somatic Mutations, Cancer Etiology, and Cancer Prevention," *Science* 355 (2017): 1330–34.
34. Joseph A. Tainter, "Problem Solving: Complexity, History, Sustainability," *Population and Environment: A Journal of Interdisciplinary Studies* 22 (2000): 3–41.
35. Rosa Brooks, *How Everything Became War and the Military Became Everything*. (New York: Simon & Schuster, 2016).
36. Joseph Campbell, "Mythologies of War and Peace," in *Myths to Live By* (New York: Viking Penguin, 1973), 174-–206.
37. Smith, *The World's Religions*, Chapter 2.

## Chapter 17: Alice in Wonderland

1. Luis W. Alvarez et al., "Extraterrestrial Cause for the Cretaceous-Tertiary Extinction," *Science* 208 (1980): 1095–108; M. G. Branstetter et al., "Dry Habitats Were Crucibles of Domestication in the Evolution of Agriculture in Ants," *Proceedings of the Royal Society B* 284 (2017): 20170095.
2. W. E. Kunin and K. Gaston, eds. *The Biology of Rarity: Causes and Consequences of Rare–Common Differences* (New York: Springer-Science+Business Media, B.V., 1997).
3. G. Ceballos et al., "Biological Annihilation via the Ongoing Sixth Mass Extinction Signaled by Vertebrate Population Losses and Declines," *Proceedings of the National Academy of Sciences* (2017).
4. M. Newman, "A Model of Mass Extinction," *Journal of Theoretical Biology* 189 (1997): 235–52; N. C. Arens and I. D. West, "Press-Pulse: A General Theory of Mass Extinction?" *Paleobiology* 34 (2008): 456–71; S. L. Pimm et al., "The Biodiversity of Species and Their Rates of Extinction, Distribution, and Protection," *Science* 344, (2014): 1246752-1-12487652-10; M. L. McCallum, "Vertebrate Biodiversity Losses Point to a Sixth Mass Extinction," *Biodiversity and Conservation*

24 (2015): 2497–519; J. M. de Vos et al., "Estimating the Normal Background Rate of Species Extinction," *Conservation Biology* 29 (2014): 452–62; J. H. Lawton and R. M. May, *Extinction Rates* (Oxford: Oxford University Press, 1995); R. Leakey and R. Lewin, *The Sixth Extinction: Patterns of Life and the Future of Humankind* (New York: Anchor Books, 1996); Edward O. Wilson, *The Future of Life* (New York: Vintage Books, 2002).

5. Yinon M. Bar-Ona et al., "The Biomass Distribution on Earth," *Proceedings of the National Academy of Sciences* (2018).
6. D. Lunney et al., "Koalas and Climate Change: A Case Study on the Liverpool Plains, North-West New South Wales," in *Wildlife and Climate Change: Towards Robust Conservation Strategies for Australian Fauna* D. Lunney and P. Hutchings eds. (New South Wales: Royal Zoological Society of New South Wales, 2012), 150–68; G. Gordon et al., "Koala Populations in Queensland: Major Limiting Factors," in *Biology of the Koala* A. K. Lee et al., eds. (New South Wales: Beatty & Sons, 1990). 85–95.
7. S. Reardon, "Faecal Transplants Could Help Preserve Vulnerable Species," *Nature* 558 (2018): 173-174.
8. C. A. Hallmann et al., "More than 75 Percent Decline over 27 Years in Total Flying Insect Biomass in Protected Areas," *PLoS ONE* 12, no. 10 (2017): e0185809.
9. T. P. Hughes et al., "Global Warming and Recurrent Mass Bleaching of Corals," *Nature* 543 (2017): 373–77.
10. L. Van Valen, "A New Evolutionary Law," *Evolutionary Theory* 1 (1973): 1–30.
11. Lewontin, *The Triple Helix*; Senge, *The Fifth Discipline*; *Immoderate Greatness*; Pielou, *After the Ice Age*; Kolbert, *The Sixth Extinction*.
12. Robert B. Reich, *Saving Capitalism: For the Many, Not the Few* (New York: Alfred A. Knopf, 2015).
13. Joseph A. Tainter, "Problem Solving: Complexity, History, Sustainability," *Population and Environment: A Journal of Interdisciplinary Studies* 22 (2000): 3–41; Tainter, *The Collapse of Complex Societies*.
14. Clive Gamble et al., *Thinking Big: How the Evolution of Social Life Shaped the Human Mind* (London: Thames & Hudson, 2013).
15. Logsdon, Gene. *Holy Shit: Managing Manure to Save Mankind* (White River Junction, VT: Chelsea Green, 2010).
16. P. Alexander et al., "Losses, Inefficiencies and Waste in the Global Food System," *Agricultural Systems* 153 (2017): 190–200.
17. Campbell and Moyers, *The Power of Myth*, 32.
18. Huston Smith, *The World's Religions: Our Great Wisdom Traditions* (New York: Harper Collins, 1991).
19. Stephen Mitchell (translation), *Tao Te Ching* (New York: Harper, 2006).
20. FAO, *The State of Food and Agriculture 2016 (SOFA): Climate Change, Agriculture and Food Security* (2016), http://www.fao.org/publications/card/en/c/18679629-67bd-4030-818c-35b206d03f34.
21. Frederick D. Provenza, "What Does It Mean to Be Locally Adapted and Who Cares Anyway?" *Journal of Animal Science* 86 (2008): E271–84; Frederick D. Provenza et al., "Our Landscapes, Our Livestock, Ourselves: Restoring Broken Linkages among Plants, Herbivores, and Humans with Diets That Nourish and Satiate," *Appetite* 95 (2015): 500–19; Frederick D. Provenza et al., "Complex Creative Systems: Principles, Processes, and Practices of Transformation," *Rangelands* 35 (2013): 6–13.
22. Courtney White, *2% Solutions for the Planet: 50 Low-Cost, Low-Tech, Nature-Based Practices for Combating Hunger, Drought, and Climate Change* (White River Junction, VT: Chelsea Green Publishing, 2015).
23. Oren Shelef et al., "The Value of Native Plants and Local Production in an Era of Global Agriculture," *Frontiers in Plant Science* (2017); C. C. Hinrichs, "Embeddedness and Local Food Systems: Notes on Two Types of Direct Agricultural Market," *Journal of Rural Studies* 16 (2000): 295–303.

24. J. W. Smith et al., "Beyond Milk, Meat, and Eggs: Role of Livestock in Food and Nutrition Security," *Animal Frontiers* 3 (2013): 6-13; M.C. Eisler et al., "Steps to Sustainable Livestock," *Nature* 507 (2014): 32–34.
25. D. Tilman and M. Clark, "Global Diets Link Environmental Sustainability and Human Health," *Nature* 515 (2014): 518–22.

## Chapter 18: The Mystery of Being

1. Campbell and Moyers, *The Power of Myth*.
2. David Lindley, *Uncertainty: Einstein, Heisenberg, Bohr, and the Struggle for the Soul of Science* (New York: Random House, 2007).
3. Capra, *The Tao of Physics*, 141.
4. Brian Greene, *The Elegant Universe* (New York: Vintage Books, 1999).
5. Clark, *Einstein*, 754–55.
6. Joseph Campbell, "The Impact of Science on Myth," in *Myths to Live By* (New York: Viking Penguin, 1973), 1–18.
7. Smith, *The World's Religions*, 5.
8. Huxley, *The Perennial Philosophy*.
9. Campbell and Moyers, *The Power of Myth*, 56–57.
10. Dalai Lama and Tutu, *The Book of Joy*, 297.
11. Schrödinger, *What Is Life?*, chapter 6.
12. Luis W. Alvarez et al., "Extraterrestrial Cause for the Cretaceous-Tertiary Extinction," *Science* 208 (1980): 1095–108.
13. Nassim N. Taleb, *Fooled by Randomness: The Hidden Role of Chance in the Markets and in Life* (New York: Texere, 2001); Nassim N. Taleb, *The Black Swan: The Impact of the Highly Improbable* (New York: Random House, 2007).

## Dining on Earth: A Visitor's Reflections

1. J. B. Tucker, "Ian Stevenson and Cases of the Reincarnation Type," *Journal of Scientific Exploration* 22 (2008): 36–43; Prophet and Prophet, *Reincarnation*.
2. Campbell and Moyers, *The Power of Myth*, 154.
3. Schrödinger, *My View of the World*, 20–22.

# BIBLIOGRAPHY

Adler, Shelley R. *Sleep Paralysis: Night-mares, Nocebos, and the Mind-Body Connection*. Piscataway, NJ: Rutgers University Press, 2011.

Alexander, Eben. *Proof of Heaven: A Neurosurgeon's Journey into the Afterlife*. New York: Simon & Schuster, 2012.

Arbesman, Samuel. *The Half-Life of Facts: Why Everything We Know Has an Expiration Date*. New York: Penguin Group, 2013.

Ausubel, Kenny, and J. P. Harpignies, eds. *Ecological Medicine: Healing the Earth, Healing Ourselves*. San Francisco: Sierra Club Books, 2004.

Barrow, John. *The Origins of the Universe*. London: Phoenix, 2014.

Baskin, Yvonne. *Underground: How Creatures of Mud and Dirt Shape Our World*. Washington, DC: Island Press, 2005.

Bell, Rob. *Love Wins: A Book about Heaven, Hell, and the Fate of Every Person Who Ever Lived*. New York: HarperCollins, 2011.

Brooks, Rosa. *How Everything Became War and the Military Became Everything: Tales from the Pentagon*. New York: Simon & Schuster, 2016.

Burrows, George E., and Ronald J. Tyrl. *Toxic Plants of North America*. Ames, IA: Iowa State University Press, 2001.

Brown, Alan. *The Normal Child: Its Care and Feeding*. Toronto: McClelland, 1926

Campbell, Joseph. *The Masks of God: Creative Mythology*. New York: Penguin Books, 1976.

———. *Myths to Live By*. New York: Penguin Books, 1972.

Campbell, Joseph, and Bill Moyers. *The Power of Myth*. New York: Doubleday, 1988.

Capra, Fritjof. *The Tao of Physics: An Exploration of the Parallels between Modern Physics and Eastern Mysticism*. Boston: Shambhala, 1991.

———. *The Turning Point: Science, Society, and the Rising Culture*. New York: Simon & Schuster, 1982.

Carr, Nicholas. *The Shallows: What the Internet Is Doing to Our Brains*. New York: W. W. Norton & Company, 2010.

Carson, Rachel. *Silent Spring*. New York: Houghton Mifflin Harcourt, 1962.

Castle, Jill, and Maryanne Jacobsen. *Fearless Feeding: How to Raise Healthy Eaters from High Chair to High School*. San Francisco: Jossey-Bass, 2013.

Chamovitz, Daniel. *What a Plant Knows: A Field Guide to the Senses*. New York: Scientific American/ Farrar, Straus and Giroux, 2012.

Cheeke, Peter R. *Natural Toxicants in Feeds, Forages, and Poisonous Plants*. Danville, IL: Interstate Publishing, 1998.

Chesterton, Gilbert K. *Orthodoxy*. New York: John Land Company, 1908

Clark, Ronald W. *Einstein: The Life and Times*. New York: Avon Books, 1971.

Crozier, Alan, Michael N. Clifford, and Hiroshi Ashihara. *Plant Secondary Metabolites: Occurrence, Structure and Role in the Human Diet*. Oxford: Blackwell Publishing, 2007.

Damasio, Antonio. *Descartes' Error: Emotion, Reason, and the Human Brain*. New York: Penguin Putnam, 1994.

Dark Mountain Project. *Walking on Lava: Selected Works for Uncivilized Times.* White River Junction, VT: Chelsea Green Publishing, 2017.

Decker, Bert. *You've Got to Be Believed to Be Heard: The Complete Book of Speaking – In Business and in Life!.* New York: St. Martin's Press, 2008.

Diamond, Jared. *Collapse: How Societies Choose to Fail or Succeed.* New York: Viking, 2005.

Diamond, Jared. *Guns, Germs, and Steel: The Fates of Human Societies.* New York: W. W. Norton & Company, 1999.

Doidge, Norman. *The Brain That Changes Itself: Stories of Personal Triumph from the Frontiers of Brain Science.* London: Penguin Books, 2007.

Duke, James A. *The Green Pharmacy Guide to Healing Foods: Proven Natural Remedies to Treat and Prevent More Than 80 Common Health Concerns.* Emmaus, PA: Rodale Inc., 2008.

Eadie, Betty J. *Embraced by the Light.* New York: Bantam, 1994.

Ehrman, Bart D. *Misquoting Jesus: The Story Behind Who Changed the Bible and Why.* New York: HarperCollins, 2005.

Eisenstein, Charles. *The Yoga of Eating: Transcending Diets and Dogma to Nourish the Natural Self.* Washington, DC: New Trends Publishing, Inc., 2003.

Engel, Cindy. *Wild Health: How Animals Keep Themselves Well and What We Can Learn from Them.* New York: Houghton Mifflin, 2002.

Ferris, Warren A. *Life in the Rocky Mountains: A Diary of Wanderings on the Sources of the Rivers Missouri, Columbia, and Colorado, 1830–1835.* New York: Western Literary Messenger, 2012.

Forbes, J. M. *Voluntary Food Intake and Diet Selection in Farm Animals.* Wallingford, UK: CAB International, 2007.

Fuller, et al. *I Seem to Be a Verb.* Berkeley: Gingko Press, 2015.

Gamble, Clive, et al. *Thinking Big: How the Evolution of Social Life Shaped the Human Mind.* London: Thames & Hudson, 2013.

Gerrish, Jim. *Management-Intensive Grazing: The Grassroots of Grass Farming.* Ridgeland, MS: Green Park Press, 2004.

Gershon, Michael. *The Second Brain: A Groundbreaking New Understanding of Nervous Disorders of the Stomach and Intestine.* New York: HarperCollins, 1999.

Gibran, Kahlil. *The Prophet.* New York: Alfred A. Knopf, 1990.

Gillam, Carey. *Whitewash: The Story of a Weed Killer, Cancer, and the Corruption of Science.* Washington, DC: Island Press, 2017.

Goreau, Thomas J., et al., eds. *Geotherapy: Innovative Methods of Soil Fertility Restoration, Carbon Sequestration, and Reversing $CO_2$ Increase.* Boca Raton: CRC Press, 2015.

Greene, Brian. *The Elegant Universe: Superstrings, Hidden Dimensions, and the Quest for the Ultimate Theory.* New York: Vintage Books, 1999.

Guiliano, Mireille. *French Women Don't Get Fat: The Secret of Eating for Pleasure.* New York: Vintage Books, 2007.

Gunderson, Lance H., et al., eds. *Barriers and Bridges to the Renewal of Ecosystems and Institutions.* New York: Columbia University Press, 1995.

Holliday, Richard J., and Jim Helfter. *A Holistic Vet's Prescription for a Healthy Herd: A Guide to Livestock Nutrition, Free-Choice Minerals, and Holistic Cattle Care.* Austin: Acres U.S.A., 2014.

Howell, Jim. *For the Love of Land: Global Case Studies of Grazing in Nature's Image.* Charleston, NC: Booksurge Publishing, 2008.

Hoyle, Fred. *The Intelligent Universe.* London: M. Joseph, 1983.

Hoyle, Fred, and Chandra Wickramasinghe. *Evolution from Space: A Theory of Cosmic Creation.* London: J. M. Dent, 1981.

———. *Our Place in the Cosmos: The Unfinished Revolution.* London: J. M. Dent, 1993.

Humphries, Suzanne, and Roman Bystrianyk. *Dissolving Illusions: Disease, Vaccines, and the Forgotten History*. Self-Published, CreateSpace, 2015.

Hungate, Robert E. *The Rumen and Its Microbes*. New York: Academic Press, 1966.

Huxley, Aldous. *The Perennial Philosophy*. New York: Harper & Row, 2004.

Illich, Ivan. *Limits to Medicine: Medical Nemesis, the Expropriation of Health*. London: Marion Boyars Publishers Ltd., 2010.

Jastrow, Robert. *God and the Astronomers*. New York: W. W. Norton & Company, 1992.

Johns, Timothy. *With Bitter Herbs They Shall Eat It: Chemical Ecology and the Origins of Human Diet and Medicine*. Tucson: The University of Arizona Press, 1990.

Kauffman, Stuart A. *At Home in the Universe: The Search for the Laws of Self-Organization and Complexity*. New York: Oxford University Press, 1996.

———. *Investigations*. New York: Oxford University Press, 2000.

Kleppel, Gary. *The Emergent Agriculture: Farming, Sustainability, and the Return of the Local Economy*. Gabriola Island, Canada: New Society Publishers, 2014.

Kolbert, Elizabeth. *The Sixth Extinction: An Unnatural History*. New York: Henry Holt and Company, 2014.

Kuhn, Thomas S. *The Structure of Scientific Revolutions*, 3rd ed. Chicago: University of Chicago Press, 1996.

Kunstler, James H. *The Long Emergency: Surviving the End of Oil, Climate Change, and Other Converging Catastrophes of the Twenty-First Century*. New York: Grove Press, 2005.

Lancy, David F. *The Anthropology of Childhood: Cherubs, Chattel, Changelings*. Cambridge, UK: Cambridge University Press, 2008.

Lane, Nick. *Oxygen: The Molecule That Made the World*. Oxford: Oxford University Press, 2003.

Laozi. *Tao Te Ching*. Translated by Stephen Mitchell. New York: Harper, 2006.

LeDoux, Joseph. *Synaptic Self: How Our Brains Become Who We Are*. New York: Penguin, 2002.

Leopold, Aldo. *A Sand County Almanac*. New York: Oxford University Press, 1949.

Levenstein, Harvey. *Paradox of Plenty: A Social History of Eating in Modern America*. Berkeley and Los Angeles: University of California Press, 2003.

———. *Revolution at the Table: The Transformation of the American Diet*. Berkeley and Los Angeles: University of California Press, 2003.

Levin, Janna. *A Madman Dreams of Turing Machines*. New York: Alfred A. Knopf, 2006.

Levinovitz, Alan. *The Gluten Lie: And Other Myths about What You Eat*. New York: Regan Arts, 2015.

Lewontin, Richard. *The Triple Helix: Gene, Organism, Environment*. Cambridge, MA: Harvard University Press, 2000.

Lindley, David. *Uncertainty: Einstein, Heisenberg, Bohr, and the Struggle for the Soul of Science*. New York: Random House, 2007.

Lipton, Bruce. *The Biology of Belief: Unleashing the Power of Consciousness, Matter and Miracles*. Carlsbad, CA: Hay House, 2005.

Logsdon, Gene. *Holy Shit: Managing Manure to Save Mankind*. White River Junction, VT: Chelsea Green Publishing, 2010..

Marris, Emma. *Rambunctious Garden: Saving Nature in a Post-Wild World*. New York: Bloomsbury, 2011.

Marsh, George P. *The Earth as Modified by Human Actions*. Chestnut Hill, MA: Adamant Media Corporation, Elibron Classics, 2006.

Massy, Charles. *Call of the Reed Warbler: A New Agriculture, A New Earth*. St. Lucia, Australia: University of Queensland Press, 2017.

Mayer, Frank H., and Charles B. Roth. *The Buffalo Harvest*. Saint Paul: Pioneer Press, 1995.

Meuret, Michel, and Frederick D. Provenza. *The Art and Science of Shepherding: Tapping the Wisdom of French Herders*. Austin: Acres U.S.A., 2014.

Moore, David S. *The Developing Genome: An Introduction to Behavioral Epigenetics*. Oxford: Oxford University Press, 2015.

Moorjani, Anita. *Dying to Be Me: My Journey from Cancer to Near Death to True Healing*. New York: Hay House, 2012.

Morgan, Edward P. *This I Believe: The Personal Philosophies of One Hundred Thoughtful Men and Women in All Walks of Life*. New York: Simon & Schuster, 1952.

Mowat, Farley. *People of the Deer*. New York: Carroll & Graf Publishers, 2005.

Mukherjee, Siddhartha. *The Emperor of All Maladies: A Biography of Cancer*. New York: Simon & Schuster, 2010.

Muller, Richard A. *Now: The Physics of Time*. New York: W. W. Norton & Company, 2016.

Nabhan, Gary P. *Why Some Like It Hot: Foods, Genes, and Cultural Diversity*. Washington, DC: Island Press, 2004.

Nestle, Marion. *Food Politics: How the Food Industry Influences Nutrition and Health*. Berkeley: University of California Press, 2013.

Ophuls, William. *Immoderate Greatness: Why Civilizations Fail*. Self-Published, CreateSpace, 2012.

Overman, Dean L. *A Case Against Accident and Self-Organization*. New York: Rowman & Littlefield, 1997.

Pert, Candace B. *Molecules of Emotion: Why You Feel the Way You Feel*. New York: Scribner, 1997.

Pielou, E. C. *After the Ice Age: The Return of Life to Glaciated North America*. Chicago: The University of Chicago Press, 1991.

Pierce, William D., and Carl D. Cheney. *Behavior Analysis and Learning: A Biobehavioral Approach*. New York: Taylor & Francis, 2017.

Plotkin, Mark J. *Medicine Quest: In Search of Nature's Healing Secrets*. New York: Penguin, 2000.

Pollan, Michael. *In Defense of Food: An Eater's Manifesto*. New York: Penguin Press, 2008.

———. *The Omnivore's Dilemma: A Natural History of Four Meals*. New York: Penguin Press, 2006.

Prescott, Lansing M., et al. *Microbiology*, Sixth Ed., New York: McGraw-Hill, 2005.

Price, Weston A. *Nutrition and Physical Degeneration: A Comparison of Primitive and Modern Diets and Their Effects*. New York: Harper & Brothers, 1939.

Prigogine, Ilya. *The End of Certainty: Time, Chaos, and the New Laws of Nature*. New York: The Free Press, 1996.

Prigogine, Ilya, and Isabelle Stengers. *Order Out of Chaos: Man's New Dialogue with Nature*. New York: Bantam, 1984.

Prophet, Elizabeth C., and Erin L. Prophet. *Reincarnation: The Missing Link in Christianity*. Gardiner, MT: Summit University Press, 1997.

Prophet, Mark L., and Elizabeth C. Prophet. *The Lost Teachings of Jesus 1–4*. Gardiner, MT: Summit University Press, 1986.

Provenza, Frederick D. *Foraging Behavior: Managing to Survive in a World of Change*. Logan, UT: Utah State University Press, 2003.

Rankin, Lissa. *Mind over Medicine: Scientific Proof That You Can Heal Yourself*. Carlsbad, CA: Hay House, 2013.

Reich, Robert B. *Saving Capitalism: For the Many, Not the Few*. New York: Alfred A. Knopf, 2015.

Robinson, Jo. *Eating on the Wild Side: The Missing Link to Optimum Health*. New York: Little, Brown and Company, 2013.

Rolls, Barbara J., and Mindy G. Hermann. *The Ultimate Volumetrics Diet: Smart, Simple, Science-Based Strategies for Losing Weight and Keeping It Off*. New York: Harper Collins, 2012.

Rosenthal, Gerald A., and May Berenbaum, eds. *Herbivores: Their Interactions with Secondary Plant Metabolites*, 2nd ed. New York: Academic Press, 1992.

Rosenthal, Gerald A., and Daniel H. Janzen, eds. *Herbivores: Their Interaction with Secondary Plant Metabolites*, 1st ed. New York: Academic Press, 1979.

Rosling, Hans and Anna Rosling Ronnlund. *Factfulness: Ten Reasons We're Wrong about the World – And Why Things Are Better Than You Think*. New York: Flatiron Books, 2018.
Savory, Allen, and Jodi Butterfield. *Holistic Management: A New Framework for Decision-Making*. Washington, DC: Island Press, 1999.
Schatzker, Mark. *The Dorito Effect: The Surprising New Truth about Food and Flavor*. New York: Simon & Schuster, 2015.
———. *Steak: One Man's Search for the World's Tastiest Piece of Beef*. New York: Viking, 2010.
Schrödinger, Erwin. *What Is Life? And Mind and Matter*. Cambridge, UK: Cambridge University Press, 1944.
Selhub, Eva M., and Alan C. Logan. *Your Brain on Nature: The Science of Nature's Influence on Your Health, Happiness, and Vitality*. Mississauga, Canada: John Wiley & Sons, 2012.
Senge, Peter M. *The Fifth Discipline: The Art and Practice of the Learning Organization*. New York: Currency Doubleday, 1994.
Shepard, Mark. *Restoration Agriculture: Real-World Permaculture for Farmers*. Austin: Acres U.S.A., 2013.
Simoons, Frederick. *Eat Not This Flesh: Food Avoidances from Prehistory to the Present*. Madison: University of Wisconsin Press, 1994.
Smith, Huston. *The World's Religions: Our Great Wisdom Traditions*. New York: Harper Collins, 1991.
Sylvia, Claire, with William Novak. *A Change of Heart*. New York: Little, Brown and Company, 1997.
Tainter, Joseph A. *The Collapse of Complex Societies*. Cambridge, UK: Cambridge University Press, 1988.
Taleb, Nassim N. *Fooled by Randomness: The Hidden Role of Chance in the Markets and in Life*. New York: Texere, 2001.
———. *The Black Swan: The Impact of the Highly Improbable*. New York: Random House, 2007.
Taubes, Gary. *Good Calories, Bad Calories: Challenging the Conventional Wisdom on Diet, Weight Control, and Disease*. New York: Alfred A. Knopf, 2007.
Taylor, Jill Bolte. *My Stroke of Insight*. New York: Viking, 2008.
Teicholz, Nina. *The Big Fat Surprise: Why Butter, Meat, and Cheese Belong in a Healthy Diet*. New York: Simon & Schuster, 2014.
Trewavas, Anthony. *Plant Behaviour and Intelligence*. Oxford: Oxford University Press, 2014.
Voth, Kathy. *Cows Eat Weeds: How to Turn Your Cows into Weed Managers*. Loveland, CO: Livestock for Landscapes, 2010.
Wansink, Brian. *Mindless Eating: Why We Eat More Than We Think*. New York: Bantam Books, 2007.
———. *Slim by Design: Mindless Eating Solutions for Everyday Life*. New York: HarperCollins, 2014.
Weiss, Brian L. *Many Lives, Many Masters*. New York: Simon & Schuster, 1988.
Wheatley, Margaret J. *Leadership and the New Science: Discovering Order in a Chaotic World*. San Francisco: Berrett-Koehler Publishers, 2006.
White, Courtney. *Revolution on the Range: The Rise of a New Ranch in the American West*. Washington, DC: Island Press, 2008.
———. *Grass, Soil, Hope: A Journey through Carbon Country*. White River Junction, VT: Chelsea Green Publishing, 2014.
———. *The Age of Consequences: A Chronicle of Concern and Hope*. Berkeley: Counterpoint Press, 2015.
———. *Two Percent Solutions for the Planet: 50 Low-Cost, Low-Tech, Nature-Based Practices for Combating Hunger, Drought, and Climate Change*. White River Junction, VT: Chelsea Green Publishing, 2015.
Wilber, Ken. *The Marriage of Sense and Soul: Integrating Science and Religion*. New York: Broadway Books, 1999.

Williams, Florence. *The Nature Fix: Why Nature Makes Us Happier, Healthier, and More Creative*. New York: W. W. Norton & Company, 2017.
Williams, Roger J. *Biochemical Individuality*. New Canaan, CT: Keats Publishing, 1988.
Wilson, Edward O. *The Future of Life*. New York: Vintage Books, 2002.
Zimmer, Carl. *Microcosm:* E. coli *and the New Science of Life*. New York: Vintage Books, 2009.

# INDEX

# ABOUT THE AUTHOR

Frederick D. Provenza is professor emeritus of behavioral ecology in the Department of Wildland Resources at Utah State University. At Utah State Provenza directed an award-winning research group that pioneered an understanding of how learning influences foraging behavior and how behavior links soils and plants with herbivores and humans. Provenza is one of the founders of BEHAVE, an international network of scientists and land managers committed to integrating behavioral principles with local knowledge to enhance environmental, economic, and cultural values of rural and urban communities. He is also the author of *Foraging Behavior* and the coauthor of *The Art and Science of Shepherding*.

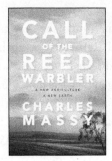

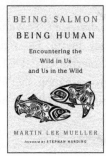

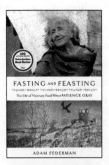

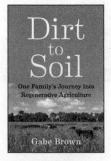